D1339718

This book is to be returned on or before
the last date stamped below.

METALLURGICAL APPLICATIONS *of* BACTERIAL LEACHING *and* RELATED MICROBIOLOGICAL PHENOMENA

METALLURGICAL APPLICATIONS *of* BACTERIAL LEACHING *and* RELATED MICROBIOLOGICAL PHENOMENA

Edited by

LAWRENCE E. MURR
ARPAD E. TORMA

John D. Sullivan Center for In-Situ Mining Research,
Department of Metallurgical and Materials Engineering

JAMES A. BRIERLEY

Department of Biology

New Mexico Institute of Mining and Technology
Socorro, New Mexico

ACADEMIC PRESS   New York   San Francisco   London   1978
A Subsidiary of Harcourt Brace Jovanovich, Publishers

The material was prepared with the support of National Science Foundation Grant No. AER77-12221. However, any opinions, findings, conclusions, or recommendations expressed herein are those of the author(s) and do not necessarily reflect the views of NSF.

ACADEMIC PRESS, INC.
111 Fifth Avenue, New York, New York 10003

United Kingdom Edition published by
ACADEMIC PRESS, INC. (LONDON) LTD.
24/28 Oval Road, London NW1 7DX
Library of Congress Cataloging in Publication Data

Main entry under title:

Metallurgical applications of bacterial leaching and
    related microbiological phenomena.

    Proceedings of an international symposium which took
place at the New Mexico Institute of Mining & Technology,
Socorro, N. M., August 3-5, 1977.
    Includes bibliographical references and index.
    1. Leaching—Congresses.   2. Metallurgy—Congresses.
3. Industrial microbiology—Congresses.   I. Murr,
Lawrence Eugene.   II. Torma, Arpad E.   III. Brierley,
James A.   IV. New Mexico. Institute of Mining and
Technology, Socorro. [DNLM: 1. Metallurgy—Congresses.
2. Microbiology—Congresses. QW75 M587 1977]
TN673.M47   622'.75   78-693
ISBN 0-12-511150-9
PRINTED IN THE UNITED STATES OF AMERICA

Dedicated to

Marguerite M. and John D. Sullivan

*microbiologist and leaching pioneer respectively,
whose professionalism and marriage
are the epitome of the objectives of this book*

Contents

**Chapter 25   Microbial Leaching of Copper at Ambient and
Elevated Temperatures   477**

James A. Brierley and Corale L. Brierley

**Chapter 26   The Use of Large-Scale Test Facilities in Studies
of the Role of Microorganisms in Commercial
Leaching Operations   491**

L. E. Murr and James A. Brierley

Contributors

Numbers in parentheses indicate the pages on which the authors' contributions begin.

WILLIAM A. APEL (45, 223), Department of Microbiology, The Ohio State University, Columbus, Ohio

A. S. ATKINS (403), Virginia Polytechnic Institute and State University, Blacksburg, Virginia

H. BALAKRISHNAN (427), National Chemical Laboratory, Poona, India

V. K. BERRY (103), Department of Metallurgical and Materials Engineering, New Mexico Institute of Mining and Technology, Socorro, New Mexico

K. BOSECKER (389), Bundesanstalt für Geowissenschaften und Rohstoffe, Hannover, Federal Republic of Germany

CORALE L. BRIERLEY (345, 477), New Mexico Bureau of Mines and Mineral Resources, Socorro, New Mexico

JAMES A. BRIERLEY (477, 491), Department of Biology, New Mexico Institute of Mining and Technology, Socorro, New Mexico

A. BRUYNESTEYN (441), British Columbia Research, Vancouver, British Columbia, Canada

A. M. CHAKRABARTY (137), General Electric Company, Corporate Research and Development, Schenectady, New York

DOUGLAS J. CORK (207), Department of Chemistry, University of Arizona, Tucson, Arizona

MICHAEL A. CUSANOVICH (207), Department of Chemistry, University of Arizona, Tucson, Arizona

CRAWFORD S. DOW (61), University of Warwick, Coventry, England

PATRICK R. DUGAN (45, 223), Department of Microbiology, The Ohio State University, Columbus, Ohio

HANS GEORG EBNER (195), University of Dortmund, Dortmund, Federal Republic of Germany

MARTIN ECCLESTON (61), University of Warwick, Coventry, England

STEFAN S. GAIDARJIEV (253), Higher Institute of Mining and Geology, Sofia—Darvenitza, Bulgaria

FRATIO N. GENCHEV (253), Higher Institute of Mining and Geology, Sofia—Darvenitza, Bulgaria

STOYAN N. GROUDEV (253), Higher Institute of Mining and Geology, Sofia—Darvenitza, Bulgaria

KAZUTAMI IMAI (275), Faculty of Agriculture, Okayama University, Tsuchima, Okayama–shi, Japan

C. A. JONES (19), Glaxo Research Ltd., Stoke Poges, Bucks, England

DONOVAN P. KELLY (19, 61, 83), University of Warwick, Coventry, England

V. S. KRISHNAMACHAR (427), National Chemical Laboratory, Poona, India

NORMAN W. LE ROUX (167, 463), Warren Spring Laboratory, Stevenage, Herts, England

DONALD LUNDGREN (151), Syracuse University, Syracuse, New York

R. O. McELROY (441), British Columbia Research, Vancouver, British Columbia, Canada

K. B. MEHTA (463), Warren Spring Laboratory, Stevenage, Herts, England

L. E. MURR (103, 491), Department of Metallurgical and Materials Engineering, New Mexico Institute of Mining and Technology, Socorro, New Mexico

D. NEUSCHÜTZ (389), Friedrich Krupp GmbH, Krupp Forschungsinstitut, Essen, Federal Republic of Germany

P. R. NORRIS (83), University of Warwick, Coventry, England

S. G. PATIL (427), National Chemical Laboratory, Poona, India

VIVIAN F. PERRY (167), Warren Spring Laboratory, Stevenage, Herts, England

P. N. RANGACHARI (427), National Chemical Laboratory, Poona, India

GIOVANNI ROSSI (297), University of Cagliari, Sardinia, Italy

M. N. SAINANI (427), National Chemical Laboratory, Poona, India

U. SCHEFFLER (389), Friedrich Krupp GmbH, Krupp Forschungsinstitut, Essen, Federal Republic of Germany

MARVIN SILVER (3), Université Laval, Quebec, Canada

TATSUO TANO (151), Okayama University, Okayama, Japan

NOBORU TOMIZUKA (321), Fermentation Research Institute, Chiba, Japan

A. E. TORMA (375), Department of Metallurgical and Materials Engineering, New Mexico Institute of Mining and Technology, Socorro, New Mexico

OLLI H. TUOVINEN (61), University of Helsinki, Helsinki, Finland

H. M. TSUCHIYA (365), University of Minnesota, Minneapolis, Minnesota

DON S. WAKERLEY (167), Warren Spring Laboratory, Stevenage, Herts, England

MITSUO YAGISAWA (321), Fermentation Research Institute, Chiba, Japan

Preface

The synergistic aspects of science and engineering and of their many disciplines and subdisciplines stand out as perhaps the major influence on the bulk of industrial and technical advances at least in the latter part of the twentieth century. Indeed, the complexity of the world society and its concomitant and interconnected social, economic, and industrial problems have been syndetic to science and engineering, with the outcome being the interdisciplinary and multidisciplinary approaches to the development of and solutions to the technological innovations necessary to sustain and advance the standard of living in many parts of the industrialized world.

In many respects, this book reflects the synergistic aspects of microbiology in metallurgy—particularly hydrometallurgy—culminating in or converging to a syndetic subdiscipline: biohydrometallurgy. The chapters comprising this volume represent the invited and contributed papers that composed an international symposium having the same title as the book. The theme of the symposium, which is so obviously reflected in these chapters, addresses an attempt to provide some basic understanding of the role of bacteria in leaching processes and other metallurgical applications, particularly hydrometallurgical. The book emphasizes the role played by microorganisms in the kinetics of leaching and similar metallurgical processes, and this was achieved by stimulating a strong interaction—a priori—between microbiologists and metallurgists. In many respects, this book represents a very successful attempt to bridge the gap between those involved in the basic study of microorganisms, in the strictly microbiological aspects of metal extraction and attendant conversion kinetics, and in the practical, engineering aspects of extraction. In this respect, it should be of interest to a wide range of students, researchers, and practitioners in microbiology, biophysics, biochemistry, mineral processing and preparation, extractive and/or chemical metallurgy, mining engineering, and many related disciplines including chemical engineering and bioengineering. A particular aim of the book, like the symposium from which it originated, is to discuss and ascertain the projected role of microbiological applications in areas of mineral extraction, especially from low-grade, nonrenewable waste-ore deposits, and the arrangement of the chapters addresses both the fundamental and practical aspects of such applications. This is accomplished somewhat syndetically by the arrangement of the chapters into three major sections:

I. Basic Microbial Studies Applied to Leaching, which also addresses fundamental microbial phenomena;

II. Waste Treatment and Environmental Considerations;

III. Bioextractive Applications and Optimization.

The major objectives of the symposium, which are forcefully reflected in these collected works, were

(1) to establish a strong interaction between the fundamentalists and practitioners in the area of bacterial leaching in its broadest sense, to have industry input, and to discuss industrial problems in bacterial leaching;

(2) to review the latest developments in optimizing bacterial activity, and in understanding the role of microorganisms in large-scale metallurgical applications;

(3) to address and discuss new metallurgical applications of bacterial leaching and the attendant economics;

(4) to establish the optimum conditions for conventional bacterial leaching of metal values and recommend appropriate research efforts to address leach dump optimization; and

(5) to establish a priority for research in the area of bacterial leaching with particular emphasis on the determination of directions of future research in the development of existing microorganisms or the search for new microorganisms.

It is not unlikely that this book would serve as a text/reference in advanced courses in biometallurgy, extractive metallurgy, hydrometallurgy, and applied or industrial microbiology. In fact, the symposium was offered for graduate credit and assignments were structured to direct the student toward an assessment of various concepts and processes after having been exposed to the topics presented herein. The arrangement of this book into distinct but interconnected regimes involving both the fundamental and practical (applied) aspects of bacterial leaching and related phenomena is indeed conducive to such usage.

The editors wish to thank those who participated in the symposium and who contributed to this volume. We are especially grateful for the support of this effort through a grant (No. AER77-12221) from the National Science Foundation's Research Applied to National Needs (RANN) Division. Finally, we are especially grateful for the patient and competent typing of most of the chapters and the discussions of the sections, as well as editorial help from Lorraine Valencia and Elizabeth Fraissinet.

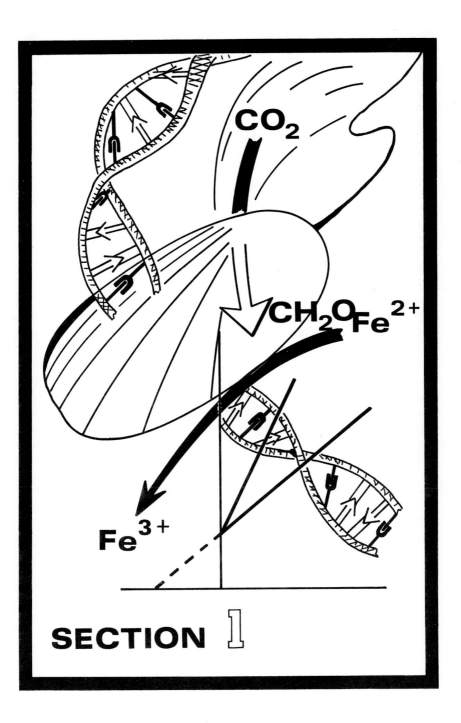

CO_2

CH_2O Fe^{2+}

Fe^{3+}

SECTION 1

I   BASIC MICROBIAL STUDIES APPLIED TO LEACHING

An understanding of the underlying principles controlling the function of thiobacilli and their reaction to environment will ultimately provide the information necessary for their successful use in solving leaching problems. Thus, this symposium included papers dealing strictly with basic cellular phenomena.

Chapter 1 of this section presents a brief review of sulfur, iron, and carbon dioxide metabolism in thiobacilli. Three chapters follow presenting new data concerning the oxidation of iron by *Thiobacillus ferrooxidans*. The kinetics of biological iron oxidation are greatly affected by the ferric to ferrous ratio, an aspect of importance in considering use of microbes for generating lixiviant for metal sulfide oxidation. The requirement for hydrogen ion by *T. ferrooxidans* for ferrous iron oxidation is also a factor affecting cell function. Chapter 8 reports development of a procedure for study of the iron oxidizing system using isolated cell membranes. Chapter 5 deals with the affect of silver and other metals toxicity on ferrobacilli. This chapter indicates that these microbes may also have a role in concentration of various metals by cellular accumulation. Some basic characteristics of cell structure and the phenomenon of apparent interspecies transition of acidophilic thiobacilli was presented. Chapter 7 speculates on the potential for alteration of thiobacilli characteristics by methods of mutation and plasmid transfer. Chapters 6 and 9 consider the microbes involved in the leaching process; one through direct observation of the microbes; the other discussing the use of microbes other than thiobacilli, an aspect certainly to receive much attention in the future.

Future symposia concerned with bacterial leaching will undoubtedly include more chapters of basic cellular studies. This is a field in which basic research has been greatly stimulated by a study of applications of leaching procedures as discussed in the chapters of the following sections.

METABOLIC MECHANISMS

OF IRON-OXIDIZING THIOBACILLI

Marvin Silver

Université Laval
Québec, Québec, Canada

The iron-oxidizing thiobacilli are remarkable for the range of inorganic compounds that are acted upon. The principal if not sole source of carbon, CO_2, is assimilated by two methods simultaneously: the Calvin-Benson or reductive pentose phosphate cycle and the carboxylation of PEP (phospho-enol-pyruvate) with the formation of oxalo-acetate by the enzyme PEP-carboxylase.

Similar to other thiobacilli, elemental sulfur and reduced inorganic sulfur compounds are used for the generation of energy; several studies have demonstrated a number of pathways, which may not be mutually exclusive, and some of the enzymes catalyzing individual reactions have been isolated and characterized. Unique amongst the thiobacilli is the ability of these organisms to oxidize ferrous iron for the generation of chemical energy. Metal sulfide mineral ores also serve this function, resulting in the oxidation of the sulfide moiety to sulfate and the dramatic modification of the initial substrate, in many cases resulting in the solubilization of the metallic entity.

The mechanisms, both known and postulated, of the above reactions are described and discussed. In addition, evidence is presented indicating that the organism known as Thiobacillus ferrooxidans *may not be one distinct bacterium, but rather a group of metabolically similar microbes.*

I. INTRODUCTION TO THE IRON OXIDIZING THIOBACILLI

Since the first of the thiobacilli, *Thiobacillus thioparus*, was discovered by Nathansohn in 1902 (1), many more sulfur-oxidizing autotrophs have been isolated and identified. Among these are the iron oxidizing bacteria. The first of these was named *T. ferrooxidans* by Temple and Colmer (2), and was followed shortly afterwards by *Ferrobacillus ferrooxidans* (3) and *F. sulfooxidans* (4). Nutritional and taxonomic studies (5-16) have all indicated that there are no major differences between these three bacteria, and thus should all be regarded as *T. ferrooxidans*. More recently, however, it was demonstrated that the iron oxidizing bacteria adapted to and grown on different substrates showed both quantitative and qualitative differences in their abilities to oxidize metal sulfide minerals, and to use some of these oxidations for the assimilation of CO_2 (17). Furthermore, these bacteria when grown on different substrates were found to contain deoxyribonucleic acid (DNA) of different base compositions (Table 1) (18,19). The DNA of *Escherichia coli* and *Rhodospirillum rubrum* were included in this study as standards of known guanine plus cytosine (G + C) content. Thus, the question is raised: is the organism currently known as *T. ferrooxidans* one distinct species, or a series of metabolically similar bacteria?

Adaptation to growth of *T. ferrooxidans* on organic compounds have been reported (20,21) accompanied by physiological and morphological alterations (22,23). Guay and Silver (24) repeated these procedures, and obtained a microorganism similar if not identical to those already described. As reported by Shafia and Wilkinson (20), the organism lost its ability to oxidize or grow on ferrous iron after a number of transfers on ferrous iron-free glucose medium. Analysis and comparison of the DNA of both the bacteria of the original iron-oxidizing culture and the new isolate, renamed *T. acidophilus*, showed great differences in the gua-

TABLE I   DNA Base Composition of Iron-Oxidizing Bacteria and Thiobacillus acidophilus Grown on Different Substrates

Bacteria	Growth Substrate	Guanine plus cytosine content of DNA, Methods:			
		T_m	CsCl density	absorbance ratio	other methods
T. ferrooxidans	$FeSO_4$	56.5[a]	57.1[a]	56.2[a]	55.0[b,d]
T. ferrooxidans	PbS	54.5[a]	54.4[a]	54.2[a]	
T. ferrooxidans	CuFeS	59.9[a]	60.2[a]	60.1[a]	
T. acidophilus	S^o	62.8[b]	64.3[b]	62.7[b]	60.0[b,d]
T. acidophilus	glucose	62.8[b]	66.3[b]	60.5[b]	62.3[b,d]
E. coli	glucose	51.1[c]	50.4[c]	51.5[a]	51.8[b,d]
					51.2[c,e]
R. rubrum	malate	62.8[c]	65.0[c]		63.5[b,d]
					62.4[c,e]

[a], data of Guay, Silver and Torma (19).   [b], data of Guay and Silver (24).
[c], data of Silver, et al. (71).   [d], calculated by the normal probability method (cf 24).
[e], calculated by the O.D. 260/O.D. 280 ratio method (cf 71).

5

nine plus cytosine content (24) (Table 1). *T. acidophilus* is similar to both the glucose-adapted strain of Shafia and Wilkinson (20), and the glucose-grown organism of Tabita and Lundgren (21) with regard to metabolic properties and enzyme content (24, D.P. Kelly, personal communication). It may also be similar to *T. organoparus* (25). As not all cultures of *T. ferrooxidans* can be adapted to growth on glucose (20, B.J. Ralph, personal communication), and those that are adaptable to growth on substrates other than ferrous iron manifest fundamentally different properties of taxonomic importance (17,19,24,26), the hypothesis formulated earlier (12) that cultures of iron-oxidizing autotrophs may be originally heterogeneous is supported.

II.   OXIDATION OF FERROUS IRON

The oxidation of ferrous iron by *T. ferrooxidans* can be described by the equation:

$$Fe^{2+} + H^+ + \tfrac{1}{4}O_2 \longrightarrow Fe^{3+} + \tfrac{1}{2}H_2O$$

for calculations of the free energy made available to the bacteria between pH 1.5 and 3.0 (27). The free energy of the reaction at physiological concentrations (ΔG) was found to be dependent upon the pH of the environment of the reaction; Tuovinen and Kelly (23) have calculated these values between pH 1.5 and 3.0 to be from 7.8 to 5.9 kcal/mole Fe^{2+} oxidized. Therefore, insufficient energy would seem to be available for the phosphorylation of ADP to form ATP, which requires between 8.9 and 14 kcal. The possibility exists that ferrous iron might be complexed with an organic molecule in order that the potential of the Fe^{2+}/ Fe^{3+} couple be lowered from 0.77 volt to around 0 (28). This has been confirmed by the polarographic and Fe^{59} uptake studies of Dugan and Lundgren (29). Then, assuming that electrons are required to be transferred to oxygen via the cytochrome transport chain in pairs, about 14 kcal/2 Fe^{2+} oxidized should be available for ATP formation.

Cytochromes of the types a and c have been detected in iron-oxidizing bacteria (30,31). Coenzyme Q has been detected by Dugan and Lundgren (32). Results of these studies indicate that the electron transport scheme for the oxidation of ferrous iron in these bacteria is as follows:

Fe^{2+} -cytochrome c
   oxidoreductase $ADP + P_i \rightarrow ATP$

Fe^{2+}⤸cytochrome c (ox.)⤸cytochrome a (red.)⤸$\frac{1}{4}O_2 + H^+$
Fe^{3+}⤴cytochrome c (red.)⤴cytochrome a (ox.)⤴$\frac{1}{2}H_2O$

with coenzyme Q possibly acting as an intermediary electron carrier between a ferrous iron-sulfate-organic complex associated with the cellular envelope. That sulfate is required for the oxidation of ferrous iron is supported by the findings of Lazaroff (33) and Lees, Kwok and Suzuki (27).

Iron-cytochrome c-oxidoreductase has been purified and characterized by Yates and Nason (34) and Din, Suzuki and Lees (31). The mechanism of the enzyme reaction has been shown by Din and Suzuki (35) to be of the type known as Ping Pong Bi Bi -- that is, each substrate is bound and modified sequentially by the enzyme with a corresponding change in the enzyme. This is shown diagramatically below:

$$Fe^{2+} \qquad\qquad Fe^{3+}$$
$$\downarrow \qquad\qquad\qquad \uparrow$$
$$E(Fe^{3+}) \rightarrow E(Fe^{3+})Fe^{2+} \rightarrow E(Fe^{2+})Fe^{3+} \rightarrow E(Fe^{2+}) \rightarrow$$
$$\text{cyt. } c \ (Fe^{3+}) \qquad\qquad \text{cyt. } c \ (Fe^{2+})$$
$$\downarrow \qquad\qquad\qquad\qquad \uparrow$$
$$E(Fe^{2+}) \qquad E(Fe^{3+}) \qquad \rightarrow E(Fe^{3+})$$
$$| \qquad\qquad | $$
$$\text{cyt. } c \ (Fe^{3+}) \quad \text{cyt. } c \ (Fe^{2+})$$

The enzyme containing one atom of ferric iron binds one atom of ferrous iron, which reduces the enzyme bound iron and is then released. Then one molecule of oxidized cytochrome c is bound by the enzyme, whose ferrous iron reduces the ferric iron of the cytochrome c. The iron of the enzyme is thus oxidized, and the reduced cytochrome c is then released. Although this mechanism is known to work on soluble ferrous salts, there is no reason to assume that it does not occur regardless of the form of iron, that is, iron in insoluble sulfide, oxide or carbonate minerals, as long as the iron is in the 2^+ valency state.

III.   OXIDATION OF ELEMENTAL SULFUR AND REDUCED SULFUR COMPOUNDS

By definition, for a bacterial species to be a member of the genus *Thiobacillus*, it must be demonstrated to be able to use the oxidation of inorganic sulfur compounds for its energy requirements. All strains of *T. ferrooxidans* do this, although with quantitative differences noted with different strains (8,16), or with the same strain grown on different substrates (12).

Most of the research on the oxidation of elemental sulfur and reduced compounds of sulfur have been carried out with thiobacilli other than *T. ferrooxidans*. However, it is not unreasonable to assume that results obtained from the other thiobacilli apply at least in part to the iron-oxidizing members of this genus.

A generalized scheme for the oxidation of elemental sulfur and reduced sulfur compounds by the thiobacilli is shown in Fig. 1. The oxygenation of elemental sulfur with the formation of sulfite is catalyzed by the sulfur-oxidizing system. Sulfide is oxidized by a particulate system to the oxidation level of elemental sulfur, which then might be oxygenated, possibly by the sulfur-oxidizing enzyme, to sulfite. Thiosulfate can either be split with the formation of sulfite and sulfide, or it may be oxidized to tetrathionate, which then yields sulfite and trithionate, which in turn is converted to sulfite and thiosulfate. Sulfite can then be oxidized to sulfate either by a cytochrome-c-mediated oxidation, or via adenosine phosphosulfate.

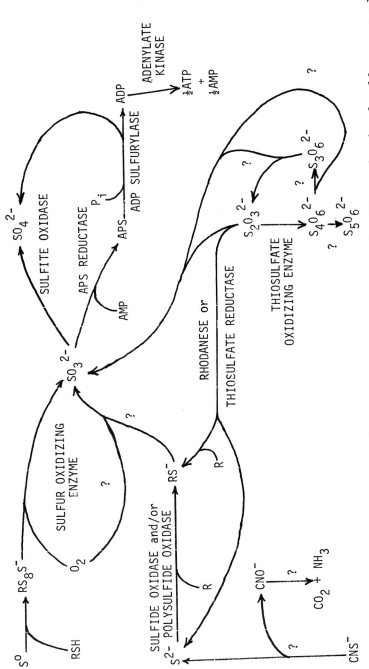

Fig. 1. Generalized scheme for the oxidation of elemental sulfur and reduced sulfur compounds by the thiobacilli.

The reactions involved in the oxidation of elemental sulfur are:

$$S_8 + GSH \xrightarrow{\hspace{4cm}} GS_8SH$$

$$GS_8SH + O_2 \xrightarrow{\text{sulfur oxidizing enzyme}} GS_8SO_2H$$

$$GS_8SO_2H + H_2O \xrightarrow{\hspace{3cm}} GS_7SH + H_2SO_3$$

Elemental sulfur, which exists most commonly in the form of a circular eight atom molecule is attacked by a sulfhydryl containing agent resulting in an organic polysulfide. The sulfur oxidizing enzyme catalyzes the oxidation of the terminal atom which is then removed hydrolytically as sulfite. The organic polysulfide, minus one molecule of sulfur is then able to be acted upon again by the sulfur oxidizing enzyme. *In vitro*, the sulfite is acted upon by the elemental sulfur with the formation of thiosulfate; whether this reaction is important *in vivo* is not known. In the presence of a suitable sulfite trapping reagent, such as formaldehyde (36, 37), or an active sulfite-oxidizing system (38), sulfite has been shown to be the true end product of this reaction. Besides *T. ferrooxidans* (37), this enzyme has been partially purified and characterized from *T. thiooxidans* (39), and *T. thioparus* (36) and its presence has been confirmed in *T. novellus* (40) and *T. neapolitanus* (41). Takakuwa (42) has resolved the sulfur-oxidizing enzyme of *T. thiooxidans* into two protein components, one of molecular weight of 120,000 and the other of molecular weight of 23,000 daltons.

The mechanism of the entry of elemental sulfur into the cell has not as yet been satisfactorily explained. It is known that when grown on elemental sulfur, the ends of the cells of *T. ferrooxidans* are more pointed in form than when grown on ferrous iron (22), and small evaginations appear on the external membrane which, when stained with silver ion, show the presence of sulfhydryl groups (43).

The initial reactions of the oxidation of sulfide have been
investigated by Moriarty and Nicholas (44,46) in $T.$ $concretivorus$,
$T.$ $thiooxidans$ and $T.$ $thioparus$, and by Aminuddin and Nicholas
(47) in $T.$ $denitrificans$. They demonstrated a cell-membrane associated enzymic process which oxidized sulfide to the oxidation
level of elemental sulfur. Actually, a linear polymeric chain of
sulfur atoms bound to a lipoprotein membrane fraction was formed,
possibly dependent on more than one enzyme; this process was
mediated by a copper-containing protein, cytochromes and ubiqui-
none and was dependent upon oxygen in $T.$ $concretivorus$, $T.$ $thio-$
$parus$ and $T.$ $thiooxidans$ and upon nitrite in $T.$ $denitrificans$.

$$H_2S \xrightarrow{\text{sulfide oxidase}} S_nSH \xrightarrow{\text{polysulfide oxidase}} XS_nSX$$

$$
\begin{array}{cc}
\downarrow & \downarrow \\
(Cu) & (flavin?) \\
\downarrow & \downarrow \\
(flavin?) & ubiquinone \\
\downarrow & \downarrow \\
cyt.\ b & cyt.\ b \\
\downarrow & \downarrow \\
cyt.\ c & cyt.\ c \\
\downarrow & \downarrow \\
\text{cyt. } d & \text{cyt. } a_1 \\
\end{array}
$$

$NO_2 \rightarrow N_2 + NO + N_2O \qquad O_2 \rightarrow H_2O \qquad (NO_3^-?) \rightarrow (?)$

Two distincly different mechanisms, which may not be mutually exclusive have been proposed for the oxidation of thiosulfate
based primarily on the existance of two enzymes, rhodanese and
the thiosulfate oxidizing enzyme. Rhodanese has been partially
purified from $T.$ $denitrificans$ (48), $T.$ $ferrooxidans$ (49) and
$Thiobacillus$ $A2$ (50), and has been shown to exist in $T.$ $novellus$
(40) and $T.$ $neapolitanus$ (51). This enzyme has been shown to
catalyze the scission of thiosulfate to sulfite and a sulfide
moiety bound to either cyanide, dihydrolipoate or dihydrolipoam-
ide (50,52).

$$S\text{-}SO_3^{2-} + SCN^- \dashrightarrow SO_3^{2-} + SCN^-$$

$$S\text{-}SO_3^{2-} + \text{dihydrolipoate} \rightarrow SO_3^{2-} + \text{dihydrolipoate persulfide}$$

$$S\text{-}SO_3^{2-} + \text{dihydrolipoamide} \rightarrow$$
$$SO_3^{2-} + \text{dihydrolipoamide persulfide}$$

An enzyme with similar function, thiosulfate reductase using reduced glutathione, has been isolated from *T. thioparus* (53), and catalyzes the following reaction:

$$S\text{-}SO_3^{2-} + 2GSH \dashrightarrow SO_3^{2-} + H_2S + GSSG$$

The thiosulfate-oxidizing enzyme, which catalyzes the oxidative condensation of two molecules of thiosulfate to form tetrathionate:

$$2H_2S_2O_3 \dashrightarrow H_2S_4O_6 + 2H^+$$

has been partially purified and characterized from *T. neapolitanus* (54), *T. novellus* (55), *T. ferrooxidans* (13) and *T. thioparus* (56). Ferricyanide and cytochrome *c* have been found to act as electron acceptors in this reaction. Based on the detection of polythionates in the medium of thiobacilli growing on thiosulfate (57), tetrathionate has been postulated to be metabolized as follows (58,13):

$$S_4O_6^{2-} \dashrightarrow S_3O_6^{2-} \dashrightarrow S_2O_3^{2-}$$
$$\qquad\quad \downarrow SO_3^{2-} \qquad\qquad\quad \downarrow SO_3^{2-}$$

Sulfite is the central intermediate in sulfur oxidation; it is through this compound that all pathways pass in the formation of sulfate and ATP. There are two principal methods for the oxidation of sulfite in the thiobacilli; via adenosine phosphosulfate (57) or by a cytochrome *c*-mediated oxidation (38). The first of these pathways:

$$SO_3^{2-} + AMP \xrightarrow{\text{APS reductase}} APS$$

$$APS \xrightarrow{\text{ADP sulfurylase}} SO_4^{2-} + ADP$$

$$ADP \xrightarrow{\text{adenylate kinase}} \tfrac{1}{2}AMP + \tfrac{1}{2}ATP$$

contains three reactions: these are catalyzed by APS reductase, causing the formation of APS from AMP and sulfite, ADP sulfurylase, which replaces the sulfate moiety by a phosphoric acid group resulting in the formation of ADP and the liberation of sulfate, and adenylate kinase, which transfers a phosphoric acid group from one ADP molecule to another, resulting in the formation of one ATP and one AMP. That this pathway is important in at least some of the thiobacilli is emphasized by the fact that in *T. thioparus*, APS reductase comprises 3 to 4% of the cellular protein (59), and 4 to 5% in *T. denitrificans* (60). It is only from these two thiobacilli that this enzyme has been purified, although it has been reported to exist in *T. thiooxidans* (61), and *T. denitrificans* (47).

A cytochrome-mediated AMP-independent sulfite oxidizing system,

$$SO_3^{2-} + \text{cytochrome } c$$
$$\text{oxidoreductase} \qquad ADP + P_i \rightarrow ATP$$

$$SO_3^{2-} \diagdown \text{cytochrome } c \text{ (ox.)} \diagup\diagdown \text{cytochrome } a \text{ (red.)} \diagdown O_2$$
$$SO_4^{2-} \diagup \text{cytochrome } c \text{ (red.)} \diagdown\diagup \text{cytochrome } a \text{ (ox.)} \diagup \tfrac{1}{2}H_2O$$

has been studied in *T. novellus* (38), *T. thioparus* (62), *T. thiooxidans* (63), *T. ferrooxidans* (64), and *Thiobacillus A2* (Silver, unpublished observations). A sulfite oxidizing enzyme has been isolated from *T. neapolitanus* in which cytochromes are not involved in the oxidation (65); sulfite is oxidized directly by oxygen. Although this oxidation is not dependent on AMP, it is stimulated by this compound, suggesting the formation of an enzyme-sulfite intermediate which might react with AMP to form APS or with water to form sulfite.

IV. FIXATION OF CARBON DIOXIDE

As with other autotrophic bacteria (66), the iron-oxidizing thiobacilli use the reductive pentose phosphate (Calvin-Benson) cycle and the carboxylation of phospho-enol-pyruvate for the assimilation of CO_2 (67,68) with ferrous iron serving as the sole energy source. Elemental sulfur, reduced inorganic sulfur compounds and metal sulfide minerals (12,14,17) could also support CO_2 fixation.

Each molecule of CO_2 converted to organic carbon by the Calvin-Benson cycle requires three molecules of ATP and two molecules of reduced pyridine nucleotide. The ATP is derived directly from the oxidation of the growth substrates, whereas pyridine nucleotides are reduced by the reversal of the electron transport chain. That sulfur compounds and ferrous iron can be used for this process is documented by Aleem (69) and Aleem, Lees and Nicholas (70) respectively.

V. REFERENCES

1. Nathansohn A., *Mitt. Zool. Sta. Neapel.*, 15, 655 (1902).

2. Temple, K.L., and Colmer, A.R., *J. Bacteriol.*, 62, 605 (1951).

3. Leathen, W.W., Kinsel, N.A., and Braley, S.A., *J. Bacteriol.*, 72, 700 (1956).

4. Kinsel, N.A., *J. Bacteriol.*, 80, 628 (1960).

5. Silverman, M.P., and Lundgren, D.G., *J. Bacteriol.*, 77, 642 (1959).

6. Silverman, M.P., and Lundgren, D.G., *J. Bacteriol.*, 78, 326 (1959).

7. Beck, J.V., *J. Bacteriol.*, 79, 502 (1960).

8. Unz, R.F., and Lundgren, D.G., *Soil Sci.*, 92, 302 (1961).

9. Ivanov, V.I., and Lyalikova, N.N., *Mikrobiologiya*, 31, 468 (1962).

10. Hutchinson, M., Johnstone, K.I., and White, D., *J. Gen. Microbiol.*, 44, 373 (1966).

11. Hutchinson, M., Johnstone, K.I., and White, D., *J. Gen. Microbiol.*, 57, 397 (1969).

12. Margalith, P., Silver, M., and Lundgren, D.G., *J. Bacteriol.*, 92, 1706 (1966).

13. Silver, M., and Lundgren, D.G., *Can. J. Biochem.*, 46, 1215 (1968).

14. Silver, M., *Can. J. Microbiol.*, 16, 845 (1970).

15. Kelly, D.P., and Tuovinen, O.H., *Int. J. Syst. Bacteriol.*, 22, 170 (1972).

16. Bounds, H.C., and Colmer, A.R., *Can. J. Microbiol.*, 18, 735 (1972).

17. Silver, M., and Torma, A.E., *Can. J. Microbiol.*, 20, 141 (1974).

18. Guay, R., Silver, M., and Torma, A.E., *IRCS Med. Sci.: Biochem. Microbiol. Parasitol. Inf. Dis.*, 3, 417 (1975).

19. Guay, R., Silver, M., and Torma, A.E., *Rev. Can. Biol.*, 35, 61 (1976).

20. Shafia, F., and Wilkinson, R.F., *J. Bacteriol.*, 97, 256 (1969).

21. Tabita, F.R., and Lundgren, D.G., *J. Bacteriol.*, 108, 328 (1970).

22. Lundgren, D.G., Vestal, J.R., and Tabita, F.R., in "Water Pollution Microbiology", (R. Mitchell, Ed.) p. 69. Wiley-Interscience, New York, 1972.

23. Tuovinen, O.H., and Kelly, D.P., *Z. Allg. Mikrobiol.*, 12, 311 (1972).

24. Guay, R., and Silver, M., *Can. J. Microbiol.*, 21, 281 (1975).

25. Markosyan, G.E., *Dokl. Akad. Nayuk SSSR*, 211, 1205 (1973).

26. Tuovinen, O.H., Kelley, B.C., and Nicholas, D.J.D., *Can. J. J. Microbiol.*, 22, 109 (1976).

27. Lees, H., Kwok, S.C., and Suzuki, I., *Can. J. Microbiol.*, 15, 43 (1969).

28. Blaylock, B.A., and Nason, A., *J. Biol. Chem.*, 238 3453 (1963).

29. Dugan, P.R., and Lundgren, D.G., *J. Bacteriol.*, 89, 825 (1965).

30. Vernon, L.P., Mangum, J.H., Beck, J.V., and Shafia, F.M., *Arch. Biochem. Biophys.*, 88, 227 (1960).

31. Din, G.A., Suzuki, I., and Lees, H., *Can. J. Biochem.*, 45 1523 (1967).

32. Dugan, P.R., and Lundgren, D.G., *Anal. Biochem.*, 8, 312 (1964).

33. Lazaroff, N., *J. Bacteriol.*, 85, 78 (1963).

34. Yates, M.G., and Nason, A., *J. Biol. Chem.*, 241, 4872 (1966).

35. Din, G.A., and Suzuki, I., *Can. J. Biochem.*, 45, 1547 (1967).

36. Suzuki, I., and Silver, M., *Biochem. Biophys. Acta*, 122, 22 (1966).

37. Silver, M., and Lundgren, D.G., *Can. J. Biochem.*, 46, 457 (1968).

38. Charles, A.M., and Suzuki, I., *Biochem. Biophys. Acta*, 128, 522 (1966).

39. Suzuki, I., *Biochim. Biophys. Acta*, 104, 359 (1965).

40. Charles, A.M., and Suzuki, I., *Biochim. Biophys. Acta*, 128, 510 (1966).

41. Taylor, B.F., *Biochim. Biophys. Acta*, 170, 112 (1968).

42. Takakuwa, S., *Plant and Cell Physiol.*, 16, 1027 (1975).

43. Karavaiko, G.I., Lyalikova, N.N., and Pivovarova, in "Ecological and Geochemical Activity of Microorganisms", (M.V. Ivanov, Ed.) p. 25. Science Centre of the Biological Institute, USSR, 1976.

44. Moriarty, D.J.W., and Nicholas, D.J.D., *Biochem. Biophys. Acta*, 184, 114(1969).

45. Moriarty, D.J.W., and Nicholas, D.J.D., *Biochim. Biophys. Acta*, 197, 143 (1970).

46. Moriarty, D.J.W., and Nicholas, D.J.D., *Biochim. Biophys. Acta*, 216, 130 (1970).

47. Aminuddin, M., and Nicholas, D.J.D., *Biochim. Biophys. Acta*, 325, 81 (1973).

48. Bowen, T.J., Butler, P.J., and Happold, F.C., *Biochem. J.*, 97, 651 (1965).

49. Tabita, R., Silver, M., and Lundgren, D.G., *Can. J. Biochem.*, 47, 1141 (1969).

50. Silver, M., and Kelly, D.P., *J. Gen. Microbiol.*, 97, 277 (1976).

51. Kelly, D.P., and Tuovinen, O.H., *Plant and Soil*, 43, 77 (1975).

52. Silver, M., Howarth, O.W., and Kelly, D.P., *J. Gen. Microbiol.*, 97, 285 (1976).

53. Peck, H.D., and Fisher, E., *J. Biol. Chem.*, 237, 190 (1962).

54. Trudinger, P.A., *Biochem. J.*, 78, 680 (1961).

55. Aleem, M.I.H., *J. Bacteriol.*, 90, 95 (1965).

56. Lyric, R.M., and Suzuki, I., *Can. J. Biochem.*, 48, 355 (1970).

57. Roy, A.B., and Trudinger, P.A., "The Biochemistry of Inorganic Compounds of Sulfur", p. 207. Cambridge University Press, London, 1970.

58. Sinha, D.B., and Walden, C.C., *Can. J. Microbil.*, 12, 1041 (1966).

59. Lyric, R.M., and Suzuki, I., *Can. J. Biochem.*, 48, 344 (1970).

60. Bowen, T.J., Happold, F.C., and Taylor, B.F., *Biochim. Biophys. Acta*, 118, 566 (1966).

61. Peck, H.D., *J. Bacteriol.*, 82, 933 (1961).

62. Lyric, R.M., and Suzuki, I., *Can. J. Biochem.*, 48, 334 (1970).

63. Kodama, A., Kodama, T., and Mori, T., *Plant and Cell Physiol.* 11, 701 (1970).

64. Vestal, J.R., and Lundgren, D.G., *Can. J. Biochem.*, 49, 1125 (1971).

65. Hempfling, W.P., Trudinger, P.A., and Vishniac, W., *Arch. Mikrobiol.*, 59, 149 (1967).

66. Kelly, D.P., *Ann. Rev. Microbiol.*, 25, 177 (1971).

67. Maciag, W.J., and Lundgren, D.G., *Biochem. Biophys. Res. Commun.*, 17, 603 (1964).

68. Din, G.A., Suzuki, I., and Lees, H., *Can. J. Microbiol.*, 13, 1413 (1967).

69. Aleem, M.I.H., *Symp. Soc. Gen. Microgiol.*, 27, 351 (1977).

70. Aleem, M.I.H., Lees, H., and Nicholas, D.J.D., *Nature*, 200 759 (1963).

71. Silver, M., Friedman, S., Guay, R., Couture, J., and Tanguay, R., *J. Bacteriol.*, 107, 368 (1971).

FACTORS AFFECTING METABOLISM AND FERROUS IRON
OXIDATION IN SUSPENSIONS AND BATCH CULTURES OF
THIOBACILLUS FERROOXIDANS: RELEVANCE TO FERRIC IRON
LEACH SOLUTION REGENERATION

D. P. Kelly

University of Warwick, Coventry, England

C. A. Jones

Glaxo Research Ltd., Stoke Poges, Buck, England

Ferric iron in acid solution is the essential reagent for the bacterially-assisted solubilization of uranium from ores such as uraninite, and is also significant in the dissolution of metal sulphides. The regeneration of ferric iron by bacterial reoxidation of ferrous sulphate is an essential step in the recycling of leach liquors in heap or other percolation leach systems. Consequently it is of use to know the factors likely to affect bacterial development and activity. The principal factor limiting growth in batch culture of Thiobacillus ferrooxidans *on ferrous iron is ferric iron product inhibition of the rate of ferrous iron oxidation. Ferric iron inhibits oxidation competitively, lowering substrate saturation coefficient (K_s) without affecting maximum attainable specific growth rate (μ_{max}). In normal batch culture, μ_{max} was $0.143h^{-1}$, minimum observed K_s was 36mM ferrous sulphate,* *and growth yield coefficient was 0.35g dry wt/g atom iron oxidized. Growth rates were exponential, measured as iron oxidation and fixation of labelled carbon dioxide. Growth rate was not limited by carbon dioxide concentration, but exhaustion of carbon dioxide in sealed systems caused a switch from exponential to linear iron oxidation, indicating growth-uncoupled oxidation of the residual iron by the static population. High potassium concentrations partially relieved ferric iron inhibition. With non-growing cell suspensions, oxygen electrode methods indicated multiple K_m values for ferrous iron oxidation: 0.7mM $FeSO_4$ between 0.4-43mM; 20-123mM between 100-400mM; and unmeasurably high between 400-700mM. Above 0.7M ferrous sulphate became an inhibitory substrate. Increased H^+ or uranyl sulphate inhibited non-competitively. Efficiency of coupling of carbon dioxide fixation to Fe oxidation was decreased at low (25mM) and high (500mM) ferrous sulphate concentrations and was maximal in the 100-300mM range. Ferric sulphate did not inhibit carbon dioxide fixation and so affected metabolism and growth only by its effect on ferrous iron oxidation.*

I.   INTRODUCTION

Thiobacillus ferrooxidans is one of the most important organisms in the leaching of metal sulphides (1,2), although recent work indicates that it is not uniquely responsible for this phenomenon and its activity may be enhanced by the presence of other organisms (3-5; Norris and Kelly, this volume; Tuovinen, et al., this volume). In many leaching systems, the oxidation of pyrite or of ferrous iron plays a central role, in which ferric iron oxidizes and consequently solubilizes other metals, such as uranium from its oxides (2,6-10). The regeneration of soluble ferric iron from ferrous sulphate is also important in any economic process with liquor recycling (1,11,12). Consequently it is of some importance to understand the physiology of *T. ferrooxidans* and in particular to be able to define the factors limiting the rate and extent of its development. Among factors limiting development could be nutrient supply (especially carbon dioxide, which is the organism's main carbon source), availability of oxygen and oxidizable iron, possible toxicity of dissolved metals and the innate capacities of the bacteria (i.e. maximum possible growth and oxidation rates and biomass yields). In this paper we are concerned with the limitations placed on growth and iron oxidation by the availability of carbon dioxide and the concentrations of ferrous and ferric ions to which the bacteria are exposed in batch culture or non-growing suspensions.

Previously (13) we have demonstrated that *Thiobacillus ferrooxidans* in continuous chemostat culture is capable of specific growth rates (μ) on ferrous iron at pH 1.6 of at least μ = 1.33 h^{-1} with a possible μ_{max} = 1.78 h^{-1}. In chemostat culture its true substrate saturation constant (K_s) for $FeSO_4$ was as low as 0.7mM, and its true growth yield was 1.33g dry wt/g atom Fe^{2+} oxidized (13,14). Such figures are in marked contrast to previously published batch culture values for μ_{max} of 0.145 and 0.2 at pH values above pH 1.9 or μ of 0.109 at pH 1.3, and to K_s values between 7.2 and 35.9mM (15-17). Apparent Michaelis constants (K_m) for $FeSO_4$ oxidation by cell suspensions are reported to lie in the range 0.568-9.4mM (18-22); the K_m for a partially separated iron oxidase was 0.59mM (23). Growth yields (Y) in batch culture are also below those possible in the chemostat and fall in the range 0.2-0.5g dry wt/g atom Fe^{2+} oxidized. The most reliable figures available to date for batch culture Y seem to be 0.35 and 0.392 (17; A. Caglar, personal communications, 1974). Limitation of growth rate or efficiency have in the past variously been attributed to limitation of available oxygen or carbon dioxide: a view we contend is not universally tenable. Our chemostat data indicate the usual values for batch cultures to be gross underestimates of the maximum potential of *T. ferrooxidans*, but also showed growth-linked iron oxidation to be subject in different steady state experiments to competitive or non-competitive inhibition by the ferric iron produced by the chemostat culture (13). The former state was associated with a low yield of 0.36-0.38 (13,24), possibly indicating a physiological similarity between batch cultures and chemostat steady states subject to predominantly competitive product inhibition. A consequence of competitive inhibition of iron oxidation by ferric ions would be an apparently poor affinity for ferrous iron, reflected as a low growth rate and high effective K_s.

These experiments provide data for an interpretation of the relatively poor values commonly found for μ, K_s and Y; and set out to establish (a) whether competitive inhibition by ferric iron is found during iron oxidation by batch cultures and suspensions; (b) whether K_s values deduced from the chemostat are valid for suspensions oxidizing iron in the virtual absence of ferric ion product; (c) whether ferrous iron at high concentrations is inhibitory to its own oxidation; (d) whether ferric ions or high levels of ferrous ions can uncouple carbon dioxide fixation from iron oxidation.

II. EXPERIMENTAL METHODS

A. Organism, Cultural Conditions and Analytical Procedures

The strain of *Thiobacillus ferrooxidans* used and culture methods are described in detail elsewhere (17, 24). All cultures were grown at 30°C in media adjusted to pH 1.6 with H_2SO_4. For use in manometry or oxygen electrode cell experiments, organisms were harvested from chemostat cultures or from 1 or 2ℓ of batch cultures grown in 1ℓ amounts shaken in 2ℓ Erlenmeyer flasks until near the end of the growth phase. Batch culture medium contained (g/ℓ): K_2HPO_4, 0.4; $(NH_4)_2SO_4$, 0.4; $MgSO_4.7H_2O$, 0.4; $FeSO_4$, 27.8 or as indicated. For chemostat culture, and batch experiments in which the effect of potassium on iron oxidation was tested, the medium contained (g/ℓ): KH_2PO_4, 0.054; $(NH_4)_2SO_4$, 0.36; $MgSO_4.7H_2O$, 0.015; and $FeSO_4.7H_2O$ as required. Harvested organisms were washed with water at pH 1.6 and suspended in pH 1.6 water or chemostat medium lacking iron. Bacterial protein in suspensions or cultures was determined after collecting organisms on membrane filters and dissolving them in 0.5N NaOH (24). Ferrous iron oxidation in cultures was monitored by titration with ceric sulphate.

1. *Fixation of ^{14}C-Carbon Dioxide*

   a. *Growing cultures.* Culture medium (50 ml) in 250 ml flasks sealed with Suba-seals received $NaH^{14}CO_3$ or CO_2 by injection and was shaken at 30°C for one hour before injecting *T. ferrooxidans.* Duplicate samples (1 ml) were taken with sterile syringes for iron estimation; further samples were filtered through 25 mm Millipore membranes (0.45 μm pore size), washed with 10 ml 0.01N H_2SO_4 and 2 x 10 ml distilled water, then dried in 20 ml scintillation vials over silica gel.

   b. *Suspensions.* Double-armed Warburg flasks (25 ml) were used with 2.0-3.5 ml liquid volumes at pH 1.6. $NaH^{14}CO_3$ (30 mM, 10μc/ml) was tipped into the main chamber, which contained $FeSO_4$ (and ferric sulphate in some cases) in pH 1.6 H_2SO_4, and CO_2 release measured for 10-15 min before tipping a suspension of *T. ferrooxidans* from the second side bulb. After measurement of oxygen consumption, samples (1 ml) were removed to measure ^{14}C fixed into filtered bacteria as described above. To measure total $^{14}CO_2$ fixed, samples (1 ml) were mixed with 1 ml 10% acetic acid in ethanol. Samples of this were dried on Whatman GF/A fibre glass discs in scintillation vials or the whole samples were diluted to 10 ml with water and 1 ml aliquots counted for ^{14}C.

   c. *Measurement of ^{14}C.* Membrane filters and glass fibre discs were immersed in 5 ml 0.5% (v/v) butyl-PBD in toluene. Aqueous samples (1 ml) were mixed with 15 ml of a mixture of 8 g butyl-PBD, 500 ml Triton X-100 and 1000 ml toluene. Vials were counted in a Philips Liquid Scintillation Analyzer programmed to compute absolute dpm rates for ^{14}C using predetermined channels ratio equations.

2. *Removal of Carbon Dioxide from the Medium in "CO_2-free"*
   Growth Experiments

Mineral salts medium (minus $FeSO_4$) with a surplus of 10 ml
water was boiled in the culture flasks until the volume was
reduced by 10 ml. $FeSO_4$ solution, similarly boiled, was added
hot to the salts solution and the flasks sealed with rubber
stoppers having inlet and outlet tubes. The medium was allowed
to cool in a stream of air freed of CO_2 by passage through
2N NaOH and saturated $Ba(OH)_2$. Outgoing air passed through a
further NaOH trap. The tubes were then sealed with vaccine caps.
Some cultures were also grown in sealed "Katz flasks" (300 ml)
with centre wells containing a roll of Whatman No. 40 filter
paper and 1 mℓ 40% (w/v) KOH.

3. *Measurement of $FeSO_4$ Oxidation by Suspensions of* Thiobacillus
   ferrooxidans

   a. *Standard Warburg manometry.* Flasks (20 ml) normally
contained 2 ml liquid at pH 1.6, 0.5-0.8 mg *T. ferrooxidans*
protein, and $FeSO_4$ (tipped from a side bulb) between 10-800 mM.
Flasks were air-filled and shaken at $30^{\circ}C$ at 150 ± 10 strokes/
min.

   b. *Oxygen electrodes.* Water-jacketed perspex cells (Rank
Bros, Bottisham, Cambs) with Clark oxygen electrodes were used
at $30^{\circ}C$ with remote magnetic stirring of 2 ml or 3 ml reaction
solutions. Oxygen consumption in air-saturated solution pH 1.6
was recorded on chart recorders (Kipp and Zonen, Switzerland) and
the systems calibrated to measure dissolved oxygen in nmoles/ml.
Generally, *T. ferrooxidans* suspensions (25-300 µg protein in
0.1-0.2 ml) were injected into appropriate pH 1.6 $FeSO_4$ solutions.
No endogenous oxygen consumption by the organisms was detectable.
Oxygen consumption due to non biological autooxidation of ferrous
iron was only significant (10-15% of the total rate) with very
high concentrations, and, where necessary, corrections for

autooxidation were always made. For inhibition experiments, organisms were injected into ferrous-ferric mixtures. Appropriate control tests showed that only biological reactions were recorded in these experiments, and there was no interference from chemical interaction of the iron species or reaction with the oxygen electrode system.

III. RESULTS AND DISCUSSION

A. Effect of Carbon Dioxide Availability on the Growth of Batch Cultures

Batch cultures in unsealed flasks effected exponential oxidation of $FeSO_4$ (Fig. 1) with a specific growth rate (μ_{Fe}) of 0.11-0.12 h^{-1}. Using sealed flasks containing normal air, initially exponential $FeSO_4$ oxidation became linear after 20-30% of the iron was consumed (Fig. 1). Removal of CO_2 into KOH in sealed flasks containing CO_2-free media resulted in entirely linear oxidation kinetics, indicating that no *growth* of organisms was occurring (Fig. 1). Autooxidation in sterile controls was

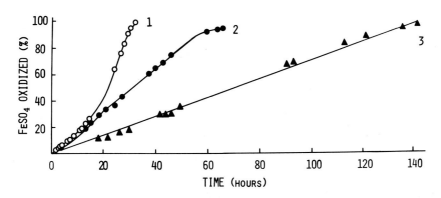

Fig. 1. Effect of CO_2-limitation on kinetics of $FeSO_4$ oxidation by Thiobacillus ferrooxidans cultures (50 ml). Curve 1: unsealed flask; Curve 2: sealed flask with 260 ml air; Curve 3: sealed Katz flask, CO_2 removed by KOH. Initial $FeSO_4$ concentration, 82-88mM.

These results are consistent with exponential $FeSO_4$ oxidation being a measure of increase in bacterial biomass and numbers, while linear oxidation is due to growth-uncoupled-oxidation by a standing population of bacteria. Our sealed flasks contained a free air space of 260 ml and 50 ml of medium. From isotope dilution data in our $^{14}CO_2$ experiments we calculated the air space to contain 0.036% (v/v) CO_2. Consequently, while the flasks contained ample oxygen (about 52 ml) to allow complete $FeSO_4$ oxidation in our experiments, they contained only enough CO_2 (about 0.1 ml) to allow growth of about 120 µg dry wt of new organisms.

The absolute correlation of exponential growth and CO_2-fixation was proved in cultures in sealed flasks, supplemented with $^{14}CO_2$. Exponential iron oxidation became linear after 30% of the available $FeSO_4$ was oxidized. $^{14}CO_2$-fixation was exactly proportional to exponential iron oxidation and $^{14}CO_2$ was all consumed at exactly the same time as the abrupt change to linear oxidation rate. This indicated that rate of growth and efficiency of coupling of CO_2-fixation to iron oxidation did not decline until carbon dioxide was essentially exhausted.

Comparison was made of the exponential rate of iron oxidation and $^{14}CO_2$-fixation with limiting and non-limiting supplies of CO_2, and of growth yield with excess CO_2 available. Specific growth rates (measured as μ_{Fe} or μ_{CO_2}) were essentially identical (Table 1); exponential growth rate was essentially unaffected by CO_2 concentration and $^{14}CO_2$-fixation was proportional to iron oxidation at all concentrations. Viable numbers of organisms produced were governed solely by the quantity of iron available in CO_2-unlimited cultures and by the amount of CO_2 available in CO_2-limited cultures. The growth yield coefficient was independent of absolute CO_2 concentration over the range 0.036% to 8% v/v (Table 2).

TABLE I   Effect of Carbon Dioxide Supply on Growth Rates of Batch Cultures on 84mM $FeSO_4$ in Sealed and Unsealed Flasks.

Volume of free air space and CO_2 supplied	μ_{Fe}[a] (h^{-1})	μ_{CO_2}[a] (h^{-1})	Increase in Viable count at 100% $FeSO_4$ oxidation (% of control)[b]
Experiment 1			
Unsealed (control)	0.110	–	100
260 ml + KOH	Linear	–	1
260 ml	0.107	–	33[c]
470 ml	0.107	–	72[d]
Experiment 2			
260 ml + 8 μl $^{14}CO_2$	0.071	0.069	32
260 ml + 0.5 ml $^{14}CO_2$	0.067	0.067	92
260 ml + 1.5 ml $^{14}CO_2$	0.069	0.069	100
Unsealed (control)	–	–	100

[a] μ_{Fe} and μ_{CO_2} are specific growth rates calculated from exponential rates of $FeSO_4$ oxidation and $^{14}CO_2$ fixation as 0.693/ doubling time (h).

[b] Control viable counts were 2.48 x 10^8 and 2.39 x 10^8 viable organisms/ml determined by colony formation on membrane filters in 20 days (17). Initial count in all flasks, 2.5 x 10^7/ml.

[c,d] $FeSO_4$ oxidation rate became linear after oxidation of 30% and 70% respectively.

Since specific growth rate was essentially constant at 0.11-0.12 h^{-1} for $FeSO_4$ concentrations of 77-165 mM, it is reasonable to assume that this represents the μ_{max} possible for batch culture at pH 1.6 and 30^0C, with a mean growth yield coefficient (on 82 mM $FeSO_4$) of 0.35 g dry wt/ g atom Fe, ranging from 0.33-0.39 and equal to about 3 x 10^{12} viable organisms.

TABLE II Effect of carbon dioxide concentration on μ_{Fe} and growth yield of 50 ml cultures on 84–88mM $FeSO_4$ in sealed flasks with a 260 ml air space. Biomass was determined as protein and dry weight calculated as protein $\times$ 1.67, after correcting for initial 2.5 µg dry wt/ml.

Carbon dioxide supplied (ml)	μ_{Fe}	Final Biomass (µg/ml)	Yield (mg dry wt/g mole $FeSO_4$)
Unsealed control	0.113	29.4	350
1.5	0.105	29.0	333
5	0.112	31.0	354
10	0.115	31.0	352
20	0.124	33.0	386

B. Effect of $FeSO_4$ Supply on Batch Growth Kinetics and Inhibition by Ferric Iron

Specific growth rates ($\mu = \mu_{Fe}$) increased in response to initial $FeSO_4$ concentration at least over the range 10-60 mM and a μ_{max} of 0.143 h^{-1} was indicated (Fig. 2). This result suggests a high K_s for $FeSO_4$, but as accurate measurement of exponential iron oxidation rate was possible only after about 15% of the $FeSO_4$ had been oxidized, significant Fe^{3+} was always present. That the apparently poor K_s for $FeSO_4$ was due to competitive inhibition by Fe^{3+} was shown by adding ferric sulphate to batch cultures with different $FeSO_4$ concentrations. Data analysis by the method of Lineweaver and Burk (25) plotting $1/\mu \times 1/FeSO_4$ for exponential rates with and without 5, 10 and 57 mM Fe^{3+} showed competitive inhibition of μ_{Fe} (Fig. 2) and an increase of K_s from 36 mM $FeSO_4$ for the control (5-8 mM Fe^{3+}) to 67 mM with 57 mM Fe^{3+}. Added ferric ions did not alter μ_{max}, indicating a purely competitive effect on iron oxidation (Fig. 2)

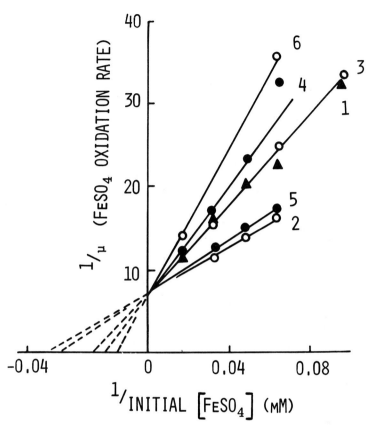

Fig. 2. Lineweaver-Burk analysis of exponential $FeSO_4$ oxidation in cultures with different initial $FeSO_4$ levels between 10-60 mM. μ_{Fe} was measured over periods where the amount of ferric iron produced was equivalent in all flasks. Flasks contained the following supplements: Curve 1: control series; Curve 2: 25mM K_2SO_4; Curve 3: 5mM Fe^{3+}; Curve 4: 10mM Fe^{3+}; Curve 5: 10mM Fe^{3+}, 25mM K_2SO_4; Curve 6: 57mM Fe^{3+}. Cultures in curves 1, 3 and 5 contained 0.37 mM potassium ion.

C. Effect of Potassium Ion on Iron Oxidation Kinetics in
    Batch Culture

Growth on 30mM $FeSO_4$ in the low-potassium medium (0.37mM K^+) was stimulated from μ_{Fe} 0.114 to 0.138 and 0.15 h^{-1} respectively by raising K^+ (as K_2SO_4) to 7.47 and 50.87mM. High K^+ partially alleviated the competitive inhibition by ferric ions, resulting in a lessened effect by added ferric sulphate and lowered the control K_s to 19.6mM $FeSO_4$ in the absence of added ferric ion. K_s with 10mM added Fe^{3+} was 46.5mM with 0.37mM K^+ but 22.7mM with 50.87mM K^+ (Fig. 2). The observed μ_{max} was unaffected by K^+ concentration (Fig. 2).

D. Effect of High Ferrous and Ferric Sulphate Levels on Iron
    Oxidation in Batch Culture

Lag before growth was increased and rate of development decreased as initial $FeSO_4$ concentration was increased from 62 to 249mM (Fig. 3).

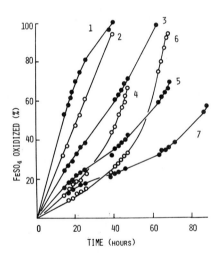

Fig. 3. Effect of high ferrous and ferric sulphate levels on $FeSO_4$ oxidation in growing cultures. Initial concentrations of ferrous + ferric ion (mM) were: Curve 1: 62 + 0; Curve 2: 59 + 120; Curve 3: 132; Curve 4: 126 + 120; Curve 5: 187; Curve 6: 184 + 120; Curve 7: 249.

Addition of ferric ions depressed growth further but subsequently appeared stimulatory at high ferrous concentrations (Fig. 3). These results indicate that $FeSO_4$ could be an inhibitory substrate at high concentration.

E. Kinetics of Iron Oxidation by Suspensions of *Thiobacillus ferrooxidans*

1. *Warburg Manometry*

The rate of $FeSO_4$ oxidation increased up to about 100mM $FeSO_4$ but then decreased at higher concentrations, indicating substrate inhibition (Table 3).

TABLE III   Kinetics of $FeSO_4$ oxidation in Warburg flasks. Oxygen uptake rates were estimated 5-30 min after tipping $FeSO_4$ to T. ferrooxidans at pH 1.7.   Constants were calculated from Lineweaver and Burk (25) plots.

s $FeSO_4$ concentration (mM)	v O_2 uptake rate $(\mu l/h)$	$\dfrac{1}{s}$ (x 10^4)	$\dfrac{1}{v}$ (x 10^5)
8	391	1250	256
10	432	1000	232
16	536	625	187
24	611	417	164
25	588	400	170
32	657	313	152
40	688	250	145
50	708	200	141
60	735	167	136
80	761	125	131
100	800	100	125
250	680	40	147
500	590	20	170
800	280	12.5	357

V_{max} = 850 $\mu l/hour$

K_m = 9.7mM $FeSO_4$

The apparent K_m for $FeSO_4$ was 9.7mM, but Warburg manometry is of questionable accuracy in measuring rates at very low $FeSO_4$ concentrations and in all cases rates were inevitably determined over a time period during which significant ferric iron had accumulated. Consequently the K_m value was probably raised by ferric inhibition. In all cases the oxygen consumption supported the equation for normal iron oxidation:

$$4FeSO_4 + O_2 + 2H_2SO_4 = Fe_2(SO_4)_3 + 2H_2O$$

2. *Oxygen Electrode Experiments*

a. *Normal kinetics.* Using $FeSO_4$ concentrations between 0.4-900mM, oxygen consumption always showed an acceleration phase of 1-5 minutes before measurement of maximum oxidation rate was possible. The presence of inhibitory amounts of ferric iron produced longer acceleration phases before a constant rate was established. It was found that organisms harvested from late exponential phase batch cultures increased 1.5-2.5 fold in specific rate of $FeSO_4$ oxidation if stored for 1-3 days at 20°C in water at pH 1.6 before use. Activity was retained, though progressively declined, in organisms kept at 20°C for 1-3 weeks. Storage at 4°C resulted in considerable extension of the acceleration phase in oxidation, although maximum rates were relatively unaltered.

Numerous experiments using 0.4-100mM $FeSO_4$ indicated K_m for $FeSO_4$ to lie with high reproducibility between 0.43-0.9mM (Fig. 4a, b), although extreme values of 1.58, 2.0 and 3.26 were obtained in individual experiments. Recently a critical study has been made of the various graphical calculation methods of deriving K_m values from kinetic data (26). While we have presented data in Fig. 4 by the method of Lineweaver and Burk (25) we have used a variety of other procedures such as $^s/_v$ versus s or v versus $^v/_s$ plots (27) and a computer programme on a CDC 6600 computer (University of London) designed to test the fit of our data to the Michaelis-Menten equation. Data from six separate experiments with different cultures of bacteria using 0.4-43mM $FeSO_4$ indicated a K_m of 0.70 ± 0.14mM $FeSO_4$. It was noted that at higher $FeSO_4$ concentrations the oxidation rate seemed further accelerated, suggesting a second, higher value for K_m (Fig. 4).

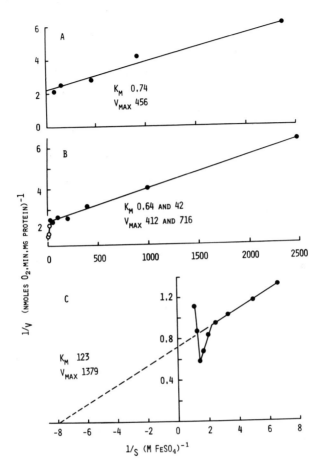

Fig. 4.   Kinetics of FeSO₄ oxidation measured in oxygen electrode cells. A, 98 µg protein/ml; 0.43–11.7mM; B, 29 µg protein/ml; 0.4–140mM, showing biphasic response to concentration; C, 76.4 µg protein/ml; 150–900mM FeSO₄. V$_{max}$ is given as nmoles O₂/min/mg protein; K$_m$ as mM FeSO₄.

b. *Multiple K$_m$ values and substrate inhibition dependent on FeSO₄ concentration.* With FeSO₄ between 100-700mM, oxidation rates accelerated further with increasing substrate concentration; Lineweaver-Burk plots indicated (Fig. 4c) a biphasic relationship of rate and concentration with apparent K$_m$ values of 123mM and infinity for the two quite distinct phases obtained. The same organisms in this experiment gave a K$_m$ of 0.43mM for 1-9mM FeSO₄.

In this experiment, $FeSO_4$ above 700mM produced substrate
inhibition (Fig. 4c). The absolute significance of these higher
K_m values is questionable as we have obtained values of 20-40mM
in some other experiments, calculated by the $^1/_v \times {}^1/_s$ or $^s/_v \times s$
methods.

    c. Effect of ferric ions on $FeSO_4$ oxidation. With organisms
freshly harvested from batch culture, the addition of low
concentrations of ferric sulphate *stimulated* oxygen uptake in the
presence of 5-80mM $FeSO_4$ (Fig. 5), but with increasing Fe^{3+}
concentration, oxidation was progressively inhibited (Fig. 5b).

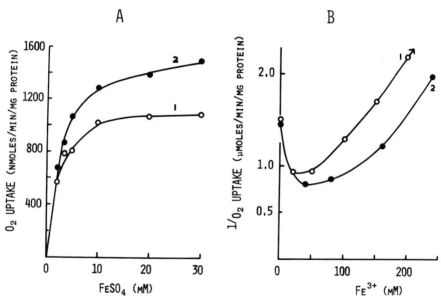

    *Fig. 5. Stimulation of $FeSO_4$ oxidation by low concentrations
of ferric iron using freshly harvested* T. ferrooxidans *A,
Oxidation by 104 µg bacterial protein/ml; Curve 1: control series
Curve 2: with 20mM* Fe^{3+}. *B, Oxidation of 10mM (Curve 1) or 20mM
$FeSO_4$ (Curve 2) by 110 µg bacterial protein/ml with various
amounts of* Fe^{3+}.

Although low Fe^{3+} concentrations stimulated the rate, they
did not affect the acceleration phase in oxidation. Storage of
the organisms at 20°C resulted in increased oxidation rates over
1-2 days, but the stimulation by Fe^{3+} did not increase in

proportion. Thus Q_{O_2} values (nmoles O_2 consumed/min/mg
T. ferrooxidans protein) with 10mM $FeSO_4$ ($\pm$ 20mM Fe^{3+}) were 785
(1009) after harvesting, 1258 (1240) after 1 day at 20°C, and 1642
(1288) after 2 days. Thus as the activity of the organisms
increased, the stimulatory effect of Fe^{3+} was completely lost.
For further study of *inhibitory* effects of Fe^{3+}, organisms
subjected to prior storage at 20°C were used, as K_m values were
not apparently altered during this period. Attempts to employ
freshly harvested organisms to estimate K_i for inhibitory
concentrations of Fe^{3+} by means of the graphical $1/v$ versus
inhibitor concentration method (28) produced useless non-linear
plots (Fig. 5b).

    d. Competitive inhibition of $FeSO_4$ oxidation by Fe^{3+}. At
all inhibitory concentrations tested, using 0.4-140mM $FeSO_4$,
ferric sulphate was demonstrated to act as a purely *competitive*
inhibitor of ferrous sulphate oxidation. The competitive mode
for inhibition was consistently indicated by five graphical
procedures for a number of experiments (Fig. 6-10). Moreover,
computer fitting of experimental inhibition data to the
Michaelis-Menten equation was consistent with no significant
alteration of maximum oxidation rate (V_{max}) but increase of K_m
(i.e. K_p) in response to Fe^{3+} concentration. While K_m in the
absence of Fe^{3+} was typically about 1 mM $FeSO_4$, K_p values
observed with Fe^{3+} ranged from 4.5-63 depending on concentration.
The inhibitor constant (K_i) for Fe^{3+} was estimated by several
procedures (27) and figures ranging from 2.5mM Fe^{3+} to 28mM using
high Fe^{3+} levels were obtained.

    Values of 12mM or 20mM were obtained by the graphical
methods of Dixon (28; Fig. 9) and Hunter and Davis (29; Fig. 10)
respectively. Calculations using the K_m and K_p values (27)
suggest a value for K_i of about 8-10mM Fe^{3+}.

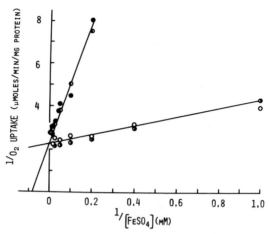

Fig. 6.   *Lineweaver-Burk plot (1/v × 1/s) of oxidation rates for 1-100mM FeSO₄ with (●,◐) and without (○,◑) 100mM Fe³⁺ by T. ferrooxidans (29 μg protein/ml). The lines are computer fits, giving identical Vₘₐₓ intercepts and Kₘ = 1.06, Kₚ = 9.6mM FeSO₄.*

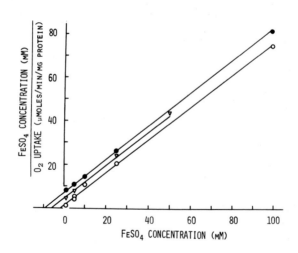

Fig. 7.   *Dixon and Webb plot (s/v × s) for oxidation of 1-100mM FeSO₄ with ferric iron at 0mM (○), 10mM (▽), or 25mM (●).*

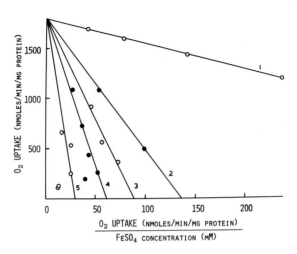

Fig. 8. Dixon and Webb plot (v x v/s) for oxidation of 5-40mM FeSO₄ (54.7 μg bacterial protein) with Fe³⁺ at OmM (1); 10 (2); 15 (3); 20 (4); or 25mM (5).

e. *Effect of pH and uranyl ion on FeSO₄ oxidation.* In contrast to the purely competitive inhibitory effect of ferric ions on $FeSO_4$ oxidation, UO_2^{2+} or increased H^+ (pH 1.2) were clearly non-competitive inhibitors of oxidation, altering V_{max} (control 1183, others 824) without affecting K_m (0.9 mM) for $FeSO_4$ (Fig. 11).

F. Effect of Ferrous and Ferric Iron Concentration on ^{14}C-Carbon Dioxide Fixation

Using bacterial suspensions in Warburg flasks supplied with enough $NaH^{14}CO_3$ to ensure that fixation was independent of carbon dioxide availability or rate of $FeSO_4$ oxidation (Kelly, unpublished data), the amounts of $^{14}CO_2$ fixed were found to be dependent on $FeSO_4$ *concentration* as well as on absolute quantity oxidized. Thus in one experiment the amount of $^{14}CO_2$ (nmoles) fixed per μmole $FeSO_4$ oxidized was 2.24 with 25mM $FeSO_4$,

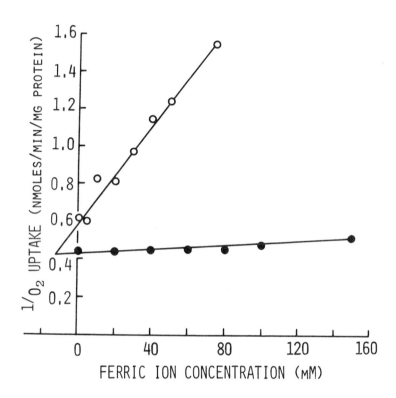

Fig. 9. Estimation of K_i for competitive ferric inhibition of $FeSO_4$ oxidation by the method of Dixon (28). $FeSO_4$ concentrations were 5mM (O) and 100mM (●). K_i is given by the intercept.

3.26 with 150mM and 3.36 with 300mM. Fixation efficiency was more than doubled when 86mM $FeSO_4$ was supplied, compared with 8.6mM (Table 4). Very high concentrations of $FeSO_4$ depressed CO_2 fixation; thus 500mM depressed fixation efficiency to about 70% of that with 100mM (Table 4).

Added ferric iron did not inhibit fixation at the concentrations tested, even at 57mM which depressed $FeSO_4$ oxidation to 45% of the control rate (Table 4). In fact, the *efficiency* of fixation was apparently increased by ferric ions.

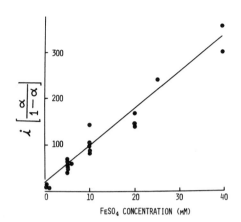

Fig. 10. Estimation of K_i for competitive ferric inhibition of $FeSO_4$ oxidation by the method of Hunter and Downs (29), using data from a number of experiments. i = inhibitor concentration (mM Fe^{3+}); $\alpha = v_i/v$, where v = nmoles O_2 consumed/min/mg bacterial protein in the absence of i, and v_i is v in the presence of i. The intercept on $i \dfrac{\alpha}{1-\alpha}$ is K_i.

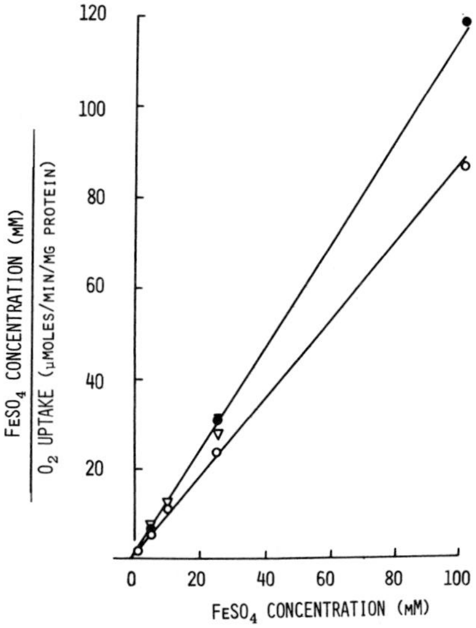

Fig. 11. Non-competitive inhibition of $FeSO_4$ oxidation by increased H^+ and by uranyl ion. s/v x s plots are given for bacteria (29 µg protein/ml) oxidizing $FeSO_4$ at pH 1.7 (O), pH 1.2 (∇) and pH 1.7 with 10mM uranyl sulphate (●). (Contrast with purely competitive ferric effect in Fig. 7).

TABLE IV. *Effect of ferrous and ferric ion concentration on $^{14}CO_2$ fixation by Thiobacillus ferrooxidans*

(a) *Total flask contents, 3.5 ml (940 µg T. ferrooxidans protein).*

FeSO₄ conc (mM)	Fe³⁺ conc (mM)	Incubation time (min)	Oxygen uptake Maximum Rate (µl/hr)	Oxygen uptake Total (µmoles)	FeSO₄ oxidised (µmoles)	$^{14}CO_2$ fixed[a] nmoles/flask Total fixed	In filtered bacteria	$^{14}CO_2$ fixed[a] nmoles/µmole FeSO₄ oxidised Total fixed	In filtered bacteria
8.6	0	65	256	6.41	25.6	82.80	69.83	3.23	2.72
8.6	28.6	75	256	7.05	28.2	102.29	83.25	3.59	2.95
8.6	57.0	85	114	5.36	21.4	89.88	65.40	4.20	3.06
86.0	0	45	720	17.32	69.3	537.53	414.21	7.76	5.98
86.0	0	60	720	25.71	102.8	777.84	667.10	7.57	6.49

(b) *Total flask contents, 2ml (744 µg T. ferrooxidans protein).*

FeSO₄ conc (mM)	Fe³⁺ conc (mM)	Incubation time (Min)	Oxygen Uptake (µmoles)	FeSO₄ Oxidised (µmoles)	$^{14}CO_2$ fixed dpm/flask	nmoles/µmole FeSO₄ oxidised
100	0	45	20.4	81.6	182,300	3.003
500	0	45	18.1	72.5	111,752	2.072
100	50	55	20.8	83.2	213,900	3.456

[a] Specific activity, 744 dpm/nmole $^{14}CO_2$

IV. GENERAL DISCUSSION AND CONCLUSIONS

This study has revealed a number of fundamental properties of *Thiobacillus ferrooxidans* that enable prediction of how the organism will behave in a variety of environments of significance to mineral leaching.

First, the efficiency with which *growing cultures* can couple energy from iron oxidation is essentially independent of the concentration of carbon dioxide available over the range < 0.001-8% and in cultures in which the $FeSO_4$ concentration decreases from at least 165mM to virtually zero, with a parallel increase in ferric iron. In some contrast, non-growing suspensions exhibited dependence both on CO_2 concentration and $FeSO_4$ concentration for maximum efficiency of coupling of oxidation to fixation. Growing cultures, whose growth ceases because of CO_2 exhaustion, are still capable of oxidizing $FeSO_4$ at a high rate for long periods. Non-growing suspensions are similarly capable of high rates of $FeSO_4$ oxidation in the absence of carbon dioxide. This is biochemically interesting, since it indicates that oxidation can be "uncoupled" from assimilatory carbon metabolism and that the organisms have a mechanism by which energy from iron oxidation can be dissipated. Moreover, it indicates that for the operation of fixed film, tank reactors or open oxidation ponds for the oxidation of iron (11, 30, 31) an abundant supply of CO_2 is not essential, since non-growing bacteria can be active for long periods, and iron oxidation is not growth-dependent.

Secondly, the true K_s for $FeSO_4$ in batch cultures and suspensions is of the order of 1 mM, as was determined from chemostat cultures (13, 24), but in batch cultures the effective K_s is much higher (and growth rate consequently lowered) because of competitive product inhibition by ferric iron. Thus in all batch cultures and iron oxidation systems employing non-growing organisms, the rate of iron oxidation is depressed by ferric ions and the observed rate will be dependent on the ferrous:ferric ratio. The ferric inhibition was purely competitive, in contrast to our demonstration of non-competitive inhibition in chemostat culture (13, 24). We are thus in disagreement with Wong, et al. (22), who claimed non-competitive inhibition in a batch system, but whose data are in any case consistent with at least an element of competitive inhibition.

Thirdly, high concentrations of $FeSO_4$ were autoinhibitory to oxidation by the bacteria in growing cultures or non-growing suspensions. Concentrations of 0.5 M and above considerably depressed $FeSO_4$-dependent CO_2-fixation, possibly indicating uncoupling as reported for dinitrophenol, uranium, copper and nickel (2, 18). Ferric iron only depressed oxidation rates, without uncoupling CO_2-fixation. Thus both substrate-ferrous and product-ferric iron must be regarded as potentially toxic metals to *T. ferrooxidans* when attempting to predict the behaviour of the bacteria under a known set of conditions. Even so, the organism has the ability to oxidize very high concentrations of $FeSO_4$, and apparently has oxidation mechanisms with three distinct sets of kinetic parameters dependent on $FeSO_4$, concentration. We do not know if these represent distinct oxidase systems or multiple binding or transport mechanisms.

In conclusion we can state that *T. ferrooxidans* exhibits iron oxidation kinetics in chemostat and batch culture, as well as in non-growing suspensions, that are far more complex than had previously been presumed. The organism has a very high potential for the turnover of ferrous iron in both growing and non-growing states, and the capacity to develop in media of low carbon dioxide and oxygen supply. Many of the important systems for the regeneration of ferric leach liquors are essentially batch or suspension systems and will be governed by the phenomena described in this paper. Chemostat kinetics (13, 24) may apply to some natural open culture systems such as heaps and drainage waters.

V. REFERENCES

1. Kelly, D.P., in "Microbial Energy Conversion," (H.G. Schlegel and J. Barnea, Eds), p. 329. E. Goltze KG, Göttingen, 1976.

2. Tuovinen, O.H., and Kelly, D.P., *Z. allg. Mikrobiol.*, 12, 311 (1972).

3. Balashova, V.V., Vedenina, I.Y., Markosyan, G.E., and Zavarzin, G.A., *Mikrobiologiya*, 43, 581 (1974).

4. Tsuchiya, H.M., Trivedi, N.C., and Schuler, M.L., *Biotechnol. Bioeng.*, 16, 991 (1974).

5. Bosecker, K., Abstracts, "Fifth International Fermentation Symposium," (H. Dellweg, Ed), p. 451. Berlin (1976).

6. Duncan, D.W., and Brynesteyn, A., "New Mexico State Bureau of Mines and Mineral Resources," Circular 118, 55 (1971).

7. Tomizuka, N., Yagisawa, M., Someya, J., and Takahara, Y., *Agr. Biol. Chem.*, 40, 1019 (1976).

8. Guay, R., Silver, M., and Torma, A.E., *European J. Appl. Microbiol.*, 3, 157 (1976).

9. Le Roux, N.W., North, A.A., and Wilson, J.C., "Tenth International Mineral Processing Conference," Paper 45. London (1973).

10. Gow, W.A., McCreedy, H.H., Ritcey, G.M., McNamara, V.M., Harrison, V.F., and Lucas, B.H., "The Recovery of Uranium," p. 195. IAEA, Vienna, 1970.

11. Derry, R., Garrett, K.H., Le Roux, N.W., and Smith, S.E., "Geology, Mining and Extractive Processing of Uranium," (M.L. Jones, Ed.), p. 56. Institution of Mining and Metallurgy, London, 1977.

12. McCreedy, H.H., CANMET Mineral Sciences Laboratories Report No. MRP/MSL 76-342 OP.

13. Jones, C.A., and Kelly, D.P., Abstracts, "Fifth International Fermentation Symposium," (H. Dellweg, Ed.), p. 126. Berlin, 1976.

14. Kelly, D.P., Eccleston, M., and Jones, C.A., "GBF Monograph Series No. 4," (W. Schwartz, Ed.), Verlag Chemie, Weinheim, 1977.

15. Lacey, D.T., and Lawson, F., *Biotechnol. Bioeng.*, 12, 29 (1970).

16. MacDonald, D.G., and Clark, R.H., *Can. J. Chem. Eng.*, 48, 669 (1970).

17. Tuovinen, O.H., and Kelly, D.P., *Arch. Mikrobiol.*, 88, 285 (1973).

18. Tuovinen, O.H., and Kelly, D.P., *Arch. Microbiol.*, 95, 153, 165 (1974).

19. Schnaitman, C.A., Korczynski, M.S., and Lundgren, D.G., *J. Bacteriol.*, 99, 552 (1969).

20. Bodo, C., and Lundgren, D.G., *Can. J. Microbiol.*, 20, 1647 (1974).

21. Steiner, M., and Lazaroff, N., *Appl. Microbiol.*, 28, 872 (1974).

22. Wong, C.H., Sharer, J.M., and Riley, P.M., M.Sc. Thesis, Univeristy of Waterloo (1973).

23. Din, G.A., and Suzuki, I., *Can. J. Biochem.*, 45, 1547 (1967).

24. Jones, C.A., Ph.D. Thesis, University of London (1974).

25. Lineweaver, H., and Burk, D., *J. Amer. Chem. Soc.*, 56, 658 (1934).

26. Atkins, G.L., and Nimmo, I.A., *Biochem. J.*, 149, 775 (1975).

27. Dixon, M., and Webb, E.C., "The Enzymes," Chapters 2 and 4. Longmans, London, 1958.

28. Dixon, M., *Biochem. J.*, 55, 170 (1953).

29. Hunter, A., and Downs, C.E., *J. Biol. Chem.*, 157, 427 (1945).

30. Livesey-Goldblatt, E., Tunley, T.H., and Nagy, I.F., "GBF Monograph Series No. 4," (W. Schwartz, Ed.) Verlag Chemie, Weinheim, 1977.

31. Mehta, K.B., and Le Roux, N.W., *Biotechnol. Bioeng.*, 16, 559 (1974).

HYDROGEN ION UTILIZATION BY

IRON-GROWN *THIOBACILLUS FERROOXIDANS*

William A. Apel
Patrick R. Dugan

Department of Microbiology
The Ohio State University
Columbus, Ohio, U.S.A.

The initial pH of aerobic suspensions of iron-grown T. *ferrooxidans was shown to increase with the addition of* $FeSO_4$. *The pH increase was proportional to the* Fe^{+2} *concentration in the 2 mM - 35 mM range and appears to be due to* H^+ *uptake by the cells.* H^+ *removal also varied directly with initial pH of the suspension in the range of 2.0 to 3.3. Initial pH increase was followed in time by a net pH decrease; leading to the postulation that the cells require* H^+ *in proportion to the amount of electrons produced from the oxidation of* Fe^{+2} *and that the net acid production resulted from a slower abiotic hydrolysis of* Fe^{+3} *(*$Fe^{+3} + 3H_2O \longrightarrow Fe(OH)_3 + 3H^+$*). Neither controls without cells nor control suspensions of cells without substrate showed any significant pH change. Suspensions of cells at pH 2.1 in the presence of DNP (1 mM - 10 mM) or oxalacetate (80 μM - 120 μM) showed some* H^+ *uptake in the absence of added* Fe^{+2} *leading to the postulation that these compounds effected the* H^+ *permeability of the cytoplasmic membrane.*

I. INTRODUCTION

Thiobacillus ferrooxidans is an acidophilic, chemolithotrophic bacterium involved in both the leaching of metals from
sulfide minerals (20), and the formation of acid drainage in
areas in which pyritic minerals are exposed to air and water
(1,7,10). This autotrophic bacterium is capable of deriving all
of its energy from the oxidation of either reduced inorganic
sulfur or reduced iron compounds. Growth on ferrous iron (Fe^{+2})
results in the formation of ferric iron (Fe^{+3}) which undergoes
abiotic hydrolytic reactions with consequent net production of
acid as shown in equations I and II (6):

$$(I) \quad 4\ FeSO_4 + 2\ H_2SO_4 + O_2 \longrightarrow 2\ Fe_2(SO_4)_3 + 2\ H_2O$$

$$(II) \quad 2\ Fe_2(SO_4)_3 + 12\ H_2O \longrightarrow 4\ Fe(OH)_3 + 6H_2SO_4$$

In accordance with equation I, *T. ferrooxidans* growing on
Fe^{+2} consumes oxygen in the ratio of one mole O_2 per four moles
of Fe^{+2} oxidized to Fe^{+3} (12,18), energy being derived from the
four electrons liberated from Fe^{+2}. The electrons must be accompanied by protons for which there is no biological source, since
Fe^{+3} is not known to be deposited in the cell. Equation II is
a non-biological reaction which furnishes an environmental supply of H^+. When (I) and (II) are summed (equation III), the
dilemma caused by the need to balance electrons taken up by the
cell is not resolved; i.e. a source of protons is required on
the left side of the reaction. Ferric hydroxide also will
react to form hydroxy sulfate complexes (reaction IV) which have
buffering capacity and can alter equations II and III so that
less H_2O enters the reactions and less H_2SO_4 is produced.
Further, equation I does not consider the proton requirement for
the metabolic reduction of CO_2.

$$(III) \quad 4FeSO_4 + O_2 + 10H_2O \longrightarrow 4\ Fe(OH)_3 + 4\ H_2SO_4$$

$$(IV) \quad Fe(OH)_3 + 2H^+ + SO_4^= \longrightarrow Fe(OH)(SO_4) + 2H_2O$$

A number of internal enzymes from $T.$ $ferrooxidans$ have been isolated and characterized. With the exception of crude preparations of a cell envelope associated iron oxidase which had a pH optimum of 2.5 to 3.5 (5), all of the above purified enzymes had pH optima ranging from 5.0 to 9.0 (4,9,11,16,19,23), which is significantly higher than the pH of the organism's environment. From this difference in internal and external pH one may conclude that the acidophilic $Thiobacilli$ possess a cell envelope which is selectively impermeable to high concentrations of H^+, a conclusion which has been directly supported by the findings of Dewey and Beecher (8). Due to the fact that resting $T.$ $ferrooxidans$ cells do not respire and are able to tolerate long periods of storage at low pH's, it was suggested by Beck (2) that their H^+ barrier is of a passive nature.

This report presents evidence to show that $T.$ $ferrooxidans$ takes H^+ from its environment in relation to the amount of Fe^{+2} utilized and presents a model depicting H^+ uptake. The effects of respiratory uncouplers on H^+ uptake by suspensions of $T.$ $ferrooxidans$ also are examined.

II. MATERIALS AND METHODS

A. Cell Culture

$T.$ $ferrooxidans$ were grown in 5 gal. carboys under forced aeration at $23 \pm 1^{\circ}C$ in the "9K" medium of Silverman and Lundgren (17). After 4 days incubation, the cells were harvested by centrifugation at $10,000 \times g$ and the resulting cell pellet was resuspended in pH 3 H_2SO_4 solution and held overnight at $4^{\circ}C$ to allow precipitation of residual insoluble iron compounds. The cells then were carefully decanted, washed three times with pH 3 H_2SO_4 solution, and resuspended to a density of 8.0×10^{-4} g (dry wt.)/ml. Cells were stored at $4^{\circ}C$ for a maximum of 48 hr. after harvesting before being either utilized or discarded.

B. H^+ Uptake

H^+ uptake was determined at $23 \pm 1^{\circ}C$ by placing 5 ml aliquots of the cell suspension in a glass polaragraphic electrolysis vessel which contained a Corning 475060 pH electrode connected to a Corning model 12 expanded scale pH meter. The pH meter was coupled to a Heath EU-20B Servo Recorder. The gas inlet of the vessel was connected to either air, N_2, or O_2, and the vessel stopcock was adjusted to produce vigorous bubbling of the suspension. Fe^{+2}, Al^{+3}, Na^+, Mg^{+2}, Fe^{+3}, Mn^{+2}, Ni^+, and Cu^{+2} as sulfate solutions, oxalacetate, and sodium thiosulfate were added individually to the cell suspension in pH 3 H_2SO_4 solutions, while dinitrophenol (DNP) was dissolved in 95% ethanol and added to yield 1 to 10 mM concentrations of DNP. Ethanol controls showed no noticeable influence on the activity of the cells with the above concentration.

The pH of the suspension was adjusted with 5 N H_2SO_4 and allowed to come to equilibrium after which the pH was continuously monitored. H^+ uptake was calculated from the pH data.

C. X-ray Microanalysis

Specimens for X-ray microanalysis were air dryed on carbon planchets (E.F. Fullam, Inc., Schnectady, N.Y. 12301) which were attached with double sided tape to aluminum specimen stubs which had been coated with Dag 154 graphite suspension (Achison Colloids Co., Port Huron, Mi. 48060). The specimens were sputter coated with 100 $\overset{o}{A}$ of carbon and then viewed and analyzed utilizing an Hitachi model S-500 scanning electron microscope equipped with an Ortec model 7844S-455 silicon-lithium (SiLi) detector coupled to an Ortec model 6230 computer unit complete with pulse height analyzer and digital printout. All specimens were analyzed for 400 sec at 30° tilt, 15 mm working distance, condenser lens setting 500, aperture setting 2, and 30,000 V power.

III.   RESULTS

Hydrogen ion (proton) removal by cells is plotted as milli-moles (from the strip chart pH curves) in order to facilitate comparisons of data from different experiments which had different pH starting and terminating points due to differences in the pH of solutions added to the reaction vessel (pH range 1.8 to 3.5). The data are only semiquantitatively accurate in terms of millimoles H^+ uptake, because ferric and sulfur salt complexes formed as reaction products in solution have some buffering capacity in the experimental pH range.

The addition of $FeSO_4$ (50 mM) to an aerated *T. ferrooxidans* suspension at an initial pH of 2.4 was found to produce an immediate increase in pH which was interpreted as an influx of H^+ into the cells (Fig. 1). Approximately 16 min after the

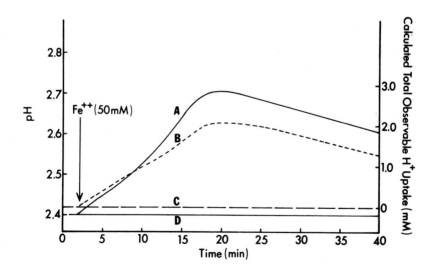

Fig. 1.   Increases in initial pH and calculated H^+ uptake in T. ferrooxidans *suspensions in the presence of Fe^{+2} at an initial pH of 2.4. (A) pH of cells with Fe^{+2}, (B) calculated H^+ uptake of cells with Fe^{+2}, (C) calculated H^+ uptake of cells without iron, and (D) pH of cells without Fe^{+2}.*

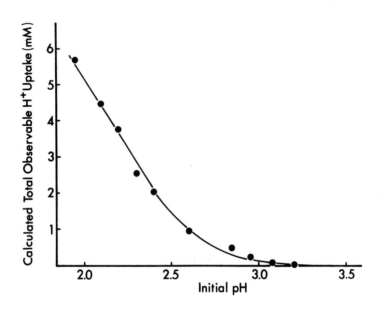

Fig. 2.   Calculated total observable changes in H^+ uptake relative to changes in initial pH of T. ferrooxidans *suspensions in the presence of Fe^{+2}.*

addition of $FeSO_4$ to the suspension, the H^+ uptake leveled off and approximately 20 min. after the addition of the $FeSO_4$, the pH began to drop. Under anaerobic conditions, when the system was purged with N_2, there was no observable H^+ uptake, although when the system was oxygenated with O_2, H^+ uptake was comparable to that attained with air.

Amounts of recordable H^+ uptake were found to be pH dependent in the range of 1.8 to 3.3 with the lower pH   producing the most noticeable uptake, while the addition of $FeSO_4$ to a suspension with a pH greater than 3.3 resulted in an immediate drop in pH (i.e. no observable H^+ uptake (Fig. 2)).

Magnitudes of H^+ uptake by the cell suspension also were found to directly correlate with the concentration of Fe^{+2}

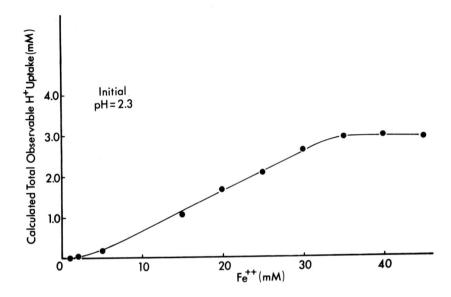

Fig. 3.   Calculated total observable changes in H^+ uptake relative to changes in Fe^{+2} concentrations in suspensions of T. ferrooxidans at an initial pH of 2.3.

present in the system in the range of 2 mM to 35 mM. At Fe^{+2} concentrations greater than 35 mM, the system appeared to be saturated, and further increases in Fe^{+2} concentrations resulted in no net H^+ uptake (Fig. 3).

The addition of 50 mM concentrations of Al^{+3}, Na^+, Mg^{+2}, Mn^{+2}, Ni^+, and Cu^{+2} as sulfates to the suspension had no observable effects on H^+ uptake under either aerobic or anaerobic conditions, and under aerobic conditions, the subsequent addition of $FeSO_4$ in the presence of the above ions resulted in normal H^+ uptake. When 50 mM concentrations of $Fe_2(SO_4)_3$ were added to the suspension there was an immediate net drop in pH, presumably due to the abiotic production of acid as previously described in equations I and II.

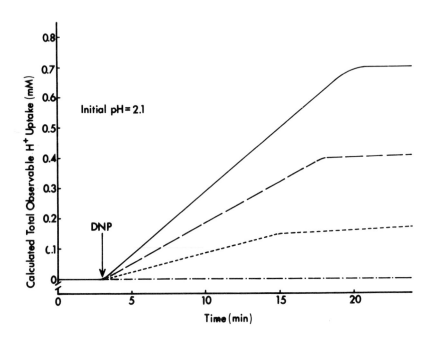

Fig. 4.   Effects of concentrations of DNP on calculated
H^+ *uptake by suspensions of* T. ferrooxidans *at an initial pH
of 2.1.   (——) cells with 10 mM DNP, (—— ——) cells with 5 mM
DNP, (----) cells with 1 mM DNP, and (—— · ——) cells with no
DNP.*

When 1 to 10 mM concentrations of DNP were added to *T.
ferrooxidans* suspensions, a net increase in pH was observed indicating H^+ uptake by the cells, with the rate of H^+ uptake
being proportional to the DNP concentration (Fig. 4).

Similar data was obtained with *T. ferrooxidans* suspensions
which had been treated with oxalacetate in the range of 80 μM
to 120 μM (Fig. 5).   Increases in pH also appeared to be proportional to concentrations of oxalacetate in the range examined.

X-ray microanalysis of the supernatants of cell suspensions
treated with 5 mM DNP and 80 μM oxalacetate revealed no increase

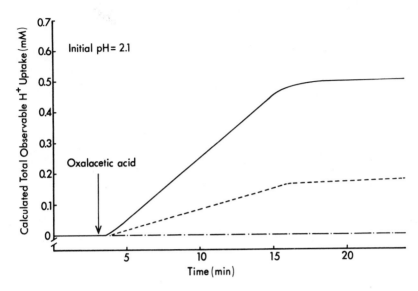

Fig. 5. *Effects of concentrations of oxalacetate on calculated H+ uptake by suspensions of* T. ferrooxidans *at an initial pH of 2.1. (——) cell with 120 μM oxalacetate, (----) cells with 80 μM oxalacetate, and (—— . ——) cells with no oxalacetate.*

in elements which are detectable with this technique and would normally be associated with cellular leakage (i.e. Na, K, P, etc.) (Fig. 6). Thus, it was concluded that increases in supernatant pH were primarily due to H^+ uptake rather than leakage of basic substances from the cells which could result in a neutralization of H^+ in the suspending solution.

Any effects of either DNP or oxalacetate in the above concentrations on H^+ uptake by *T. ferrooxidans* in the presence of Fe^{+2} were below the detection levels of the monitoring system.

IV. DISCUSSION

Iron grown *T. ferrooxidans* are unique since unlike many other chemolithotrophic bacteria (e.g. *Nitrosomonas,*

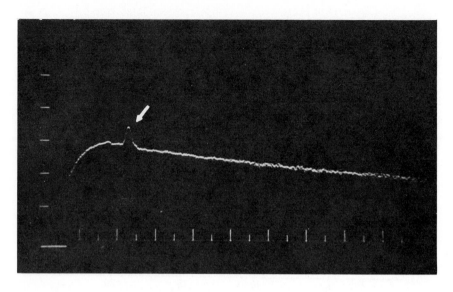

Fig. 6. Photograph showing X-ray microanalysis spectrum of supernatant containing cells treated with 80 μM oxalacetic acid. Arrow denotes sulfur peak which was present due to acidification of suspension to pH 2.1 with H_2SO_4. An identical spectrum was obtained from the supernatant of cells treated with 5 mM DNP.

Hydrogenomonas, Methanobacterium, etc.) there are no protons associated with its energy source. This may explain the obligately acidophilic nature of the bacterium in that the oxidation of Fe^{+2} to Fe^{+3} results in the removal by the cell of one electron, and in order to maintain balance of electrical charge, a proton obtained from the cell's environment may ultimately be utilized for the reduction of CO_2 and O_2. Therefore, in this context, H^+ seems to be an essential nutrient for *T. ferrooxidans.*

The data presented in this paper support this viewpoint since it appears that the addition of Fe^{+2} to a culture of *T. ferrooxidans* being maintained under conditions favorable for iron oxidation by the organism (i.e. low pH and aerobic conditions) results in a significant net increase in the pH of the organism's environment for a period of several minutes after the addition of the Fe^{+2} (Fig. 1). This pH increase is attributable to H^+ uptake by the cells. Furthermore, it seems plausible that H^+ uptake continues for as long as the cells are oxidizing Fe^{+2}. The leveling off and subsequent decrease in pH observed during these experiments is thought to be due to the build up of Fe^{+3}, resulting from the cell's metabolic activities, and the subsequent abiotic production of acid as previously discussed. Thus, even though there is a net production of acid in the system, the cells probably continue to utilize environmental H^+.

This postulation is consistent with the chemiosmotic coupling theory of Mitchell (13,14) which states that through the activities of respiratory chains, cells pump H^+ across a H^+ impermeable membrane resulting in both a H^+ and electrostatic gradient across that membrane. The cells then utilize the energy associated with this gradient in the formation of ATP. Essentially, *T. ferrooxidans* possesses a pre-formed gradient since the organism's external pH is considerably lower than its internal pH. Hence, *T. ferrooxidans* has a H^+ impermeable barrier and both pH and electrostatic gradients across that barrier, all of which are primary criteria for generation of ATP according to Mitchell's hypothesis.

Uncouplers, like DNP, are believed to adversely effect the integrity of H^+ barriers (13). This is evidently the case with *T. ferrooxidans* since the addition of DNP to cell suspensions resulted in an increase in the pH of the suspension, presumably due to H^+ uptake by the cells. Similarly, H^+ uptake was observed by Noguchi, et al.(15) when they added DNP to low pH suspensions of *T. thiooxidans* indicating that a parallel process may exist in both of these acidophilic species. Furthermore, it should be mentioned that it has been shown by Beck and Shafia (3) that levels of certain phosphates, particularly orthophosphates, are critical relative to the effects of DNP on the physiological activities of *T. ferrooxidans*. Under low phosphate conditions, DNP was found to be much more effective in inhibiting Fe^{+2} oxidation related metabolic activities of the cells than it was under high phosphate conditions. All DNP related experiments included in this report were performed utilizing suspensions of *T. ferrooxidans* containing little if any free phosphate.

Oxalacetic acid in the 10^{-4}M range may also effect *T. ferrooxidans* in a manner similar to that of DNP. It has been previously shown that oxalacetic acid inhibited iron oxidation by *T. ferrooxidans* due to the effects of the organic acid on the integrity of the cell envelope of the organism (21). When *T. ferrooxidans* was treated with 10^{-2}M concentrations of oxalacetic acid, electron micrographs revealed that the cell wall and cytoplasmic membrane were grossly disrupted, allowing leakage of cytoplasmic components into the cell's environment (22). Thus, the possibility exists that when *T. ferrooxidans* was treated with oxalacetic acid in hundred-fold lower concentrations, a more subtle disruption of the organism's cell envelope occurred, allowing leakage of environmental H^+ into the cell. This is consistent with X-ray microanalysis data which indicated that no detectable cellular elements leaked into the surrounding solution.

When considering the concentrations of DNP and oxalacetic acid necessary to result in the above described effects, one should keep in mind that concentrations may vary greatly relative to the density of the cell suspension utilized, since the reactions of the above chemical agents with the cells primarily appear to be surface related. The cell suspensions used in the experiments contained a much greater cell density and hence a greater surface area than one would normally expect to find under natural conditions; therefore, much lower concentrations of the chemical agents may accomplish effects of the same magnitude in natural cell populations. However, in this study high cell densities were essential in order to demonstrate measurable H^+ uptake.

Figure 7 summarizes schematically the possible roles of H^+ and iron in relation to cell envelope structure. Fe^{+3} is represented as bound to a receptor site at the cytoplasmic membrane surface. Fe^{+2} exchanges with and replaces Fe^{+3} at the receptor site and is immediately oxidized, liberating an electron to an acceptor (cytochrome c) in the membrane. The

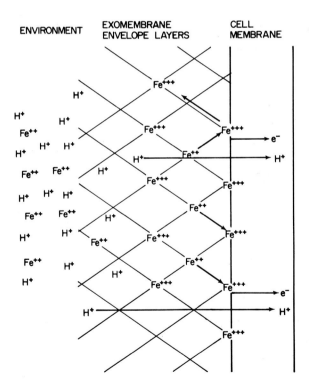

Fig. 7. Schematic diagram representing the Fe^{+2} and Fe^{+3} ions which are bound or complexed to the cell envelope-membrane area of T. ferrooxidans. *The movement of H^+ is also depicted.*

electron is accompanied by a proton from the environment and electric neutrality is maintained. Higher concentrations of H^+ and Fe^{+2} are depicted in the environment than in the envelope layers. A high concentration of bound Fe^{+3} in the envelope-membrane area contributes to the H^+ barrier due to the high electrostatic charge.

In conclusion, the observations incorporated in this report explain in part the obligate acidophilic nature of *T. ferrooxidans*. Data has been presented which indicates that H^+ is taken up during iron oxidation by the cells and it has been postulated that H^+ is an essential nutrient for the bacterium. However, more definitive experiments will be required to further elucidate the role of hydrogen ions in the metabolic processes of the acidophilic *Thiobacilli*.

V.   REFERENCES

1.   Apel, W.A., P.R. Dugan, J.A. Filppi, and M.S. Rheins, *Appl. Environ. Microbiol.*, 32, 159 (1976).

2.   Beck, J.V., *J. Bacteriol.*, 79, 502 (1960).

3.   Beck, J.V., and F.M. Shafia, *J. Bacteriol.*, 88, 850 (1964).

4.   Blaylock, B.A., and A. Nason, *J. Biol. Chem.*, 238, 3453 (1963).

5.   Bodo, C.A., and D.G. Lundgren, *Can. J. Microbiol.*, 20, 1647 (1974).

6.   Carpenter, L.V., and L.K. Henderson, Res. Bull. #10. Eng. Experim. Station, U. of West Virginia (1933).

7.   Colmer, A.R., K.L. Temple, and M.E. Hinkle, *J. Bacteriol.*, 59, 317 (1949).

8.   Dewey, D.L. and J. Beecher, *Radiation Research*, 28, 289 (1966).

9.   Din, G.A., I. Suzuki, H. Lees, *Can. J. Biochem.*, 45, 1523 (1967).

10. Dugan, P.R., *Ohio J. Sci.*, 75, 266 (1975).

11. Howard, A., and D.G. Lundgren, *Can. J. Biochem.*, 48, 1302 (1970).

12. Lees, H., S.C. Kwok, and I. Suzuki, *Can. J. Microbiol.*, 15, 43 (1969).

13. Mitchell, P., *Biol. Rev.*, 41, 445 (1966).

14. Mitchell, P., *J. Bioenergetics*, 3, 5 (1972).

15. Noguchi, A., M. Takama, T. Sekiguchi, and Y. Nosoh, *Agric. Biol. Chem.*, 41, 451 (1977).

16. Silver, M., and D.G. Lundgren, *Can. J. Biochem.*, 46, 1215 (1968).

17. Silverman, M.P., and D.G. Lundgren, *J. Bacteriol.*, 77, 642 (1959).

18. Silverman, M.P., and D.G. Lundgren, *J. Bacteriol.*, 78, 326 (1959).

19. Tabita, R., M. Silver, and D.G. Lundgren, *Can. J. Biochem.*, 47, 1141 (1969).

20. Tuovinen, O.H., and D.P. Kelly, *Int. Met. Rev.*, 19, 21 (1974).

21. Tuttle, J.H., and P.R. Dugan, *Can. J. Microbiol.*, 22, 719 (1976).

22. Tuttle, J.H., P.R. Dugan, and W.A. Apel, *Appl. Environ. Microbiol.*, 33, 459 (1977).

23. Vestal, R., and D.G. Lundgren, *Can. J. Biochem.*, 49, 1125 (1971).

METABOLIC TRANSITIONS IN CULTURES OF
ACIDOPHILIC THIOBACILLI

Olli H. Tuovinen

University of Helsinki
Helsinki, Finland

Donovan P. Kelly
Crawford S. Dow
Martin Eccleston

University of Warwick
Coventry, England

Differences in the lipopolysaccharide composition and fatty acid profiles have been reported for acidophilic thiobacilli. The studies do not generally account for the growth-phase related changes which have been seen to occur also in the phospholipid composition. There is no evidence for specific (stimulation of) phospholipid production during growth on solid substrates. In Thiobacillus ferrooxidans *a change of the inorganic source of energy may affect its iron-oxidizing capability and an autotrophy-heterotrophy transition may result in apparent loss of the iron-oxidation activity. This may be due to metabolic repression and current data also suggest that some cultures are heterogeneous on the basis of their DNA base composition depending on the growth substrate. The isolation of purely autotrophic and heterotrophic counterparts from such cultures has not been achieved although the growth conditions selectively determine the predominance of one type or another. The close association of mixed populations indicates tight metabolic coupling and interdependence.*

I. INTRODUCTION

The thiobacilli are a group of chemolithotrophic bacteria
that can obtain energy from the oxidation of inorganic sulphur
compounds. Most of the *Thiobacillus* species are also autotrophic.
During growth on an inorganic substrate, carbon dioxide is assimilated primarily via the Calvin cycle and the intermediary carbon
metabolism is predominantly biosynthetic in autotrophic bacteria.
In some thiobacilli transient mixotrophic conditions may repress
both autotrophic and heterotrophic key enzymes. The transition to
heterotrophy involves a change in the intermediary carbon metabolism so that energy and reducing power can be generated by such
mechanisms as the TCA cycle.

The thiobacilli exhibit a remarkable diversity of physiological and biochemical characteristics, different species being
able to grow at pH values between pH 1 and 10 and being obligately
chemolithotrophic, facultatively heterotrophic or even chemolitho-
trophically heterotrophic. Taxonomically they have been divided
into fairly distinct species, but analysis of their DNA base composition has shown that the G + C content ranges over nearly 20 %
between extreme values. Such a range is consistent with genetic
differences between the various species of such magnitude as to
indicate the group to be made up of distinct *genera*. Examined
critically, the only unifying feature of the thiobacilli is their
ability to oxidize inorganic sulphur compounds. Some acidophilic
thiobacilli can also oxidize ferrous-iron. The best known of
these is *Thiobacillus ferrooxidans* which grows at pH 1 - 4 and can
use iron, pyrite, soluble or mineral sulphides and other inorganic
S-compounds as energy substrates. This organism is of prime importance in many mineral leaching operations. Some work indicates
that *T. ferrooxidans* can be adapted to heterotrophic growth and
that the ability to oxidize iron can be *permanently lost* after
adaptation to growth on, for example, glucose. Such a phenomenon

is possibly unique among bacterial adaptation mechanisms and in-
cidentally deprives *T. ferrooxidans* of its key diagnostic character: the ability to oxidize ferrous-iron.

In recent years a number of observations have suggested that
some *Thiobacillus* species are capable of remarkable physiological
variation or that some supposed pure cultures are in fact closely
associated mixtures of organisms possessing diverse properties,
different members of which develop best under different environmental conditions. Apparent inter-species changes led Johnstone
et al. (1) to the conclusion that "single organisms of some
strains of thiobacilli can give rise to the range of organisms
described" in their paper.

In this paper we shall discuss various metabolic and structural aspects relating to a change of substrate and environmental
conditions for thiobacilli. We shall also review the current evidence and show additional data for the reputed heterogeneity of
cultures of acidophilic thiobacilli.

II. MORPHOLOGICAL AND STRUCTURAL FEATURES

Morphological variations have been encountered in cultures
of iron-oxidizing thiobacilli. Under forced aeration the normally
rod-shaped cells develop enlarged coccoid forms (2). These morphological alterations are reversible and dependent on the rate
of aeration. The enlarged cells have a smaller surface-to-volume
ratio and exhibit a decreased iron-oxidizing activity. These observations may suggest a regulatory mechanism, causing morphological variations, for oxygen uptake coupled with iron oxidation
in response to the oxygen tension in the environment. Morphological alterations have also been reported in cultures grown on
pyrite and ferrous sulphate (3) but these studies are not comparable because of the entirely different growth conditions. Morphologically heterogeneous *T. ferrooxidans* cultures have cellular

forms ranging from single rods to chains of several cells depending on the growth stage of the bacteria (4).

Autotrophic iron- and sulphur-oxidizers have characteristically polyhedral inclusion bodies (5, 6). These are electron dense areas in electron micrographs and have been shown (in *T. neapolitanus*) to contain high ribulose-1,5-biphosphate carboxylase activity (7 - 9) which is the key enzyme of carbon dioxide fixation via the Calvin cycle. The abundance of these inclusion bodies, also called carboxysomes, decreases when the bacteria grow mixotrophically (10). These inclusions are completely absent in heterotrophic cells. The ribulose-1,5-biphosphate carboxylase activity is also detected in the soluble cytoplasm fraction. Carboxysomes are very fragile and rupture readily during preparation and purification which may partially account for the presence of the soluble ribulose-1,5-biphosphate carboxylase activity unassociated with carboxysomes. The transition of thiobacilli from autotrophy to heterotrophy seems to result virtually in complete repression of this enzyme (11, 12). The relationship between carboxysomes and the ribulose-1,5-biphosphate carboxylase activity is a close association and their disappearance under heterotrophic growth conditions may indicate that carboxysomes are the main sites of carbon dioxide fixation in autotrophic cells.

Another readily distinguishable change takes place when thiobacilli are grown heterotrophically. The cells develop a granular appearance due to the formation of poly-β-hydroxybutyrate (PHB) inclusions which are completely absent in cells grown autotrophically on iron or sulphur compounds (5). PHB synthesis represents a deviation in the fatty acid metabolism and is normally triggered off by excess carbon in the medium. The lack of the PHB granules in autotrophic bacteria may be due to the repression of one or more of the enzymes of fatty acid metabolism involved in the PHB synthesis. Carbon dioxide fixation utilizes most of the energy derived from inorganic substrate oxidation and therefore excessive

CO_2 fixation leading eventually to PHB formation is effectively prevented because of the unavailability of energy.

III. LIPOPOLYSACCHARIDES, FATTY ACIDS, PHOSPHOLIPIDS

A lipopolysaccharide layer has been established as the main component of the cell envelope constituting about 65 % of its dry weight in iron-oxidizing thiobacilli (13). Comparative studies indicate quantitative differences in the chemical composition of lipopolysaccharide fractions isolated from autotrophic and hete-rotrophic *T. ferrooxidans* (Table I). The phosphorus content is characteristically low irrespective of the growth conditions unlike many other gram-negative heterotrophic bacteria. Although ferric iron has been detected as the major trace element in the lipopoly-saccharide fraction isolated from iron-grown bacteria (6) the neg-ligible phosphorus content established in later studies (14, 15) may cast some doubt on the involvement of LPS-phosphate in ferrous-iron oxidation at the cell envelope in contrast to a previous sug-gestion (6).

The sugar composition is quantitatively altered in response to different growth substrates (Table I). In sulphur-grown cells the hexose and heptose content of the lipopolysaccharide fraction is much lower than in iron- or glucose-grown cells whereas the methyl pentose level is elevated. 2-deoxy-sugars were practically absent in all three types of lipopolysaccharides. However, one should be cautious in concluding chemical comparisons since dif-ferent studies indicate wide variation depending on the growth conditions and analytical procedures employed. The high protein levels of the lipopolysaccharide fractions (Table I) indicate re-latively unpurified preparations. Immunological differences of lipopolysaccharides have also been reported for the various cul-tures of autotrophic and heterotrophic *T. ferrooxidans* but the immunodiffusion analysis indicated the presence of heterogeneous components in the material (14).

TABLE I   *Comparisons of available data on lipopolysaccharide fractions from autotrophic and heterotrophic* Thiobacillus ferrooxidans

| Analysis | LPS isolated from bacteria grown on | | | |
	Fe^{2+}	S^o	glucose	Reference
Sedimentation	105			6
coefficient $(S_{w,20})$	> 280	61; 9.3	128	15
Density in sucrose	1.28			
gradient	1.252	1.335; 1.176	1.319; 1.284	15
Protein (%)	2.5	18	25	15
Nitrogen (%)	1.00	1.81	1.55	14
Sugars (%)^a	40.8	56.43	32.24	14
Fatty acids (%)	14.2	8.6	13.8	14

^{a}*Sum of hexosamine, KDO, heptose and methyl pentose.*

The fatty acid composition of lipopolysaccharides has been shown to differ in *T. ferrooxidans* grown on iron, sulphur or glucose. Iron-grown bacteria accumulate relatively more C 14 β-hydroxymyristic acid and C 18:1 fatty acid in lipopolysaccharides when compared with sulphur- or glucose-grown bacteria (15). The study also indicates analytical difficulties since the major fatty acid in each lipopolysaccharide preparation remained unidentified.

Over 50 % of the fatty acid content of the chloroform-methanol extractable lipid fraction is composed of C 16:1 and C 18:1 fatty acids in each cell type (15). Analytical difficulties are apparent again as there appears to be a poor, if any, relationship between the fatty acid distribution in whole cells and extractable lipids. A prominent fatty acid is C 19 cyclopropane acid found in *T. ferrooxidans* (15), *T. thiooxidans* (16) and *T. novellus* (17).

C 19 cyclopropane acid production is enhanced in the presence of biotin (16, 17). It has been suggested (17) that biotin may be a rate-limiting compound in the fatty acid synthesis which proceeds via the carboxylation of acetyl-CoA involving carboxybiotinyl enzyme proteins.

In this context one should point out that C 19 cyclopropane acid levels are increased in stationary phase cells while the precursor (unsaturated fatty acids) levels fall down (18). Hitherto fatty acid analysis has been carried out using stationary phase thiobacilli and the data do not actually indicate the relative fatty acid profiles prevalent during growth. Variations in temperature, agitation and phosphate level had little effect on the cellular fatty acids in *T. novellus* whereas major differences occurred with respect to the autotrophic and heterotrophic type of energy and carbon metabolism (17). Increased levels of saturated fatty acids (C 16, C 18) could be found in autotrophic bacteria. It is emphasized, however, that entirely different fatty acid profiles have been presented in two studies (16, 19) for the same strain of *T. thiooxidans* which may indicate that besides the different analytical techniques employed other factors such as the age of bacteria may significantly alter the cellular fatty acid distribution. In fact, the fatty acid data presented for *T. denitrificans* grown aerobically and anaerobically (19) indicate that the fatty acid profiles are not comparable under different cultural conditions and thus may have little value in characterizing different species of thiobacilli.

Similarly, contradictory reports have been published for the phospholipid metabolism in different strains of autotrophic iron-grown *T. ferrooxidans*. Phosphatidylserine was first reported to occur in the outer membrane of the cell envelope (20) but was not detected at all in a later study of a different strain (21). The relative proportions of individual phospholipids vary during different stages of growth (22, 23) suggesting a coupling between

phospholipid synthesis and growth-related reactions. Phospholipid
compositions have been characterized in various thiobacilli but it
still remains to be elucidated how phospholipid metabolism is regulated and altered under various growth conditions involving inorganic and organic compounds as an energy and carbon source. A
comparative study (24) of *Sulfolobus* and "Ferrolobus", both thermophilic and acidophilic autotrophs, indicates somewhat similar
lipid composition in autotrophic and heterotrophic cells and
points out the presence of glycolipids which have not been described in thiobacilli. In *T. thiooxidans* the phospholipid composition changes over the growth cycle (22) which again imposes
limitations for comparative evaluations based on published data.
Similar growth-phase-related transitions in the phospholipid profiles have been reported for *T. neapolitanus* (23). Since phospholipids may have more or less specific interactions with membrane
proteins in various metabolic functions including respiratory
chain activities (25), metabolic transitions are naturally reflected in bacterial phospholipid profiles. It has been suggested
that phospholipids are involved as "wetting" agents in the oxidation of sulphur and insoluble metal sulphides, facilitating the
attachment of thiobacilli to the solid surface. Current data do
not indicate any specific phospholipid production stimulated by
a solid substrate. Phospholipids detected in the medium may in
fact represent excretion products resulting from bacterial membrane turn-over and related reactions.

IV. ASSIMILATORY METABOLISM OF NUTRIENTS

T. *intermedius* is reportedly unable to utilize sulphate as
a sulphur source but requires reduced sulphur compounds for heterotrophic growth (26). In contrast, *T. ferrooxidans* assimilates
sulphate during growth on ferrous-iron and glucose indicating that
assimilatory sulphur metabolism is unaffected by the two different
modes of growth (27). During thiosulphate-dependent growth, sul-

phate is not assimilated and reduced for cellular biosynthesis
(28). Under these conditions the sulphur requirement is primarily
met by the outer S-atom of thiosulphate after its cleavage reaction although the major portion of thiosulphate is oxidized for
energy. The enzymes mediating sulphate activation and reduction
are present in *T. ferrooxidans* irrespective of the three growth
substrates (29). This suggests a regulatory mechanism for sulphate
assimilation by S-intermediates rather than the repression of the
enzymes of the pathway. Alternatively, the cell wall may have
similar sites for binding and uptake of sulphate and thiosulphate
which would account for the preferential utilization of S from
thiosulphate in the presence of sulphate.

Nitrogen metabolism has not been studied in detail in acidophilic thiobacilli. Ammonium is the preferred nitrogen source but
growth also occurs on amino acids and other organic N-compounds
(30). Use of inorganic nitrate has not been proved conclusively
nor has an assimilatory nitrate reductase been shown. *T. neapolitanus* can use nitrate as a sole N-source. Dissimilatory nitrate
reduction is known only in *T. denitrificans* and *Thiobacillus* A2.
In the absence of suitable N-compounds some strains are able to
reduce nitrogen gas to ammonia (31, 32) which is then incorporated
into amino acids. The shift to nitrogen fixation is strongly dependent on the level of NH_4^+ in the medium. In situ studies have
not indicated any nitrogen fixation activity in heap leaching
operations (33) possibly owing to the presence of minute amounts
of NH_4^+ or because of iron-oxidizers unable to fix N_2-gas.

V. OXIDATIVE ACTIVITIES, SUBSTRATE SPECIFICITY AND DNA BASE RATIOS

A common feature of all strains of *T. ferrooxidans* is their
ability to derive energy from the oxidation of ferrous-iron. The
original isolations of autotrophic iron-oxidizers (34 - 36) indicated differences in the ability to utilize inorganic sulphur compounds for energy (*T. ferrooxidans, Ferrobacillus ferrooxidans,*

Ferrobacillus sulfooxidans). Since their isolation it has been shown that these three bacteria do in fact oxidize sulphur as well as thiosulphate and many other inorganic S-compounds also support their growth (37 - 40). Differences have been reported for growth parameters and the kinetics of thiosulphate oxidation between the subcultures of the original isolates of *T. ferrooxidans, F. ferrooxidans* and *F. sulfooxidans* (38, 39) but they do not warrant species differentiation. This view is further supported by the similarity of the DNA base ratios (Table II) which fell within the range of 52.5 - 53.6 for the three organisms (38). As a result of the reevaluation of the inorganic sulphur compound oxidizing activities of the three isolates of iron-oxidizers they are now all labelled as strains of *T. ferrooxidans* (40).

The transition of *T. ferrooxidans* to growth on thiosulphate has been shown to take place after a distinct adaptation period (41). It is possible that adaptation from iron to thiosulphate involves changes in membrane permeability for substrate translocation and enzymic induction. With the exception of the thiosulphate oxidase the enzymes mediating the oxidation of thiosulphate are present also in iron-grown *T. ferrooxidans* (29). Naturally, other changes such as those associated with electron transfer components accompany this kind of metabolic shift but have not been

TABLE II Base composition of DNA of
Thiobacillus ferrooxidans, Ferrobacillus ferrooxidans
and F. sulfooxidans[a]

Organism	*G+C %*
T. ferrooxidans	*52.5 ± 0.57*
F. ferrooxidans	*53.6 ± 0.63*
F. sulfooxidans	*52.9 ± 0.71*

[a]*Data from reference (38).*

fully characterized hitherto. Iron-grown *T. ferrooxidans* usually adapts more readily to growth on tetrathionate or elemental sulphur; their oxidation does not necessarily involve thiosulphate oxidase but does, by analogy, imply changes in membrane permeability and other phenomena related to substrate binding and uptake.

After continued growth on thiosulphate the ability to grow on iron gradually declines (42). This may indicate slight repression by thiosulphate, or its intermediates, of enzymes mediating ferrous-iron oxidation. Alternatively, while eliminating the possibility of a contaminating organism, the decline may reflect increased sensitivity of thiosulphate-grown cells to factors prevailing during growth on ferrous-iron. This phenomenon has not been observed in tetrathionate-grown cells of *T. ferrooxidans* from batch (42) or chemostat cultures.

Different strains of *T. ferrooxidans* vary in their ability to oxidize metal sulphides (43). Unfortunately, the results are not comparable since different strains, rather than only one grown with various substrates, were tested for oxygen uptake and carbon dioxide fixation activities coupled with metal sulphide oxidation. Adaptation to factors like lattice structure and metal toxicity varies in different strains of *T. ferrooxidans* and may thus influence the amenability of metal sulphides to bacterial oxidation. There is currently some evidence that *T. ferrooxidans* may oxidize selenide (44) and cuprous compounds (45 - 48) for energy. Their bacterial oxidation, particularly that of Cu^+, is likely to be associated with particular enzyme reactions to introduce electrons into the respiratory chain and transfer them to oxygen with coupled energy generation. This is indicated by the different redoxpotentials of Fe^{2+}/Fe^{3+} (+770 mV) and Cu^+/Cu^{2+} (-153 mV). The bacteria may have a mechanism of altering substrate redoxpotential for metabolic coupling by way of chelation and association of lipid material as suggested previously (49).

Acidophilic thiobacilli have been demonstrated to be capable
of growing heterotrophically on organic compounds (12, 50 - 54).
It is well established that major changes occur in the pattern of
carbon metabolism on transition to heterotrophy. These have been
summarized at length elsewhere (11, 12, 51, 52). The transition
from an inorganic substrate to an organic compound is routinely
brought about by increasing gradually or stepwise the concentration of the organic substrate at the expense of the inorganic
energy source. In *T. ferrooxidans* the transition to heterotrophy
was believed to be an irreversible process leading to complete
loss of iron-oxidizing activity (53, 54).

Various arguments have been put forward to explain this phenomenon. It may be that the derepression of iron-oxidizing activity is difficult to demonstrate after prolonged growth on glucose
(the bacterium may also lack a coupling factor under these conditions). Alternatively, the possibility has never been tested ri-
gorously that some strains of *T. ferrooxidans* may be mixed cultures. Heterotrophic variants can survive on organic excreta produced by the autotrophic strain in a medium containing ferrous-
iron (or inorganic sulphur compound) as an energy source. In such
tightly coupled mixed populations, the autotrophic bacteria may
be able to survive at low concentrations in heterotrophic cultures.
Their level may be below the detection of standard microbiological
methods such as manometric techniques. The loss of iron oxidation
may thus represent a situation where autotrophic iron-oxidizers
are gradually diluted to a very low level in the course of subculture on glucose. Two organotrophic strains of acidophilic thiobacilli have been studied in this respect.

It has been shown with the heterotrophic strain KG-4 of *T.
ferrooxidans* that transfers of this culture give rise to autotrophic iron-oxidizers (55). Colony counting after growth on glucose-
and ferrous-iron-agar indicated that glucose-grown cultures contained at most 0.1 % bacteria which could develop colonies on

iron-agar. This estimate does not take into account the possibility of the greater sensitivity of glucose-grown bacteria to the components of the autotrophic iron-containing medium as already suggested for the thiosulphate → ferrous-iron transition. Autotrophic iron oxidation was redeveloped by the bacteria also in liquid cultures. These observations contradict the original verdict indicating irreversible loss of iron oxidation in the glucose-grown culture of the KG-4 strain (53).

The iron oxidizing strain TM of *T. ferrooxidans* was converted to heterotrophy and the strain obtained renamed as *T. acidophilus* in view of the reputedly irreversible loss of iron oxidation (54). The species differentiation was further supported by the different G + C % ratios determined for the heterotrophic and original iron-grown culture. The heterotrophic strain of *T. acidophilus* has been restudied and this presentation gives an account of the work in progress on the growth and G + C % ratios of *T. acidophilus*.

The original glucose-grown culture of *T. acidophilus* was kindly provided by Dr. M. Silver. It was routinely maintained in shake flasks in the glucose medium (53) for about one and a half years before the studies described herein were carried out. The cultures were adapted to the change of the energy source in two ways. First, cells from 0.1 - 5.0 ml aliquots of cultures were collected onto membrane filters and incubated on agar media until visible growth was apparent. The membrane filters were then transferred to the respective liquid media. The media have been described previously (41, 53, 56). Alternatively, 1 ml inoculations were directly transferred to liquid media. The ferrous-iron culture was further adapted to growth at pH 1.5 to avoid ferric-iron precipitate. Bulk cultures on glucose were grown in the medium of Guay and Silver (54) or on $FeSO_4$ at pH 1.4 (56).

The following transitions were achieved:

```
    A            B              C             D             E
[GLUCOSE]→ THIOSULPHATE →[FERROUS-IRON]→ THIOSULPHATE → TETRATHIONATE
                ↓             ↓             ↓
             F              G       H
          [GLUCOSE]     [GLUCOSE]  SULPHUR
```

The DNA base composition was determined for the organisms indicated by the boxes and is given in Table III.

TABLE III   Base composition of DNA of
Thiobacillus acidophilus *grown on different
substrates (see Text)*[a]

Substrate	Buoyant density	G+C %
Glucose (A)	1.7227	64.0
Ferrous-iron (C)	1.7168	58.0
Glucose (G)	1.7227	64.0
Glucose (F)	1.7226	63.9

[a]*Bacterial cultures were harvested by centrifugation and washed in TES buffer, pH 7.1 (0.05 M Tris pH 7.2 + 0.001 M EDTA + 0.1 M NaCl). Freshly prepared lysozyme was added to a final concentration of 2 mg/ml and the culture incubated at $30^{\circ}C$ for 1 hour. The addition of sodium lauryl sarcosinate to a final concentration of 2 % (w/v) brought about lysis. Separation of DNA from cellular debris, RNA and protein was achieved by centrifuging the lysate at 120,000 g_3 for 36 hours in a linear CsCl gradient of mean density 1.71 g/cc^3. Fractions were collected and dialyzed overnight at $4^{\circ}C$ against saline citrate (0.15 M NaCl + 0.015 M sodium citrate adjusted to pH 7.0). Sample purity was determined by reading A280/A260 and the DNA concentration calculated from the extinction coefficient. If required, further purification was achieved by the addition of 1 ml redistilled, water saturated phenol per 2 ml sample, rotating slowly for 10 minutes, centrifuging and retaining the aqueous phase which was dialysed against saline citrate. Samples were stored in saline citrate at $4^{\circ}C$ over chloroform. Equilibrium sedimentation was performed in a Beckman Model E analytical ultracentrifuge for at least 22 hours at 44,000 rev/min and $25^{\circ}C$ in a cell with a 12 mm 4° Kel F centre piece. Photographs were taken using ultraviolet absorption optics and the films examined either by photographic*

enlargement or with a Joyce-Loeble recording microdensitometer.
DNA from Escherichia coli (ρ = 1.7100 g/cc³) and Micrococcus lyso-
deiklicus (ρ = 1.7310 g/cc³) were used as markers and buoyant den-
sities were calculated by the method of Mandel et al. (57). The
buoyant density of each sample in neutral CsCl was determined at
least twice.

The growth studies indicated that *T. acidophilus* is able to
grow autotrophically on thiosulphate, ferrous-iron, tetrathionate
and elemental sulphur. After repeated subculturing on Fe^{2+} or
$S_2O_3^{2-}$, the bacterium is still able to revert back to heterotrophic
growth on glucose.

The DNA base ratios for *T. acidophilus* and *T. ferrooxidans*
grown on iron are very similar, indicating possible genetic similarity of the organisms in contrast to *T. acidophilus* grown on
glucose (Table III and IV). This view is also supported by the
resemblance of the radiorespirometric pattern for glucose oxidation in *T. acidophilus* (61) and "*T. ferrooxidans*" (51).

The change of the DNA base composition of *T. acidophilus* on
the transition from heterotrophy to autotrophy (Table III) indicates culture heterogeneity. The different G + C ratios may also
indicate the involvement of plasmid DNA in the metabolic transition between heterotrophy and autotrophy but this possibility is
currently not supportable as the changes are so large and DNA heterogeneity was not observed. The probability of such a big
change in plasmid DNA is extremely unlikely although extrachromo-
somal DNA has not been reportedly studied in thiobacilli.

Mixed cultures may provide explanation for the heterotrophy -
autotrophy interconversions accompanied by G + C % changes. The
heterotrophic counterpart may survive on the organic excreta in
the autotrophic medium as already pointed out. A number of metabolic products have been identified in the spent autotrophic medium of *T. ferrooxidans* (62, 63). If there are two types of bacteria present in the heterotrophic culture of *T. acidophilus*, then
the iron-oxidizing autotrophic types must also be able to maintain

TABLE IV   *Base composition of DNA of*
Thiobacillus ferrooxidans *and*   T. acidophilus

Substrate	G+C %	Reference
A. T. FERROOXIDANS		
Fe^{2+}	56.2	58[a]
	56.6	59[a]
	57	60
	58	M. Mandel[b]
	58.3	This study[c]
$S_4O_6^{2-}$	58.7	This study[c]
PbS	53.6	58[a]
	54.4	59[a]
$CuFeS_2$	59.9	58[a]
	60.1	59[a]
Glucose	59	This study[d]
B. T. ACIDOPHILUS		
Glucose	63.0	54[a]
Sulphur	63.2	54[a]

[a] *Mean of various methods.*
[b] *Personal communication.*
[c] *Experimental details described in Footnote to Table III.*
[d] *Strain KG-4; DNA was extracted by J.F. Jackson (60) and the base composition determined by M. Mandel.*

a reasonable growth rate in the glucose medium as otherwise they would be diluted out in the heterotrophic medium in the course of subculturing. The loss of the iron-oxidizing ability in the hete-rotrophic *T. acidophilus* - previously also shown in *T. ferrooxidans* - may not solely be due to repression of the enzymes mediating the oxidation of inorganic substrates but may rather suggest the pre-dominant selection of the heterotrophic counterpart since the tran-sition involves a distinct change in the DNA base ratio. On the other hand, it is evident that the autotrophic population is ex-tremely low in the heterotrophic culture of *T. acidophilus* since

satellite bands of DNA were not observed and since the iron-oxidizing activity is below the limit of detection by manometric techniques in the glucose-grown cultures. It is not known what actually
accounts for the maintenance of the autotrophic population in the
glucose-grown culture and further studies are clearly required to
establish the cause of the tightly coupled *T. ferrooxidans/T. acidophilus* mixture.

Transitions of an autotrophic strain of *T. ferrooxidans* between tetrathionate and ferrous-iron did not affect the DNA base
composition (Table V) suggesting that these growth conditions do
not result in selective changes of the predominant population.

TABLE V   *Base composition of DNA of*
Thiobacillus ferrooxidans *grown on*
tetrathionate and ferrous-iron[a]

Substrate	Buoyant density	G+C %
A. *Tetrathionate*	1.7176	58.8
B. *Ferrous-iron (from A)*	1.7171	58.3
C. *Tetrathionate (from B)*	1.7166	57.8

[a]*DNA extracted and analyzed as described in Footnote to Table III.*

It is noteworthy that the KG-4 strain of *T. ferrooxidans* (53)
has a G + C % of 59 (Table IV) when grown on glucose, indicating
that at least some strains of *T. ferrooxidans* are indeed capable
of true adaptation to heterotrophy. The *T. ferrooxidans/T. acidophilus* mixture would always give rise to an iron-oxidizing *T. ferrooxidans* population (G + C 56 - 58 %) when grown on iron, but
would equally always give rise to a predominantly *T. acidophilus*
population (G + C 63 - 64 %) on glucose *even if* the *T. ferrooxidans*
consortant can grow on glucose, since the *T. acidophilus* organisms
must have a selective advantage due to higher specific growth rates
on the concentration of glucose supplied. Thus in batch culture

both organisms would be expected to persist during prolonged subculture. In a glucose-limited chemostat, however, *T. ferrooxidans*
would quickly be eliminated if it was solely competing with *T. acidophilus* for glucose.

We must conclude that acidophilic thiobacilli are either capable of unusual physiological adaptation with changes in the ge-
nome DNA, or more likely, that there exist mixed populations at
least in some cultures so closely interdependent that it is not
easy to separate them. Not only may there be inter-species changes
with respect to the substrate but also other factors may contribute
to the selective development of mixed populations. An example of
this kind is the pH-limit for growth often quoted for various species of thiobacilli. Our studies show that the heterotrophic culture of *T. acidophilus* is capable of adapting to pH-values ranging
from pH 1.5 to 6.5 while growing on glucose. The wide range of
substrates and pH-values suitable for growth indicates unusual metabolic and physiological flexibility which stretches the survival
of acidophilic thiobacilli in natural environments. It is evident
that further work is required to establish the nature of the organisms present in acid leaching environments and the taxonomic relationships of facultatively heterotrophic acidophilic thiobacilli.

VI. REFERENCES

1. Johnstone, K.I., Townshend, M., and White, D., *J. Gen. Microbiol.*, 24, 201 (1961).

2. Silverman, M.P., and Rogoff, M.H., *Nature*, 191, 1221 (1961).

3. Le Roux, N.W., North, A.A., and Wilson, J.C., in "Proceedings
   of 10th International Mineral Processing Congress", (M.J. Jones, Ed.), p. 1051. Institution of Mining and Metallurgy,
   London, 1974.

4. Karavaiko, G.I., and Avakyan, A.A., *Mikrobiologiya*, 39, 950
   (1970).

5. Wang, W.S., and Lundgren, D.G., *J. Bacteriol.*, 97, 947 (1969).

6. Wang, W.S., Korczynski, M.S., and Lundgren, D.G., *J. Bacteriol.*,
   104, 556 (1970).

7. Shively, J.M., Ball, F., Brown, D.H., and Saunders, R.E., *Science*, 182, 584 (1973).

8. Shively, J.M., Ball, F.L., and Kline, B.W., *J. Bacteriol.*, 116, 1405 (1973).

9. Fellman, J.K., *Abst. Ann. Meet. Am. Soc. Microbiol.*, p. 155 (1977).

10. Cannon, G.C., and Shively, J.M., *Abst. Ann. Meet. Am. Soc. Microbiol.*, p. 155 (1977).

11. Kelly, D.P., *Ann. Rev. Microbiol.*, 25, 177 (1971).

12. Tabita, R., and Lundgren, D.G., *J. Bacteriol.*, 108, 328 (1971).

13. Bodo, C., and Lundgren, D.G., *Can. J. Microbial.*, 20, 1647 (1974).

14. Vestal, J.R., Lundgren, D.G., and Milner, K.C., *Can. J. Microbiol.*, 19, 1335 (1973).

15. Hirt, W.E., and Vestal, J.R., *J. Bacteriol.*, 123, 642 (1975).

16. Levin, R.A., *J. Bacteriol.*, 108, 992 (1971).

17. Levin, R.A., *J. Bacteriol.*, 112, 903 (1972).

18. Goldfine, H., *Adv. Microbiol. Physiol.*, 8, 1 (1972).

19. Agate, A.D., and Vishniac, W., *Arch. Mikrobiol.*, 89, 257 (1973).

20. Korczynski, M.S., Agate, A.D., and Lundgren, D.G., *Biochem. Biophys. Res. Comm.*, 29, 451 (1967).

21. Short, S.A., White, D.C., and Aleem, M.I.H., *J. Bacteriol.*, 99, 142 (1969).

22. Shively, J.M., and Benson, A.A., *J. Bacteriol.*, 94, 1679 (1967).

23. Agate, A.D., and Vishniac, W., *Arch. Mikrobiol.*, 89, 247 (1973).

24. Langworthy, T.A., *J. Bacteriol.*, 130, 1326 (1977).

25. Eshafani, M., Rudkin, B.B., Cutler, C.J., and Waldron P.E., *J. Biol. Chem.*, 252, 3194 (1977).

26. Smith, D.W., and Rittenberg, S.C., *Arch. Microbiol.*, 100, 65 (1974).

27. Tuovinen, O.H., in "Conference Bacterial Leaching", (W. Schwartz, Ed.), p. 9.   Verlag Chemie, Weinheim, 1977.

28. Kelley, B.C., Tuovinen, O.H., and Nicholas, D.J.D., *Arch. Microbiol.*, 109, 205 (1976).

29. Tuovinen, O.H., Kelley, B.C., and Nicholas, D.J.D., *Can. J. Microbiol.*, 22, 109 (1976).

30. Lundgren, D.G., Andersen, K.J., Remsen, C.C., and Mahoney, R.P., *Dev. Ind. Microbiol.*, 6, 250 (1964).

31. MacIntosh, M.E., *J. Gen. Microbiol.*, 66, i (1971).

32. MacIntosh, M.E., *Proc. Soc. Gen. Microbiol.*, 4, 23 (1976).

33. Khalid, A.M., and Ralph, B.J., in "Biogeochemistry of Defined Micro-environments in Aquatic and Terrestrial Systems", (W.E. Krumbein, Ed.), in press. Ann Arbor, Michigan, 1977.

34. Temple, K.L., and Colmer, A.R., *J. Bacteriol.*, 62, 605 (1951).

35. Leathen, W.W., Kinsel, N.A., and Braley, S.A., *J. Bacteriol.*, 72, 700 (1956).

36. Kinsel, N.A., *J. Bacteriol.*, 80, 628 (1960).

37. Colmer, A.R., *J. Bacteriol.*, 83, 761 (1962).

38. Bounds, H.C., Ph.D. thesis, Louisiana State University, Louisiana (1969).

39. Bounds, H.C., and Colmer, A.R., *Can. J. Microbiol.*, 18, 735 (1972).

40. Kelly, D.P., and Tuovinen, O.H., *Int. J. Syst. Bacteriol.*, 22, 170 (1972).

41. Tuovinen, O.H., and Kelly, D.P., *Arch. Microbiol.*, 98, 351 (1974).

42. Kelly, D.P., and Tuovinen, O.H., *Plant Soil*, 43, 77 (1975).

43. Silver, M., and Torma, A.E., *Can. J. Microbiol.*, 20, 141 (1974).

44. Torma, A.E., and Habashi, F., *Can. J. Microbiol.*, 18, 1780 (1972).

45. Nielsen, A.M., and Beck, J.V., *Science*, 175, 1124 (1972).

46. Imai, K., Sakaguchi, H., Sugio, T., and Tano, T., *J. Ferm. Technol.*, 51, 865 (1973).

47. Golding, R.M., Rae, A.D., Ralph, B.J., and Sulligoi, L., *Inorg. Chem.*, 13, 2499 (1974).

48. Lewis, A.J., and Miller, J.D.A., *Can. J. Microbiol.*, 23, 319 (1977).

49. Tuovinen, O.H., and Kelly, D.P., *Z. Allg. Mikrobiol.*, 12, 311 (1972).

50. Shafia, F., and Wilkinson, R.F., *J. Bacteriol.*, 97, 256 (1969).

51. Tabita, R., and Lundgren, D.G., *J. Bacteriol.*, 108, 334 (1971).

52. Tabita, R., and Lundgren, D.G., *J. Bacteriol.*, 108, 343 (1971).

53. Shafia, F., Brinson, K.R., Heinzman, M.W., and Brady, J.M., *J. Bacteriol.*, 111, 56 (1972).

54. Guay, R., and Silver, M., *Can. J. Microbiol.*, 21, 281 (1975).

55. Tuovinen, O.H., and Nicholas, D.J.D., *Arch. Microbiol.*, in press (1977).

56. Tuovinen, O.H., and Kelly, D.P., *Arch. Mikrobiol.*, 88, 285 (1973).

57. Mandel, M., Schildkraut, C.L., and Marmur, J., in "Methods in Enzymology", (S.P. Colowick, and N.O. Kaplan, Eds.), Vol. 12B, p. 184. Academic Press, New York, 1968.

58. Guay, R., Silver, M., and Torma, A.E., *IRCS Med. Sci. Biochem. Microbiol. Parasitol. Inf. Dis.*, 3, 417 (1975).

59. Guay, R., Silver, M., and Torma, A.E., *Rev. Can. Biol.*, 35, 61 (1976).

60. Jackson, J.F., Moriarty, D.J.W., and Nicholas, D.J.D., *J. Gen. Microbiol.*, 53, 53 (1968).

61. Wood, A.P., Kelly, D.P., and Thurston, C.F., *Arch. Microbiol.*, 113, 265 (1977).

62. Schnaitman, C., and Lundgren, D.G., *Can. J. Microbiol.*, 11, 23 (1965).

63. Tuttle, J. H., Dugan, P.R., and Apel, W.A., *Appl. Environ. Microbiol.*, 33, 459 (1977).

TOXIC METALS IN LEACHING SYSTEMS [a]

P. R. Norris
D. P. Kelly

University of Warwick, Coventry, England

For most organisms, the most toxic metals released into mine waters are cadmium, silver, mercury and uranium. However, their toxicity to acidophilic bacteria involved in leaching of metals from ores ranges from the relatively non-toxic cadmium to the very toxic silver. Slight inhibition of growth of Thiobacillus ferrooxidans *in ferrous sulphate medium at pH 1.5 was observed at 10^{-9}M silver nitrate (0.1 ppb silver) with more severe inhibition as the silver concentration increased. Silver toxicity was alleviated at higher pH but not by addition of potassium, sodium or other cations to the medium. The toxicity was related to specific accumulation of silver by the cells. Serial subculturing of* T. ferrooxidans *in the presence of silver enabled selection of a silver-resistant strain with increased sensitivity to mercury, compared with organisms grown in the absence of silver. Although* T. ferrooxidans *and* T. thiooxidans *could not grow with silver sulphide as a sulphur source, the toxicity of silver nitrate or silver sulphide to the organisms in artificial leaching systems including pyrite or cadmium sulphide was greatly reduced or insignificant compared to the effect of silver on iron oxidation in a ferrous sulphate medium.*

[a] *This work was supported in part by Imperial Chemical Industries (UK) Ltd.*

83

I. INTRODUCTION

Among the factors that could affect the activity of the bacteria involved in mineral leaching is the possible toxicity of some of the metals released from ores. Metal toxicity could lead to reduced rates of leaching which would be undesirable for an industrial leach process but beneficial in reducing metal contamination of mine drainage waters.

The tolerance of acidophilic bacteria, notably *Thiobacillus ferrooxidans*, to high concentrations (greater than 10 g/ℓ) of most metal cations is well documented (1) and is essential for direct microbial leaching of ores. However, soluble uranium (2) and thallium (3) can be moderately toxic to *T. ferrooxidans*, and silver (4) can be very toxic to the organism. Prevention of growth of *T. ferrooxidans* in ferrous iron medium by 1 ppm silver has led to the suggestion that trace amounts of soluble silver would adversely affect the leaching performance of the bacteria (4).

Of metal cations that may be released in mine drainage or process waters, mercury, silver and cadmium are the most toxic to many organisms, including zooplankton (5), shellfish (6) and fish (7). Furthermore, contamination of rivers with silver and other metals from mining sources can lead to abnormal concentrations of the metals in organisms, particularly in filter-feeding and deposit-feeding shellfish (8).

Consequently, the activity of acidophilic bacteria in solubilization of these toxic metals from ores is of environmental and biochemical interest.

The initial studies described in this report chiefly involved *T. ferrooxidans*, the prevalent organism in leaching studies, and silver, the most toxic cation. Comparative silver toxicity is also described for sulphur-oxidizing *Thiobacillus thiooxidans*; for *Leptospirillum ferrooxidans*, an iron oxidizing

acidophile which, in mixed culture with the sulphur-oxidizers
T. thiooxidans, *T. acidophilus* or *T. organoparus*, can release
iron from pyrite in laboratory cultures more efficiently than
T. ferrooxidans (Norris and Kelly, unpublished); and for a
thermophilic acidophile that oxidizes sulphur, iron and pyrite.

II. MATERIALS AND METHODS

A. Organisms and Culture Conditions

The following organisms were used: (a) *Thiobacillus*
thiooxidans ATCC 8085. (b) *Thiobacillus ferrooxidans*, the Yates
and Nason strain as described earlier (9, 10). A silver-resistant strain, designated *T. ferrooxidans* Ag-R, was obtained by
serial subculture of *T. ferrooxidans* in ferrous iron medium with
increasing concentrations of silver up to $2 \cdot 10^{-7}$M $AgNO_3$.
(c) *Leptospirillum ferrooxidans* (11) kindly provided by
G. A. Zavarzin. (d) The thermophilic iron- and sulphur-oxidizing
Thiobacillus-like bacterium isolated from Icelandic thermal
springs (12).

For growth on ferrous iron, the medium contained (g/ℓ):
K_2HPO_4, 0.4; $(NH_4)_2SO_4$, 0.4; $MgSO_4 \cdot 7H_2O$, 0.4; $FeSO_4$ $7H_2O$,
27.8 (10^{-1}M). K_2HPO_4 was replaced by $(NH_4)_2HPO_4$(0.4g/ℓ) to give
a K-deficient medium. The pH was adjusted to 1.5 with H_2SO_4 for
growth of *T. ferrooxidans*, unless stated otherwise in the text.
The standard medium was diluted 1:1 with distilled water and
readjusted to pH 1.5 for growth of *L. ferrooxidans*. Yeast extract
(0.02% w/v) was added to the medium for growth of the thermophilic
bacterium. For growth of *T. ferrooxidans* on pyrite, the $FeSO_4$
was replaced by 10 g Tharsis pyrite/ℓ and the pH was adjusted
to 2.0. Sulphur medium contained (g/ℓ): KH_2PO_4, 3.0; $(NH_4)_2SO_4$,
3.0; $MgSO_4 \cdot 7H_2O$, 0.5; $CaCl_2 \cdot 2H_2O$, 0.25; $FeSO_4 \cdot 7H_2O$, 0.01;
S (flowers), 5.0. The pH was adjusted to 3.0 with H_2SO_4. Silver
sulphide (Ag_2S) and cadmium sulphide (CdS) were added (5g/ℓ) in

place of or with elemental S as required. $FeSO_4$ and pyrite media were inoculated (1% v/v) from $FeSO_4$ grown (100% Fe^{2+} oxidized) cultures. S medium was inoculated (1% v/v) from S grown cultures in which the pH had fallen from 3.0 to 1.5. *T. ferrooxidans*, *L. ferrooxidans* and *T. thiooxidans* were incubated at $30^{\circ}C$ and the thermophile at $50^{\circ}C$ in all growth and toxicity experiments using 100 ml of medium in 250 ml Erlenmeyer flasks on an orbital shaker at 200 rpm. Larger volumes of *T. ferrooxidans* cultures to provide cells for metal accumulation experiments were grown at $30^{\circ}C$ in 600 ml of medium in 2 ℓ flasks on an orbital shaker. Growth on $FeSO_4$ medium was followed as Fe^{2+} oxidation after determination of residual Fe^{2+} with ceric sulphate and 1, 10-phenanthroline-ferrous sulphate complex solution as indicator. Growth on S medium was estimated as acid production by measuring culture pH. Growth on pyrite was followed as culture pH and release of soluble Fe.

B. Silver Accumulation Experiments

T. ferrooxidans was harvested from $FeSO_4$ medium when Fe^{2+} oxidation was complete, washed by resuspension and finally resuspended in distilled water adjusted to pH 1.5 with H_2SO_4, using an optical density-cell dry weight calibration curve to give a cell suspension of 0.05 mg dry weight/ml. Suspensions (25 ml) were mixed with $FeSO_4$ (25 ml, $10^{-1}M$ at pH 1.5) or H_2SO_4 (pH 1.5) in 250 ml Erlenmeyer flasks. The flasks were placed in a shaking water bath at $30^{\circ}C$ for 30 min before adding $AgNO_3$. Other additions and $AgNO_3$ concentrations are indicated in the text. Samples (4 ml) were taken at intervals and filtered through 0.6 µm pore size membrane filters (Sartorius). Cells were washed immediately on filters with 8 ml H_2SO_4 (pH 1.5). Filters, including control filters washed with H_2SO_4, were retained for metal analyses and, where appropriate, filtrates were assayed for Fe^{2+}.

C. Metal Analyses

Filters and cells were boiled in 10 M HNO_3 to release cell-associated metals. Samples were diluted with distilled water, centrifuged to remove any undigested material, and supernatants assayed for Ag, Cu, Mg and K by atomic absorption spectrophotometry. Quantities of cell-accumulated Ag and Cu and residual cell cations, Mg and K, are expressed as nmol/ml cell suspension. Ag, Cd and Fe released from sulphides in growth experiments were also determined by atomic absorption spectrophotometry.

III. RESULTS

A. Effect of Silver and Other Metals on Fe^{2+} Oxidation

AgNO$_3$ at 10^{-9}M, $5 \cdot 10^{-9}$M and 10^{-8}M caused 1, 5 and 11 h extensions respectively to the lag phase of $T.$ $ferrooxidans$ cultures before subsequent growth proceeded almost at the control growth rate in absence of Ag with a doubling time for Fe^{2+} oxidation of 8.5 h. No growth occurred after addition of 10^{-7}M AgNO$_3$ (0.01 ppm Ag). The period of inhibition was not affected by the presence of $5 \cdot 10^{-2}$M KHSO$_4$ or NaHSO$_4$ in the medium but was reduced at higher pH. A lag phase of 15 h after addition of $5 \cdot 10^{-8}$M AgNO$_3$ at pH 1.7 was reduced to 1 h at pH 2.0. Ag toxicity to $T.$ $ferrooxidans$ and $T.$ $ferrooxidans$ Ag-R, an Ag-resistant strain, is illustrated in Fig.1.

Inhibition of growing $T.$ $ferrooxidans$ cultures was investigated with addition of AgNO$_3$ (arrowed) when 8-10% of medium Fe^{2+} had been oxidized in K^+-deficient and K^+- or Na^+- excess media (Fig.2). Growth was severely inhibited by 10^{-7}M AgNO$_3$, but resumed after 100 h in the presence of $5 \cdot 10^{-2}$M K^+ and after 120 hours in K^+-deficient media. A similar slight alleviation of growth inhibition occurred in the presence of $5 \cdot 10^{-2}$M Na^+ (data not shown).

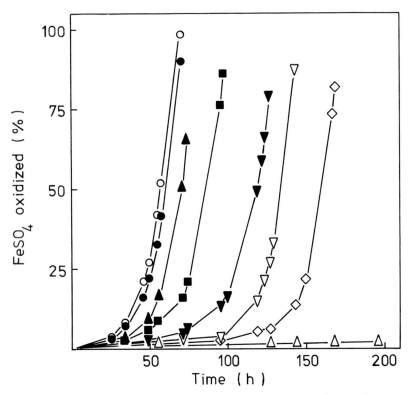

Fig. 1.   Effect of Ag on growth of T. ferrooxidans (open
symbols) and T. ferrooxidans Ag-R (closed symbols) in FeSO$_4$
medium.   Control growth (○, ●) and growth after addition of
$4 \cdot 10^{-8}$M (▽), $6 \cdot 10^{-8}$M (◇), $8 \cdot 10^{-8}$M (△, ▲), 10^{-7}M (■) and
$2 \cdot 10^{-7}$M (▼) AgNO$_3$.

Addition of 5 g Ag$_2$S/ℓ FeSO$_4$ medium at pH 1.5 prevented
growth of T. ferrooxidans.   A soluble Ag concentration of 1.3 ×
10^{-6}M (0.14 ppm) had been reached in the medium after incubation
at 30°C for 200 h.

AgNO$_3$ caused an extension of the lag phase preceding growth
of the thermophile at 50°C.   Onset of growth was delayed for 6 h
and 15 h by 10^{-7}M and 10^{-6}M AgNO$_3$ respectively;   subsequent
growth occurred as in the absence of Ag with a doubling time
for Fe^{2+} oxidation of 4 h.

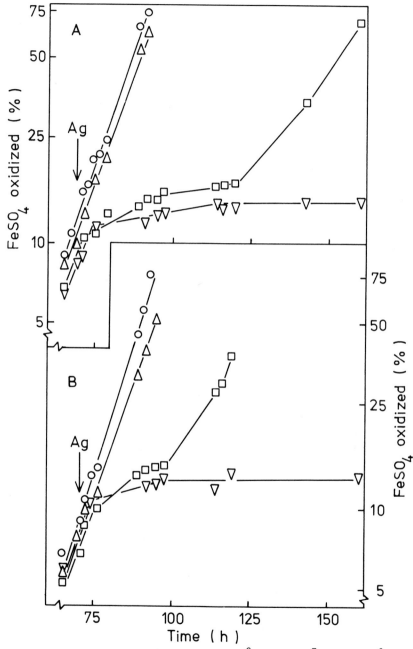

Fig. 2.   Inhibition by AgNO$_3$ (10^{-8}M, $\triangle$; 10^{-7}M, $\square$; 10^{-6}M, ∇) of T. ferrooxidans *(control growth, $\bigcirc$) growing in K-deficient FeSO$_4$ medium (A) and in FeSO$_4$ medium plus 5·10^{-2}M KHSO$_4$ (B).*

Leptospirillum ferrooxidans grew slowly in $FeSO_4$ medium (doubling time for Fe^{2+} oxidation of 24 h) and only after several days lag. Growth of the organism has not yet been fully characterized but it appeared less sensitive to Ag than *T. ferrooxidans* or the thermophile with 10^{-6}M $AgNO_3$ causing a 32 h extention of the lag phase (i.e. less than 1.5 x the generation time of the organism) before growth at the control rate.

Only mercury (Hg) and gold (Au) of other metals tested showed a level of toxicity approaching that of Ag to *T. ferrooxidans*. Au (added as sodium chloroaurate, $Na[AuCl_4].2H_2O$) in $FeSO_4$ medium caused extended lag phases before growth approximately 20% slower than growth in its absence. Maximum inhibition occurred with 10^{-7}M Au; less inhibition occurred with 10^{-8}M and 10^{-6}M and higher concentrations at which Au precipitated.   Addition of 10^{-4}M Au resulted in a lag phase shorter than that of control cultures but equivalent to that in the presence of 10^{-4}M NaCl, indicating an effect of Na^+ or Cl^- ions rather than of Au. Hg added as $HgCl_2$ prevented growth at 10^{-5}M (2 ppm Hg) in $FeSO_4$ medium. Inhibition of *T. ferrooxidans* and *T. ferrooxidans* Ag-R by Hg (Fig. 3) showed increased sensitivity of the Ag-resistant strain to the metal.

B. Effect of Silver on Growth in Sulphur Medium

$AgNO_3$ was added to S medium before inoculation with *T. thiooxidans* or *T. ferrooxidans*. Growth, estimated as acid production, was initially inhibited at the higher Ag concentrations (Fig. 4) but eventually proceeded at a similar rate and to a similar extent as control growth in the absence of Ag.

Samples of cell- and S- free medium were prepared by centrifugation after growth had occurred at all Ag concentrations tested. Only traces of Ag were detected in supernatant liquids whereas residual S in the culture medium was coloured more

intensely brown as the total Ag concentration increased,
indicating removal of soluble Ag by S.

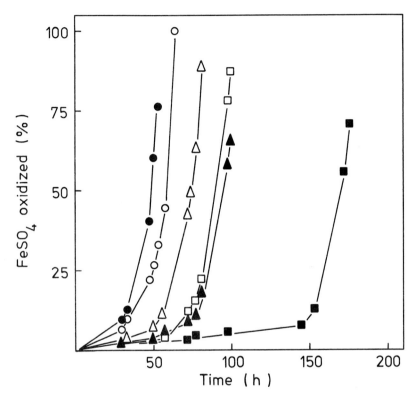

Fig. 3. Effect of Hg on growth of T. ferrooxidans *(open
symbols) and* T. ferrooxidans Ag-R *(closed symbols) in FeSO$_4$
medium. Control growth (○,●) and growth after addition of
$5 \cdot 10^{-7}$M (△,▲) and 10^{-6}M (□,■) HgCl$_2$.*

C. Release of Metals from Sulphides in the Presence of Silver
    Sulphide

Growth of *T. ferrooxidans* and *T. thiooxidans* did not occur in
mineral salts medium with Ag$_2$S as the only source of reduced S and
there was consequently no release of Ag in the medium (pH 3.0).
Growth of both organisms in medium to which Ag$_2$S and elemental S
were added was the same as on elemental S alone; pH of
stationary phase cultures after 45 days incubation at 30°C
were 1.03 (*T. thiooxidans*) and 1.22 (*T. ferrooxidans*). During the

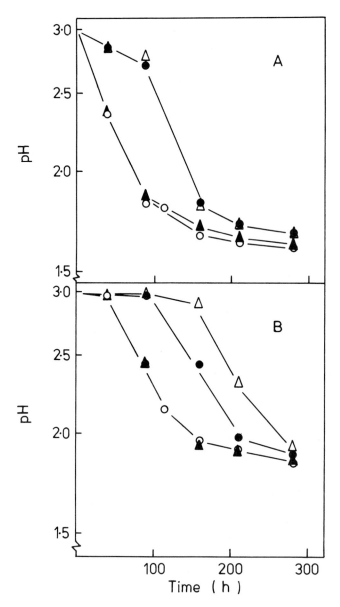

Fig. 4. *Effect of Ag on growth (culture pH) of*
T. thiooxidans *(A) and* T. ferrooxidans *(B) in S medium. Control*
growth (O) and growth after addition of 10^{-6}M (▲), $5 \cdot 10^{-6}$M (●)
and 10^{-5}M (△) AgNO$_3$.

growth of both organisms, but not in sterile medium acidified to
the same extent as the growing cultures, some Ag_2S became
progressively more finely divided. After 45 days incubation and
24 h standing, up to 1.5 ppm Ag remained in suspension.
Filtration through 0.45 μm membrane filters gave clear filtrates
which contained 0.25 μM (0.027 ppm Ag) 'soluble' Ag from a
T. thiooxidans culture and 0.95 μM (0.1 ppm Ag) 'soluble' Ag from
a *T. ferrooxidans* culture. The bulk of the suspended Ag therefore
probably included finely divided sulphide, complexes and
precipitates with medium constituents such as phosphate, and
possibly some cell-associated Ag.

The effect of the presence of Ag_2S on leaching of another
metal sulphide was tested with CdS. CdS was suitable to compare
the reaction of *T. thiooxidans* and *T. ferrooxidans* to Ag_2S
because release of Cd by both organisms follows the same pattern
and requires the same conditions, although Cd release is slightly
faster during growth of *T. thiooxidans* (Norris and Kelly,
unpublished). Growth and Cd release required added S, as
described previously for *T. thiooxidans* (13), and release of up
to 90% of Cd from 5 g CdS/ℓ through the action of both organisms
was not inhibited by 5 g Ag_2S/ℓ.

Release of soluble Fe from pyrite (FeS_2) was unaffected by
10^{-5} M $AgNO_3$ (1 ppm Ag) when $AgNO_3$ was added to growing cultures
of *T. ferrooxidans* after 10% of available Fe was released, but
was transiently inhibited by 10^{-4}M $AgNO_3$ before Fe release
occurred at the control rate in absence of Ag, and was completely
inhibited by 10^{-3}M $AgNO_3$.

D. Accumulation of Silver by *T. ferrooxidans*

Addition of $AgNO_3$ to washed cell suspensions of
T. ferrooxidans oxidizing Fe^{2+} resulted in accumulation of Ag by
the cells and inhibition of oxidation. Fe^{2+} oxidation was
inhibited to a lesser extent and less Ag was accumulated by

T. ferrooxidans Ag-R than by *T. ferrooxidans* (Fig. 5).

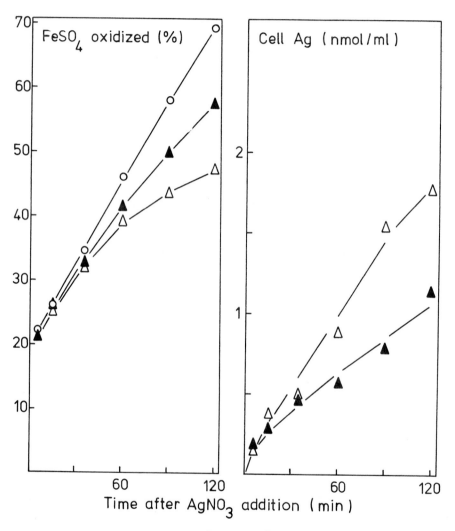

Fig. 5. Oxidation of Fe^{2+} (5.10^{-2}M $FeSO_4$) and accumulation of Ag at $30°C$ and pH 1.5 by T. ferrooxidans ($\triangle$) and T. ferrooxidans Ag-R ($\blacktriangle$) after addition of 10^{-5}M $AgNO_3$ to cell suspensions (0.05 mg dry weight/ml) and oxidation of Fe^{2+} by both cell suspensions in the absence of Ag (O).

Further experiments (data not shown) have shown that the extent of Ag accumulation and rate of inhibition of Fe^{2+} oxidation was proportional to the Ag concentration with complete inhibition occurring within 3 h of addition of $10^{-4}M$ $AgNO_3$. Addition of cysteine ($10^{-4}M$) before $AgNO_3$ ($10^{-5}M$) prevented inhibition of Fe^{2+} oxidation but its effect on Ag accumulation was not clear with retention of more Ag with cells by filters possibly resulting from collection of an Ag-cysteine precipitate. $10^{-4}M$ $CuSO_4$ (6.35 ppm Cu) did not inhibit Fe^{2+} oxidation and did not affect the accumulation of Ag from $10^{-5}M$ $AgNO_3$ (1 ppm Ag). Cu accumulation from $10^{-4}M$ $CuSO_4$ was not measurable (i.e. less than 0.1 nmol Cu/ml suspension of 0.05 mg cell dry weight/ml).

The effect of Ag accumulation by *T. ferrooxidans* on the levels of the major cell cations, Mg and K, is shown in Fig. 6. Mg loss from the cells was gradual and almost complete in 3 h while approximatley 50% of the cell K was lost rapidly, the cell K concentration then remaining almost constant.

Further experiments (data not shown) with rapid sampling have shown that the K loss occurred within 1 min of $AgNO_3$ addition. The K loss was reduced to 25% of the cell K in the presence of $5 \cdot 10^{-2}M$ $KHSO_4$ while the Ag accumulation and Mg loss were unaffected. Ag accumulation was the same in the presence or absence of $FeSO_4$ but K loss (75% of cell K) was greater and the Mg loss was more rapid in the absence of $FeSO_4$.

IV. DISCUSSION

The reported (4) high toxicity of Ag to *T. ferrooxidans* has been confirmed with inhibition of the organism at Ag concentrations several orders of magnitude less than the inhibitory concentrations of most metal cations. The thermophilic bacterium appeared less sensitive than *T. ferrooxidans* to Ag but components of yeast extract in the growth medium probably complexed some Ag and reduced its soluble concentration, in a

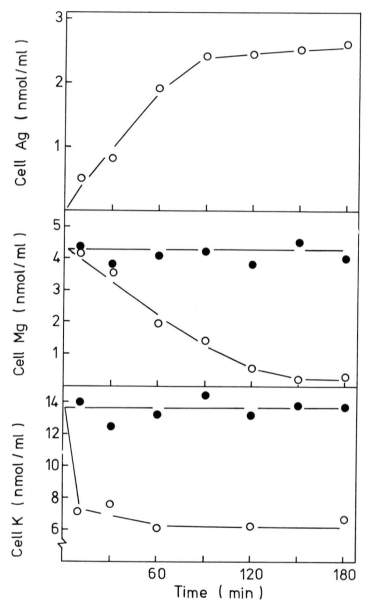

Fig. 6. Cell-associated Ag, Mg and K (open symbols) of T. ferrooxidans (0.05 mg dry weight/ml) at 30°C and pH 1.5 after addition of 10^{-5}M AgNO$_3$ in the presence of $5 \cdot 10^{-2}$M FeSO$_4$; and cell-associated Mg and K (closed symbols) in the absence of Ag.

similar but less effective manner than the cysteine which prevented Ag inhibition of Fe^{2+} oxidation by non-growing suspensions of *T. ferrooxidans*. *L. ferrooxidans* appeared less sensitive than the other organisms but some inhibition of all three organisms at an Ag concentration of $10^{-6}M$ indicated the general toxicity of the metal to Fe^{2+} oxidizing acidophiles.

Metal toxicity to microorganisms is influenced by the physiological state of the organisms, the chemical state of the metal and by the extent of their interaction as governed by their environment. The tolerance of acidophiles to most metals in low pH media probably results from effective competition by H^+ ions for negatively-charged sites at the cell surface. Thus, Cu^{2+} was not accumulated and did not affect Fe^{2+} oxidation by *T. ferrooxidans* in the present experiments, and even at the high concentrations of UO_2^{2+} and Ni^{2+} required to inhibit *T. ferrooxidans*, only small amounts of the metals were associated with the cells (14). In contrast, the high toxicity of Ag to *T. ferrooxidans* oxidizing Fe^{2+} appeared related to ready accumulation of the metal by the cells.

After accumulation of Ag by cells, there are several possibilities for disruption of normal metabolism, including the inhibition of many enzymes and reactions dependent on activity of thiol groups, as Ag^+ forms a strong complex with cysteine (15). Exposure of *Escherichia coli* to Cd^{2+} results in single strand breaks in DNA which are repaired during long lag phases before cells proliferate at a normal rate (16). The similar pattern of Ag toxicity in the growth experiments might have resulted from interaction of accumulated Ag and the DNA of *T. ferrooxidans*. Ag is known to bind to DNA *in vitro*, particularly to a nitrogen of guanosine bases (17). Inhibition of Fe^{2+} oxidation of cell suspensions of *T. ferrooxidans* by higher Ag concentrations than those inhibiting growth does not concern such possible direct interference with growth and cell division. A metal-concentration-dependent multiplicity of toxic actions has previously been suggested with UO_2^{2+} inhibition of *T. ferrooxidans* in which Fe^{2+} oxidation was severely inhibited by UO_2^{2+} concentrations greater than those required to inhibit CO_2 fixation (14). Ag inhibition of Fe^{2+} oxidation and the rapid loss of cell K after Ag accumulation might have resulted from interference with membrane cation and electron transport. A transmembrane pH gradient of *T. ferrooxidans* in acid medium, with an internal pH near neutrality to allow normal metabolism to function, could be used to conserve energy with the coupling of the gradient to ATP synthesis via a chemiosmotic ATPase reaction (18). Net entry of protons would be prevented and the pH gradient maintained by the half-reaction of H^+ ions with oxygen inside the cell matrix ($2e^- + \frac{1}{2}O_2 + 2H^+ \rightarrow 2H_2O$) with the other half-reaction ($2Fe^{2+} \rightarrow 2Fe^{3+} + 2e^-$) occurring outside the cell matrix. Inhibition of the cytochrome-mediated O_2 half-reaction by Ag could inhibit a respiratory mechanism of eliminating the protons from the cell. Ag^+ is known to inhibit the respiratory chain of *E. coli* with the most sensitive sites to inhibition between the b-cytochromes and cytochrome $\underline{a}_2$ (19). Failure of *T. ferrooxidans* to remove protons entering cells would lead to a collapse of the pH gradient and the proton motive force would fall to zero. Rapid loss of K from cells could have followed the disruption of an energized membrane state. Rapid inhibition of respiration of yeast by Hg is followed by a rapid depletion of the endogenous ATP level and a loss of cell K (20).

The observed characteristics of Ag toxicity appear specific compared with the other cations which inhibit *T. ferrooxidans*. The toxicity of Au was particularly concentration-dependent as a result of the unstable nature of the Au complex in the $FeSO_4$ medium. A similar action of Au and Ag has been shown with their inhibition of Hg^{2+} reduction to elemental Hg by *E. coli* (21), an inhibition of similar specificity to that seen with

$T.\ ferrooxidans$ in that other metals were far less inhibitory than Au or Ag. Similar actions of both metals in $T.\ ferrooxidans$ are possible, but each probably reached sensitive sites by different routes. Au in solution was probably in the form of $(AuCl_3OH)^-$ anions. Oxyanions of arsenate, selenate, tellurate and molybdate are more toxic than most metal cations to $T.\ ferrooxidans$ [1], perhaps through relative ease of access, without H^+ ion competition, to cells.

UO_2^{2+} toxicity is of several orders of magnitude less than that of Ag and in the absence of clear UO_2^{2+} accumulation, it has been suggested that the toxicity, which can be alleviated by less toxic cations such as Zn, probably involves loose binding of UO_2^{2+} at membrane sites [14].

The increased sensitivity of the Ag-resistant strain to Hg suggests different interactions of the metals in inhibition of $T.\ ferrooxidans$.

Thallium (Tl) toxicity to $T.\ ferrooxidans$ is alleviated in growth media with high K^+ concentrations [3], in keeping with Tl^+ accumulation by microorganisms in competition with K^+ uptake via a K^+ transport pathway [22]. Although some Ag and Tl compounds share similar properties, and Ag^+ is of similar ioniç radius (1.26 Å) to K^+ (1.33 Å) and Tl^+ (1.40 Å) and of the same charge - both likely criteria for effective competition with K^+ uptake in yeast and $E.\ coli$ [22] - there was no evidence to link Ag^+ toxicity with K^+ metabolism of $T.\ ferrooxidans$. In contrast to its effect on Tl^+ toxicity, K^+ (or Na^+) did not alleviate extended lag phases before growth in the presence of Ag, or alleviate toxicity of Ag added to growing cells and did not affect Ag accumulation by non-growing suspensions of Fe^{2+}- oxidizing cells.

The mechanism of Ag accumulation by $T.\ ferrooxidans$ is not known but comparison can be made with the accumulation of toxic non-essential Cd by another Gram-negative organism, *E. coli* (23). The cell Mg and cell K concentrations of *T. ferrooxidans* were similar to those of washed, non-growing cell suspensions of *E. coli* (22). Unlike uptake of Co^{2+} and Ni^{2+} which are accumulated by an energy-dependent pathway normally associated with Mg^{2+} accumulation, Cd accumulation by *E. coli* was independent of an external energy source (glucose) just as Ag accumulation by *T. ferrooxidans* was independent of $FeSO_4$, but accumulation of both cations probably depended on an energized membrane state of the cells. Cd accumulation experiments used different conditions (pH 6.5, 10 x metal salt concentrations and 10 x the biomass) than Ag accumulation experiments with *T. ferrooxidans* but the extent of metal uptake by cells (excluding cell surface binding of Cd^{2+} by *E. coli*) was similar in both cases with up to about 50 μmol Ag^+ or Cd^{2+} accumulated/g cell dry weight. Further experiments are in progress with specific inhibitors of cation transport and membrane functions in attempts to define the mechanism of accumulation of such toxic metal cations.

Interaction of Ag and *T. ferrooxidans* under conditions where most cations do not affect the cells could have been facilitated by the high affinity of Ag for thiol groups. High affinity for sulphur (log K_{sp} of Ag_2S ≈ -50) also explains the relative non-toxicity of Ag in S medium. The presence of S, Cl^- ions and the higher pH all probably contributed to allow growth and associated leaching activities of *T. ferrooxidans* and *T. thiooxidans* in the presence of greater total Ag concentrations than would be found in ore bodies or ore concentrates and provides, in comparison with Ag toxicity in $FeSO_4$ medium, an example of the influence of the environment on metal toxicity. The development of an Ag-resistant strain of *T. ferrooxidans* suggests that if bacteria active in leaching were threatened by increasing Ag concentrations, an increase in tolerance to the metal could occur. Development of *T. ferrooxidans* populations with increased tolerance to UO_2^{2+} has been attributed to selection of mutant organisms which retained resistance in the absence of UO_2^{2+}(2). *T. ferrooxidans* Ag-R has not been investigated in the same detail. Whether the acidophiles can contribute indirectly to Ag release from Ag_2S is being further studied with leaching columns. Chemical oxidation of cinnabar by Fe^{3+} in acid mine waters occurs at much greater rates than release of Hg into solution as a result of Hg binding to remaining cinnabar, but Hg release is increased by typical mine water levels of Cl^- ions that complex Hg (24). Fe^{3+}, produced by microbial action, could contribute to Ag release under similar conditions while Ag accumulation by cells could contribute in a small way to mobilization of the metal into mine drainage waters.

V. REFERENCES

1. Tuovinen, O.H., Niemelä, S.I., and Gyllenberg, H.G., *Antonie v. Leeuwenhoek*, 37, 489 (1971).

2. Tuovinen, O.H., and Kelly, D.P., *Arch. Microbiol.*, 95, 153 (1974).

3. Tuovinen, O.H., and Kelly, D.P., *Arch. Microbiol.*, 98, 167 (1974).

4. Hoffman, L.E., and Hendrix, J.L., *Biotech. Bioeng.*, 18, 1161 (1976).

5. Biesinger, K.E., and Christensen, G.M., *J. Fish Res. Bd. Canada*, 27, 1227 (1972).

6. Calabrese, A., Collier, R.S., Nelson, D.A., and Macinnes, J.R., *Mar. Biol.*, 18, 162 (1973).

7. Hale, J.G., *Bull. Environ. Contam. Toxicol.*, 17, 66 (1977).

8. Bryan, G.W., and Hummerstone, L.G., *J. Mar. Biol. Ass. U.K.*, 57, 75 (1977).

9. Tuovinen, O.H., and Kelly, D.P., *Arch. Mikrobiol.*, 88, 285 (1973).

10. Tuovinen, O.H., and Kelly, D.P., *Arch. Microbiol.*, 98, 351 (1974).

11. Balashova, V.V., Vedenina, I. Ya., Markosyan, G.E., and Zavarzin, G.A., *Microbiology*, 43, 491 (1973).

12. Le Roux, N.W., Wakerley, D.S., and Hunt, S.D., *J. Gen. Microbiol.*, 100, 197 (1977).

13. Brissette, C., Champagne, J., and Jutras, J.R., *Can. Mining. Met. Bull.*, 64, 85 (1971).

14. Tuovinen, O.H., and Kelly, D.P., *Arch. Microbiol.*, 95, 165 (1974).

15. Gruen, L.C. *Biochim. Biophys. Acta.*, 386, 270 (1975).

16. Mitra, R.S., and Bernstein, I.A., *Biochem. Biophys. Res. Commun.*, 74, 1450 (1977).

17. Tu, A.T., and Reinosa, J.A., *Biochemistry*, 5, 3375 (1966).

18. Ingledew, W.J., Cox, J.C., and Helling, P.J., *Proc. Soc. Gen. Microbiol.*, 4, 74 (1977).

19. Bragg, P.D., and Rainnie, D.J., *Can. J. Microbiol.*, 20, 883 (1974).

20. Brunker, R.L., *Appl. Env. Microbiol.*, 32, 498 (1976).

21. Summers, A.O., and Silver, S., *J. Bacteriol.*, 112, 1228 (1972).

22. Norris, P., Man, W.K., Hughes, M.N., and Kelly, D.P., *Arch. Microbiol.*, 110, 279 (1976).

23. Norris, P.R., Ph.D. thesis, University of London, (1976).

24. Burkstaller, J.E., Mcarty, P.L., and Parks, G.A., *Env. Sci. Technol.*, 9, 676 (1975).

DIRECT OBSERVATIONS OF BACTERIA AND QUANTITATIVE

STUDIES OF THEIR CATALYTIC ROLE IN THE LEACHING

OF LOW-GRADE, COPPER-BEARING WASTE

V. K. Berry
and
L. E. Murr

New Mexico Institute of Mining and Technology
Socorro, New Mexico   87801   USA

Observations have been made of Thiobacillus ferrooxidans, Thiobacillus thiooxidans, *and a high-temperature* Sulfolobus-*like microorganism attached to mineral-phase surfaces following various stages of leaching low-grade copper waste ore. Attachment in all cases is observed to be selective, that is bacteria attach only to sulfide-phase surfaces (principally CuFeS₂ and FeS₂), but attachment has been observed not to be a requirement for leaching to occur. The regularity of attachment is much less for* T. ferrooxidans *and* T. thiooxidans *than for the higher-temperature* Sulfolobus-*like microbe and this seems to be because of the absence of a flagellum in the case of the latter. Attempts to relate bacterial attachment on chalcopyrite surfaces to emergence sites of dislocations and related crystal defects which can cause local changes in surface stoichiometry (using the transmission electron microscope) are also described. Direct, systematic observations and comparison of surface degradation (corrosion) of FeS₂ and CuFeS₂ in sterile acid environments and inoculated acid environments clearly indicate a significant enhancement in leaching rate by bacteria. This bacterial catalysis is shown to be important in enhanced conversion rates by galvanic processes when*

FeS₂ and CuFeS₂ phase regions are in intimate (electrical) contact; and the implications of bacterial catalyzed galvanic conversion as it relates to sulfide dissemination in large ore bodies is briefly described.

I. INTRODUCTION

The subject of bacterial leaching has received a great deal of attention only in recent years, although bacterial leaching of sulfide minerals has been occuring for centuries (1). Microbiological leaching is a process of metal extraction in which certain microorganisms catalyze oxidation reactions. Due to the great increase in demand for metals during this century and the likelihood of this demand further increasing in the future, it is necessary to look to the lower-grade deposits as ore resources for the future. Known and accessible high-grade deposits of many minerals are now almost depleted. It is thus appropriate to search for technologically and economically feasible processes to recover greater amounts of metals from the lower grade ores. It is in this context that bacterial leaching and the role of microorganisms in altering minerals and metals in solution has assumed a role of considerable importance. The bacterial leaching process is not only economical as compared to the conventional methods but it is also essentially pollution free.

The earliest recorded mining activity which can be attributed to bacteria involved the recovery of copper from mine waters at Rio Tinto, Spain (1) in 1670. The presence of bacteria in the leaching fluids was not confirmed in that operation until 1962 (2). However, in 1922 Rudolf et al. (3,4) reported the microbiological leaching of metal sulfides by an unidentified autotrophic bacterium. The economic importance of the extraction of metals from low-grade sulfide-bearing ores was thus suggested. In 1947, Colmer and Hinkle (5) first isolated the microorganism, *Thiobacillus ferrooxidans*, from an acid drainage source which was responsible for the oxidation of sulfide minerals. Bryner et al.

(6,7) reported on the role of this microrganism in the leaching of sulfide minerals. However, it was Malouf and Prater (8) who systematically studied the role played by bacteria in the alteration of sulfide minerals.

The chemolithotrophic[1] microorganism *T. ferrooxidans* can be found wherever sulfide minerals occur in an acidic environment. In the earliest reported studies of autotrophic bacteria, Winogradsky (9) in 1888 proposed chemolithotrophy. Bacteria use the energy released from the oxidation of ferrous iron (10) or other inorganic substrates such as reduced inorganic sulfur compounds (11,12). The chemical energy is converted primarily by oxidative phosphorlylation to adeosine triphosphate (ATP) (13). ATP plays an important role in the energy metabolism of these microorganisms as discussed for example in Chap. 1.

The bacterial leaching of metal sulfides can be considered as a biochemical oxidation process which is catalyzed by microorganisms. This process is represented by the following simple equation

$$MS + 2O_2 \xrightarrow{\text{Microorganisms}} MSO_4 \qquad (1)$$

where M is ideally a bivalent metal.

Microbiological leaching processes involve complex instructions between biological (microorganisms), aqueous (nutrients in the medium) and solid mineral phases. In order to obtain a better understanding of the overall leaching process and to make better economical use of this process the interactions between each of three phases need study.

---

[1] *A process in which microorganisms synthesize their protoplasmic constituents from inorganic sources and carbon from CO_2 and derive the energy from the oxidation of inorganic compounds.*

In the past ten to fifteen years significant research has
been reported on several aspects of bacterial leaching (14-19).
Many laboratory, pilot, and field studies of microbial leaching
of copper minerals have been concerned primarily with determining
the conditions under which bacterial oxidation occurs and then
optimizing the process (20-24). There have been studies of the
mechanism of bacterial oxidation of iron oxidizing bacteria *T.
ferrooxidans* (2, 25,26) and sulfur oxidizing bacteria *Thiobacillus thiooxidans* (27,28). Some recent investigations have been
directed towards the isolation and development of bacteria able
to metabolize at high temperatures (29-31) (in excess of those
temperatures which optimize the growth of *T. ferrooxidans* and
T. thiooxidans).

One aspect of the bacterial leaching process which has not
been extensively studied is the attachment characteristics of the
bacteria to the mineral surfaces after microbe/mineral interaction.
The direct contact mechanism proposed by Silverman (32) has never
been verified by direct observations even though bacterial attachment to mineral surfaces has been described by various authors
(33-36) and observed for a number of systems (37-39). Silverman
(32) has shown that the direct contact mechanism, which requires
physical contact between the bacteria and the sulfide mineral
under aerobic conditions, is difficult to demonstrate with iron
containing sulfide minerals such as pyrite (FeS_2), bornite
(Cu_5FeS_4) and chalcopyrite ($CuFeS_2$). Thus, in the bacterial
leaching of high-grade sulfide minerals and low-grade ores, it is
important to understand the contact or attachment mechanism. In
the case of low-grade ores it is of interest to know whether bacteria attach selectively to specific mineral inclusions, and what
the mechanism and principal features of attachment include.

Another significant feature of bacterial/mineral interaction
involves the mode and character of bacterial attachment. Bacterial attachment to surfaces has been described as occurring

through a variety of holdfasts, including adhesive pili (38,40,41) as well as surface adhesion (42,43). One feature of attachment which has not been investigated with regard to mineral leaching with *T. ferrooxidans* involves the attachment location. There have been no attempts to determine if the attachment site is a result of random selection, or whether it occurs at regions of specific ion concentration, changed stoichiometry, or strained regions containing crystal defects such as dislocations. It may be of interest to determine whether crystal defects such as emergence sites of dislocations, grain boundaries, cracks, etc. influence the characteristics and degree of attachment in bacterial leaching.

Another interesting feature in the oxidation of minerals is the effect of crystal structure on the oxidation of these minerals. Silverman and Ehrlich (19) reported that the crystal structure of a given mineral is an important factor in influencing the oxidation of these minerals. They suggested that well ordered crystal lattices are less susceptible to oxidation than imperfect crystal structures. They did not elaborate on this. There has been no direct attempt to systematically study the influence of different crystallographic orientations on the rate of leaching.

The purpose of this investigation was to study the mode and characteristics of attachment of bacteria to mineral surfaces and to establish a correlation, if any, between the attachment and crystal defects (dislocations in particular) in the minerals. In the case of low-grade ores, preferential attachment of bacteria to specific mineral surfaces was studied. The effect of crystallographic orientations on the surface corrosion (corrosion features) of sulfide phases in bacterial leaching was also studied. These studies were performed using transmission and scanning electron microscopy. Specimen preparation procedures involved a combination of biological and physical science

techniques. The degree of surface corrosion with time was studied with the scanning electron microscope (SEM) and correlated
with the metals solubilized (Fe and Cu).

II. MATERIALS AND METHODS

A. Ores and Minerals

The following copper ores were obtained and used in the present experiments. i) Phelps Dodge Corporation Ajo waste ore
having chalcopyrite as the principal copper mineral, ii) Kennecott Corp., Chino Mines Division, waste copper ore having chalcocite as the principal copper mineral being chalcopyrite. Besides these ores, museum grade specimens of chalcopyrite were also
used. Waste ore from each of the three sources was crushed and
screened to give a -4 to +6 mesh size (3 to 5mm) fraction for the
leaching experiments.

B. Microorganisms: Cultivation and Harvesting

A *Sulfolobus*-like microorganism described by Brierley et al.
(30) and *T. ferrooxidans* were used in the various studies. The
Sulfolobus-like cells were cultured in 125 ml Erlenmeyer flasks
containing 75 ml of nutrient medium reported by Bryner and
Anderson (7) and 0.02% yeast-extract (see ref. 30). The initial
pH of the medium was adjusted to 2.5, and elemental sulfur was
used as an energy source. The flasks were incubated at 60°C in a
water bath. An aliquot of the culture was transferred into a new
medium every 20 days to maintain the stock culture. The bacteria
free of sulfur were harvested from the remaining part whenever
needed, otherwise these were discarded.

Bacterial cells of the *Sulfolobus*-like microorganism free of
sulfur were obtained by carefully decanting the medium containing
cells into another flask after 20 days of growth. Most of the
sulfur used as an energy source settled to the bottom of the
flask. The decanted medium containing the cells was centrifuged

at 5000 rpm in a Sorvall superspeed Model RC2-B centrifuge. The heavier sulfur sedimented and the supernatant fluid containing the bacteria were carefully removed so as not to disturb the sediment. The decanted fluid contained the cells essentially free of elemental sulfur. This fluid was stored in a refrigerator at $\approx$ 4°C. The cells remained viable for six days.

Thiobacillus ferrooxidans cells were grown in 125 ml. Erlenmeyer flasks containing 75 ml of the nutrient medium reported by Silverman and Lundgren using ferrous sulfate as the energy source. The initial pH of the medium was adjusted to 2.5, and the flasks containing the medium were incubated at room temperature (28°C). An aliquot of the culture was transferred into a fresh medium every 14 days to maintain the stock culture. To obtain cells free of iron, cultures were grown in 2 ℓ. Fernbach flasks containing 1 ℓ. of the nutrient medium (44) and ferrous sulfate as the energy source. The cells were harvested after 14 days of growth. To secure cell suspensions containing a minimum of precipitated iron the harvesting procedure described by Silverman and Lundgren (44) was followed. The contents of the Fernbach flasks were centrifuged (Sorvall superspeed Model RC2-B) at 15000 rpm. The resulting paste of cells and precipitated iron was suspended in $\approx$ 100 ml of cold distilled water (acidified to pH 2.6 with H_2SO_4) in a 500 ml Erlenmeyer flask and shaken vigorously. The flask was then allowed to stand in a refrigerator for at least 6 hours. The turbid supernatant fluid was carefully decanted from the underlying layer of precipitated iron. This procedure was repeated at least two time. The cells contained in the combined supernatant were removed by centrifugation at 15,000 rpm. The pellet was washed with distilled water (pH 2.6) two or three times and finally the pellet containing the cells was brought to a volume of 20 ~ 30 ml with distilled water (acidified to pH 3.5 with H_2SO_4). The clean cell suspensions were stored in a refrigerator and used as an inoculam of cells free of iron.

C. Experimental Procedures

1. *Preparation of Chalcopyrite Thin Sections for Specific Attachment Studies*

Unlike molybdenite which was originally used as a test material to study specific bacterial attachment (45), chalcopyrite does not have any preferential cleavage plans, although some preference for cleavage along {011} might be expected. Thus in order to obtain electron transparent thin sections, natural chalcopyrite (Quebec Canada) was crushed. An optical microscope was used to isolate small sections considered to be electron transparent. These carefully picked thin sections were then examined in a Hitachi Perkin-Elmer H.U. 200F transmission electron microscope operated at 200 kV, and fitted with a goniometer-tilt stage. The sections which were electron transparent were then used for the study. Also small sections (3 to 4 mm on one side) of the natural chalcopyrite samples were taken and polished by hand to thin flakes on a 600 grit (final polish) polishing paper. The polished sections were washed in distilled water in an ultrasonic cleaner (Bransonic 12) to remove debris and air dried. The polished thin sections were used to see bacterial attachment characteristics in the scanning electron microscope. After preparing a sufficient number of thin chalcopyrite chips having electron transparent edges, a special immersion apparatus was devised. Plexiglass discs 25 mm x 25 mm were prepared from a 5 mm thick sheet. Small shallow wells about 2 mm in diameter were drilled in the plexiglass discs. Each disc was then attached to another similar disc slightly bigger but without holes with the help of two plexiglass rods $\approx$ 3 mm in diameter and $\approx$ 12 cm long at the two diagonal ends. This arrangement was required to keep the thin sections submerged during incubation and growth of the microorganisms.

Thin sections of chalcopyrite prepared as described above were placed in the shallow wells of the immersion apparatus and

the discs were covered with plastic mesh to prevent them from
floating when immersed in the medium. Two 250 ml Phillips flasks,
each containing 100 ml of nutrient medium (7), were taken and
covered with aluminum foil. Two other similar flasks with 20 to
30 ml distilled water were kept ready. The immersion assembly
with specimens in place was suspended in these flasks. The water
level was kept at least 10 mm below the lower surface of the
plexiglass disc. These two flasks were then also covered with
aluminum foil. The four Phillips flasks (the two containing the
medium and the other two containing the immersion assemblies with
chalcopyrite specimens) were sterilized in autoclave at 120°C for
15 minutes and then cooled. On cooling 2 ml of 1% yeast extract
was added to one of the two flasks containing the medium under
asceptic conditions. This flask was then inoculated with an aliquot of stock culture of *Sulfolobus*-like microorganisms. The
immersion device containing the specimens was then transferred to
the inoculated flask. The flask was incubated at 60°C in a water
bath for two weeks.

At the conclusion of the incubation period the sample disc
arrangement was withdrawn from the medium. The samples were removed with a fine tweezer (the end of the tweezer sterilized with
alcohol) and placed in a 1% w/v solution of uranyl acetate for at
least 5 minutes to achieve negative staining of the microbes
which were presumed to have attached to the mineral surfaces. The
staining of the microbes has to be done to obtain mass-thickness
contrast (46). The samples after staining were lightly rinsed in
distilled water on a drop plate two times to wash off uranyl
acetate and then dried. The samples with attached, stained
microbes were then examined in the transmission electron microscope operated at 200 kV as described earlier.

Some of the leached chalcopyrite specimens without staining
with uranyl acetate were mounted on standard cylindrical aluminum
stubs (14 mm x 10 mm) with silver conducting paint. These were

then coated with ≈ 100 Å   60/40 Au/Pd sputtered onto the speci-
mens mounted on aluminum stubs in a residual vacuum of ≈ 500 mm
Hg (Torr) in a sputtering unit (Commonwealth Scientific Minicoater
Model CSC-100). The specimens were then examined in a Hitachi-
Perkin Elmer HHS-2R scanning electron microscope operated at 25
kV accerlerating voltage in the secondary electron emission mode.
The SEM is fitted with an ORTEC energy-dispersive X-ray micro-
analyser (Model 6200) utilizing Si(Li) detector for nondispersive
X-ray analysis of different phases in the specimen. The SEM examination was done to study the bacterial distribution on the
surface of chalcopyrite and also to study the bacterial morphologies.

2.   *Bacterial Attachment in Low-Grade Ore*

Waste-ore sections having a mesh size -4 to +6 (3 to 5 mm)
were somewhat randomly selected from Phelps Dodge Corp., Ajo waste
ore containing chalcopyrite as the main copper ore. The waste
ore sections were polished flat on one side, the final polish was
done on 600-grit polishing paper. The polished samples were
washed in distilled water in an ultrasonic cleaner to remove
debris and then air dried.

100 ml of nutrient medium (7) was taken in each of five
250 ml. Erlenmeyer flasks. The pH of the medium was adjusted to
2.3 and the flasks were covered with aluminum foil. The Erlenmeyer flasks containing the medium were sterilized in steam under
pressure (15 ~ 20 minutes, temp. ~ 120°C). The flasks were cooled and 2 ml of 1% sterilized yeast extract solution was added to
each of the flasks (final concentration 0.02%). Six to eight
polished sections were then transferred to each of the flasks
under aseptic conditions. The ore sections were not sterilized
to avoid chemical and related alterations. The flasks were then
inoculated with an aliquot from *Sulfolobus*-like stock culture and
incubated for periods ranging from 2 to 6 weeks at 60°C in a
water bath.

The ore sections were removed from the culture flasks after varying periods (2 to 6 weeks) of incubation and were rinsed lightly in distilled water. This was done to ensure that bacteria not actually adhering to the mineral surface would not remain at random locations as a result of surface tension effects upon withdrawal from the culture medium and also to rinse off the dissolved salts of the culture medium. The specimens were then dried in air in a covered container for 24 hours.

The dried specimens were mounted on standard aluminum stubs (14 mm x 10 mm) with silver paste. These were then coated with approximately 300 Å 60/40 Au/Pd in a unit operated at a residual air pressure of ~ 500 mm Hg.

The specimens were examined in a scanning electron microscope operated at 25 kV in secondary electron emission mode. Pyrite and chalcopyrite phase areas in the specimens were identified by selecting Fe-S or Cu-Fe-S regions simultaneously as regions of interests in the X-ray energy spectrum of the X-ray analyser display. The sample was then scanned at low magnifications (100 X to 500 X) while observing both the secondary electron image and the associated Fe-S or Cu-Fe-S characteristics X-ray map of the imaged region. Phase regions identified as Fe-S (FeS_2) or Cu-Fe-S ($CuFeS_2$) were then systematically studied over a range of much higher magnifications to identify the extent of bacterial attachment and to characterize the selectivity of the attachment.

The experiments described above for the *Sulfolobus*-like microorganisms were repeated using a *T. thiooxidans* in one case and *T. ferrooxidans* inoculum in the second case. The two experiments with *T. thiooxidans* and *T. ferrooxidans* were similar to that described above except that no yeast extract was added to the nutrient medium and the flasks were incubated at room temperature (~ 28°C).

3. *Surface Reactions at Sulfide Phases*

Waste ore sections having a mesh size -4 to +6 (3 to 5 mm) were randomly selected from crushed and screened Duval Corporation-Sierrita mine waste (chalcopyrite being the main copper mineral). These sections were polished flat on one side on 220 grit polishing paper, then successively to 600 grit polishing paper (final polish). The polished sections were washed in distilled water in an ultrasonic cleaner to remove debris, air dried and kept in a dessicator for later use. Also -4 to +6 mesh size ore ($\approx$ 200 g.) was given several washes in water to remove loose dust adhering onto the surface and also air dried and stored in a dissicator for later use.

A complete ore analysis of the Sierrita ore was done. Copper and iron analysis of -4 to +6 mesh size Sierrita ore was also done. The sized ore was taken, washed and dried and crushed to powder. A known weight of the powdered ore was digested by standard procedure and brought to a known volume.

Total iron in the solution (digested ore) was analyzed by titration against standardized potassium dichromate solution as titrant using diphenylamine sulfonate as indicator (47). Copper analysis was done by atomic absorption spectroscopy using a Perkin Elmer Model 303 atomic absorption spectrophotometer.

a. Flask Leaching. 250 ml. Erlenmeyer flasks containing 100 ml. nutrient medium (47) with pH adjusted to 2.0 were taken. The flasks were covered with aluminum foil and sterilized in an autoclave (120°C, 15 minutes). 10 grams of earlier washed and dried -4 to +6 mesh size (3 to 5 mm) ore including 6 to 8 polished slices were accurately weighed and transferred to each of the flasks on cooling. Two flasks were prepared for each sample, one used as a sterile control and the second was inoculated with 1 ml. of *Sulfolobus*-like microorganisms. Cells free of sulfur were used in the inoculum. The flasks were incubated at 55$\pm$2°C in a water saturated air incubator (Precision Thelco Model 6, temperature range 65°C) with no agitation. Eight sets, each set containing two flasks; one sterile and one inoculated with microbes were used in the experiment. Any loss in water due to evaporation during the course of this experiment was made up by adding sterile distilled water. Samples were taken out at weekly intervals, up to eight weeks. At the end of the weekly periods, the polished slices were removed from the medium, rinsed lightly in distilled water and air dried for at least 24 hours. The long drying period was utilized to reduce morphological distortions by surface tension effects during drying because a critical point drying apparatus was not available for this study. The medium was removed for the chemical analysis of Fe^{3+}, Fe^{2+} and Cu content in the solution and the bacterial count (MPN)[2].

The flat specimen, after drying were mounted on aluminum stubs with silver paste and coated with approximately 300 Å layer of Au/Pd in a sputtering unit. The specimens were then examined in the scanning electron microscope at 25 kV fitted with an ORTEC energy dispersive X-ray analyzer. The liquid medium was analyzed for Fe^{3+} and Fe^{2+} by titration method using standarized potassium dichromate solution as titrant and diphenylamine sulfonate as indicator (47). Cu was analyzed by atomic absorption spectroscopy. The bacterial count in the medium was done using MPN (most probable number)[2] estimates procedure (48). This experiment was repeated with the flasks incubated at 60±2°C.

The experimental procedure as outlined above was followed using *Thiobacillus ferrooxidans*. The nutrient medium used was the one reported by Bryner and Anderson (7). The initial pH of the medium was adjusted to 2.1. As before, two flasks were prepared for each sample. The flasks were inoculated with 5 ml of

[2]*This is a statistical method using various dilutions of the medium containing bacteria.*

inoculum (2.5 x 10^8 cells) of iron free cells, 5 ml of panacide
solution (0.8 g panacide/1 ℓ ethyl alcohol), which is a bacteri-
cide, and a fungicide was added to each of the sterile flasks
used as controls. The flasks were placed on a rotary shaker
(Lab-Line Junior Orbit Shaker) run at 150 rpm continuously. The
chemical analysis of the medium at weekly intervals (1 to 8 weeks),
bacterial count (MPN) and the SEM observation of the flat ore
sections were performed as described in detail above and the
experiment was duplicated under the same conditions. Another
experiment was performed with *T. ferrooxidans* using a stationary
flask technique for comparison with the stationary flask leaching
experiments utilizing the *Sulfolobus*-like microorganism.

III. RESULTS AND DISCUSSION

A. Morphology of Microorganisms Attached to Sulfide-Phase Surfaces

Figure 1 shows for comparison typical views of *Thiobacillus
thiooxidans, Thiobacillus ferrooxidans,* and the *Sulfolobus*-like,
thermophillic microorganisms as they appear in the scanning electron microscope when attached to either pyrite or chalcopyrite
surfaces. While there is a distinct morphological difference
between the *Thiobacilli* and the *Sulfolobus*-like microorganism,
there is also a rather marked difference in the density of attachment, with the *Sulfolobus*-like thermophile attaching more readily
than the *Thiobacilli*. This may be due in part to the increased
mobility of the *Thiobacilli* as a result of their flagellum [Fig.
1 (b) insert], and the fact that the cell wall of the *Sulfolobus*-
like microorganism is much less rigid, and perhaps more conducive
to adhesive attachment. The contact area associated with the
attachment of the *Sulfolobus*-like microorganisms appears to be
generally larger than that associated with the *Thiobacilli*
although there is some evidence, as apparent in the insert of
Fig. 1 (b), that the *Thiobacilli* secrete a substance of some

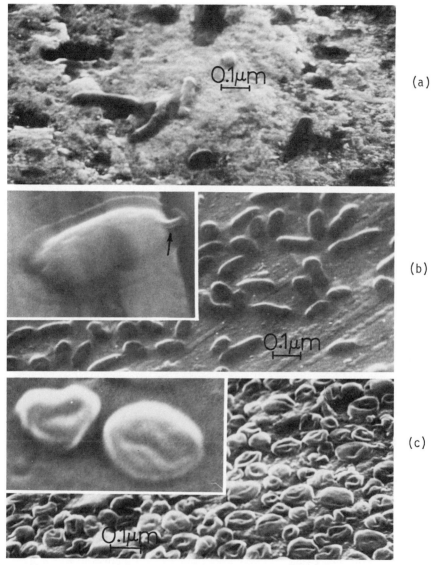

Fig. 1. Bacteria attached to pyrite surfaces. (a) Thiobacillus thiooxidans *on pyrite after incubation for 7 weeks,
(arrows). (b)* Thiobacillus ferrooxidans *after 7 weeks. The
insert shows a magnified view of a single microbe on chalcopyrite.
Flagellum disturbed by drying is visible (arrow) and the microorganism appears to be resting upon a biomat of some kind, (c)*
Sulfolobus-*like microorganisms. Insert shows magnified view. The
distortion results by drying in air; vacuum collapse of the cell wall.*

kind which might aid in their attachment. This observation is, however, not well documented at present, and the nature of the substance is also unknown. It does not, however, appear to be artifactually produced, but this will be apparent only after preparation involving critical-point drying.

B. Bacterial Attachment: Site Specificity and the Possible Role of Crystal Defects

In previous studies involving bacterial attachment to molybdenite utilizing thin electron transparent sections in which dislocations could be observed, it was difficult to associate the attachment site of *Sulflobus*-like microorganisms with specific dislocations or dense arrays of dislocations partly because in molybdenite (MoS$_2$) the thin sections cleave preferentially along (0001), and the dislocations lie mainly parallel to the surfaces (45). Only dislocations associated with surface ledges can terminate on the free surface. In an effort to circumvent this shortcoming, and to apply the technique to chalcopyrite, of interest in the present studies, crushed and carefully selected thin sections were examined and treated experimentally as described previously. While it was possible to prepare several excellent thin sections, it was not possible to induce attachment of the *Sulfolobus*-like microorganisms near the thin edges, and it was therefore not possible to associate the specific attachment site with any crystal defects, particularly dislocations. However, bacteria were observed away from the chalcopyrite specimen edges which were arranged in patterns resembling grain or subgrain boundaries, i.e. the emergence of such defects on the crystal surface. An unambiguous demonstration of this contention was, however, not possible.

Figure 2 illustrates some of the important features of the experiments outlined above, and some of the results alluded to in the preceeding paragraph. Figure 2 (a) and (b) illustrate

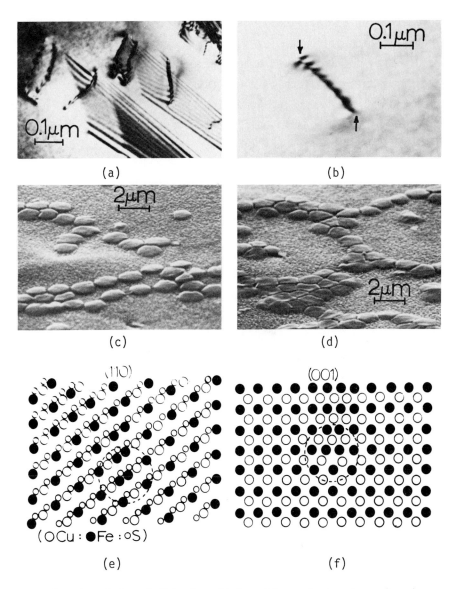

Fig. 2. (a) and (b) show dislocation substructure in electron transparent thin sections of chalcopyrite which are imaged by diffraction contrast (46). (a) shows numerous dislocations and stacking faults while (b) shows a single dislocation. (c) and (d) show cell-like attachment characteristics for Sulfolobus-like microorganisms attached to the thin section surfaces. (e) and (f) illustrate schematic views of edge dislocations intersecting surfaces having a (001) or (011) orientation respectively. The arrows in (b) show the dislocation ends which terminate on the surface.

some examples of individual dislocations and other defects in the
initial chalcopyrite samples prior to immersion in the medium.
Figure 2(c) and (d) show the arrangement of *Sulfolobus*-like microorganisms on the surface of a chalcopyrite thin section. In Fig.
2(e) and (f) are shown two idealized views of dislocations intersecting the surface having different orientations, and the local
(surface) stoichiometry changes with respect to both Fe and S.

C. Selective Bacterial Attachment to Sulfide Phase Surfaces

It has already been demonstrated in previous work that when
the *Sulfolobus*-like microorganisms attach to the surface of a polished section of a low-grade waste rock, the attachment is
specific to sulfide phase regions, viz. FeS_2 and $CuFeS_2$ surface
regions (49). These observations have been made repeatedly, and
were also supported by observations made in the present investigations. In addition, and as depicted somewhat generally in Fig.
1, the attachment of *T. ferrooxidans* and *T. thiooxidans* to low-
grade waste-rock surfaces was also specific to exposed FeS_2 and
$CuFeS_2$ regions, but the attachment was not as frequent, and the
density of attached microbes (number/unit area of exposed sulfide
phase surface) was considerably less than for the *Sulfolobus*-like
microorganism (Fig. 1).

The specifity of attachment of the *Sulfolobus*-like microorganisms is illustrated typically in Fig. 3 which shows a sharp
boundary separating a $CuFeS_2$ region on the left where microbes
are attached from quartz matrix on the right which is free of
attached microbes. Figure 4 also illustrates the specific attachment features and the identification of sulfide-phase regions
utilizing energy dispersive X-ray spectrometry and X-ray mapping.
In addition, Fig. 4 illustrates the complete surface coverage of
bacteria which has been observed to occur in some cases.

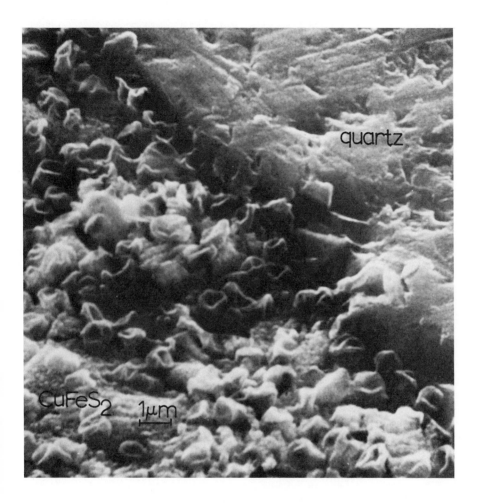

Fig. 3. Sulfolobus-*like microorganisms selectively attached
to a chalcopyrite region in a quartz host rock. Note the sharp
boundary between the CuFeS$_2$ on the left and the quartz on the
right. Bacteria appear distorted as a result of cell wall collapse during drying. This could be avoided by drying at the
critical point, but this process might then alter the mineral
substrate.*

While it has been observed that profuse attachment usually
requires 3 weeks or more of incubation, attachment is not always
observed after this time. In many experiments, essentially no
attachment was observed even after 8 weeks incubation, even though

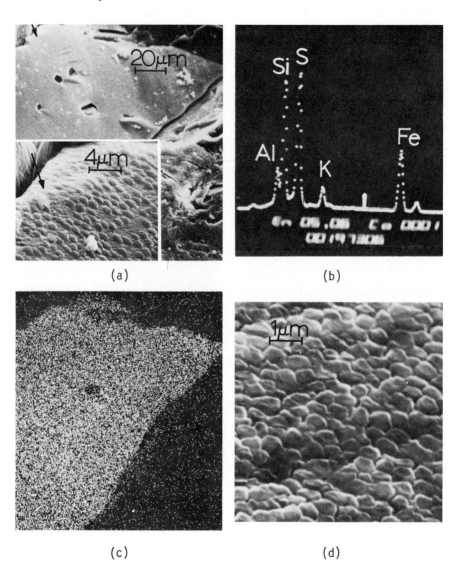

(a)   (b)

(c)   (d)

Fig. 4. Selective attachment of Sulfolobus-*like microorganisms on pyrite. (a) Pyrite grain with bacteria attached. The insert shows a magnified view of the bacteria, (b) Energy-dispersive X-ray spectrum of the entire area in (a), including the matrix region which consists mainly of K-Al silicates; (c) X-ray map using Fe and region on which the bacteria are selectively attached, (d) Dense (continuous) coverage of a pyrite surface by* Sulfolobus-*like microorganisms.*

there was a marked difference between the copper solubilized in the inoculated case when compared to the uninoculated (sterile) control situation.

It is of course not unexpected that bacteria would attach to the sulfide regions because these regions (FeS_2 and $CuFeS_2$) are, after all, sources of energy in the metabolic processes. However attachment does not seem necessary for enhanced solubilization of copper, and as a consequence metabolism does not seem to be predicated upon direct attachment. Attachment could, nonetheless, be important and could enhance the leaching kinetics, but measurements of leaching kinetics related to bacterial attachment have been neither consistent nor systematic. Such measurements do not exist at present.

D. Surface Reactions at FeS_2 and $CuFeS_2$ and Their Relation to Bacterial Leaching

While bacterial attachment may not be necessary for leaching to be stimulated as a consequence of the catalytic role played by the bacteria, leaching must involve the degradation, i.e. reaction or corrosion of the surface of the exposed pyrite and chalcopyrite phase inclusions. In the case of pyrite inclusions, the following prominent reactions are taking place:

$$FeS_2 + 3.5O_2 + H_2O \longrightarrow FeSO_4 + H_2SO_4 \qquad [1]$$

$$2FeSO_4 + 0.5\ O_2 + H_2SO_4 \xrightarrow{bacteria} Fe_2(SO_4)_3 + H_2O \qquad [2]$$

$$FeS_2 + Fe_2(SO_4)_3 \longrightarrow 3FeSO_4 + 2S \qquad [3]$$

$$2S + 3O_2 + 2H_2O \xrightarrow{bacteria} 2H_2SO_4 \qquad [4]$$

In the case of chalcopyrite regions, the following reactions are taking place:

$$2CuFeS_2 + 8.5O_2 + H_2SO_4 \xrightarrow{bacteria} 2CuSO_4 + Fe_2(SO_4)_3 + H_2O \qquad [5]$$

$$CuFeS_2 + 2Fe_2(SO_4)_3 \longrightarrow CuSO_4 + 5FeSO_4 + 2S \qquad [6]$$

where the sulfur formed in Eq. [6] is converted to sulfuric acid according to Eq. [4]. While bacterial attachment to FeS_2 and $CuFeS_2$ might accelerate or optimize the catalytic activity indicated in Eqs. [2], [4] and [5], above, such activity would also be expected simply by the presence of sufficient numbers of microorganisms in the solution proximate to the reacting surface.

Figures 5 and 6 clearly illustrate the catalytic role played by bacteria in the leaching of low-grade, copper-bearing waste rock. The catalytic activity is especially graphic on comparing the Fe^{3+}/Fe^{2+} ratio in Fig. 6. If this is taken as a quantitative measure of bacterial catalysis, then it is readily apparent that the enhancement is as much as a factor 10^3. It should be noted in Fig. 5 that the solubilization of copper is very small in stationary leaching with $T.\ ferrooxidans$. This is due in large part to the formation of hydronium jarosites which prevent direct exposure of the sulfide surfaces to the leach solution. The decline and valley in the Fe^{3+}/Fe^{2+} response noted in Fig. 6 is also due to this effect. In addition, differences in the leaching rates for the $T.\ ferrooxidans$ runs (stationary and agitated flask leaching experiments) and the $Sulfolobus$-like microorganism studies (which were also stationary flask runs) are due to the temperature difference and not because of a difference in the catalytic activity being better for the $Sulfolobus$-like microorganisms. This feature is treated in detail in Chap. 25.

Figures 7 and 8 show the corresponding surface corrosion on FeS_2 and $CuFeS_2$ which results from the reactions indicated in Eqs. [1] - [6], and the qualitative surface features corresponding to the quantitative leaching responses noted in Figs. 5 and 6. There are several interesting features to note in Figs. 7 and 8. First it should be apparent that there is in general a dearth of attached microorganisms. While one might expect that in the heavily corroded surface regions bacteria might remain tenaciously attached or entrapped within the surface pits and

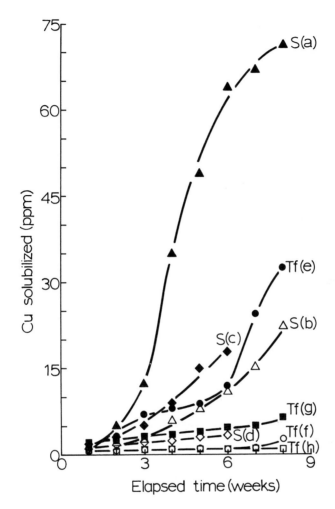

Fig. 5. Comparison of copper leaching with T. ferrooxidans *and* Sulfolobus-*like microorganisms.* S(a) *and* S(b) *denote the inoculated (a) and control or sterile (b) condition for the* Sulfo-lobus-*like stationary flask run at 60°C. The initial pH was 2.1 while the final inoculated (a) and control (b) pH was 2.5 and 2.8 respectively.* S(c) *and* S(d) *denote the inoculated (c) and control (d) condition for the* Sulfolobus-*like stationary flask run at 55°C. The intial pH was 2, and the final pH was 2.8 and 3 for (c) and (d) respectively.* Tf (e) *and* Tf (f) *denote the inoculated (e) and control (f) condition for* T. ferrooxidans *in a shake flask run at 28°C. The initial pH was 2.1 while the final pH was 2.6 and 2.8 for (e) and (f) respectively.* Tf (g) *and* Tf (h) *denote the inoculated (g) and control (h) condition for* T. ferrooxidans *in a stationary flask run at 28°C. The initial pH was 2.3 while the final Ph was 2.8 and 3.0 for (g) and (h) respectively.*

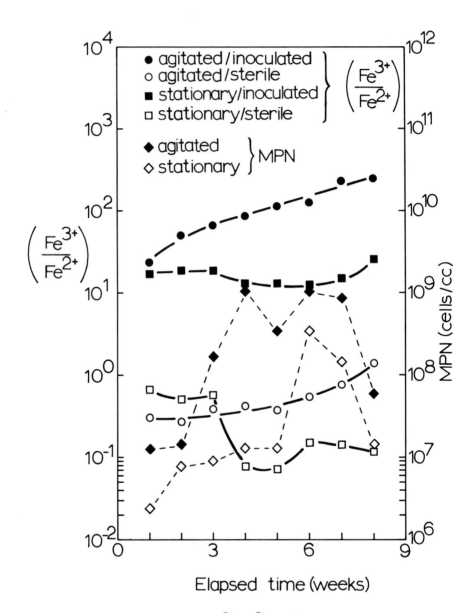

Fig. 6. Comparison of Fe^{3+}/Fe^{2+} ratios for stationary and agitated flask leaching of Duval-Sierrita waste using T. ferrooxidans. The most probable number (MPN) of microorganisms is also shown corresponding to the leaching times. All experiments were performed at 28°C, and correspond to those recorded in Fig. 5.

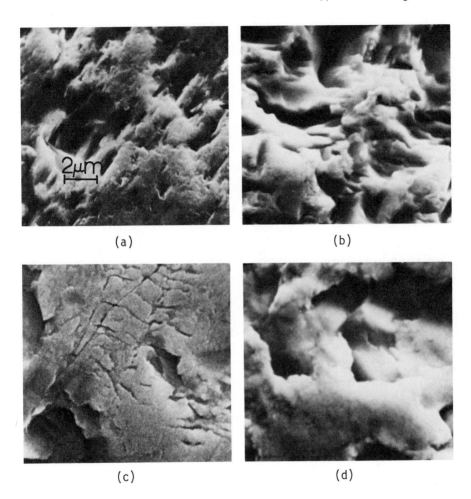

(a)   (b)

(c)   (d)

*Fig. 7. Corrosion of pyrite and chalcopyrite phase regions
in low-grade waste after 8 weeks in agitated flasks. (a)sterile
(control) pyrite surface, (b) pyrite surface exposed to medium
inoculated with* T. ferrooxidans, *(c) Sterile (control) chalcopyrite surface, (d) chalcopyrite surface exposed to medium inoculated with* T. ferrooxidans. *All runs at 28°C. All magnifications
are given in (a).*

and related corrosion morphology, there was no evidence of this.
There simply was no prominent attachment of bacteria even though,
as noted in Figs. 5 and 6, leaching in the presence of bacteria
was markedly enhanced. There were however, instances where

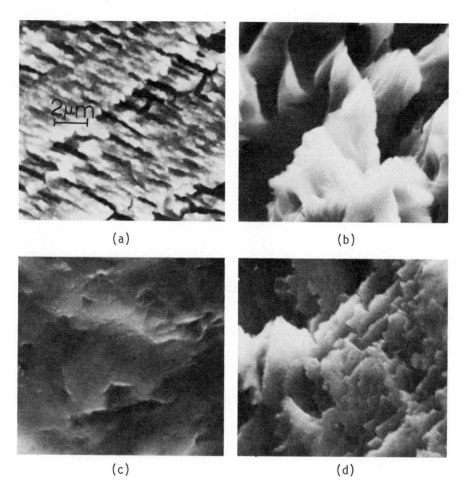

(a)   (b)

(c)   (d)

Fig. 8. Corrosion of pyrite and chalcopyrite phase regions low-grade waste after 8 weeks in stationary flasks. (a) sterile (control) pyrite surface, (b) pyrite surface exposed to medium inoculated with Sulfolobus-*like microorganisms (c) sterile (control) chalcopyrite surface, (d) chalcopyrite surface exposed to medium inoculated with the* Sulfolobus-*like microorganisms. All runs at 60°C. All magnifications are given in (a).*

bacteria were observed to attach to the corroded surface, and this is illustrated in Fig. 9. In addition, it was apparent that the corrosion texture in FeS_2 or $CuFeS_2$ was not the same over the waste particle surface treated in an identical way, and this particular feature was attributed to differences in the crystallographic

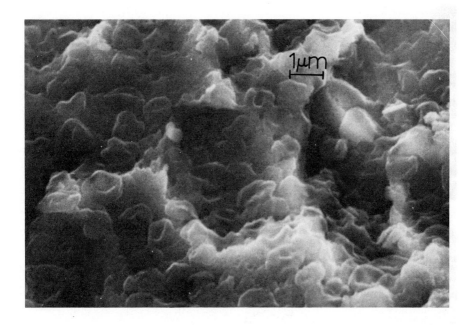

Fig. 9. Sulfolobus-*like microorganisms attached to the corroded surface of a pyrite inclusion after incubation at 60°C for 4 weeks.*

surface orientation exposed to the leach solution. This feature is shown in Fig. 10 (a) - (c), and unambiguously characterized as differences in grain surface orientation in Fig. 10 (d) and (e) which show different grains separated by grain boundaries, with the degree and texture of surface corrosion changing across the boundaries. This occurs simply because the rates of reaction are different for different crystallographic surface orientations just as rates of crystal growth are different along different crystallographic directions.

E. Galvanic Interaction Between Chalcopyrite and Pyrite in
    Intimate (Electrical) Contact

It can be observed in Figs. 7 and 8 that when the FeS_2 and $CuFeS_2$ phase inclusions leach separately, there is a greater rate of reacation on the FeS_2. The chalcopyrite, in fact, leaches

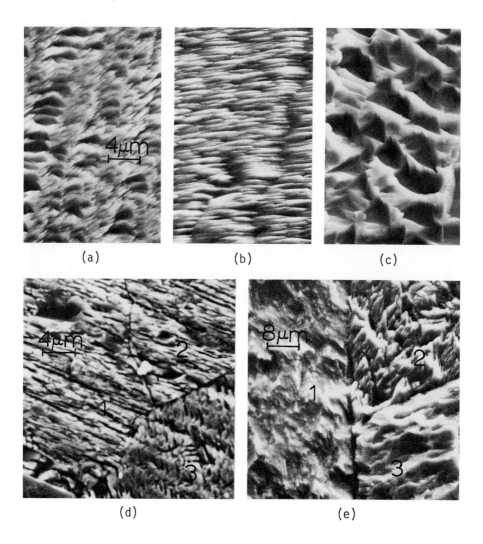

Fig. 10. *Variations in the degree (texture) and rate of corrosion for different phase (grain) surfaces exposed to identical leaching conditions; illustrating the effects of grain surface (crystallographic) orientation. (a) - (c) show three different pyrite surface areas in the same waste particle leached in the presence of* Sulfolobus-*like microorganisms for 6 weeks at 60°C. (d) shows the junction of three grains of pyrite leached for 4 weeks and showing different corrosion on each grain surface, (e) junction of three chalcopyrite grains showing differences in corrosion after 6 weeks. (d) and (e) in medium inoculated with* Sulfolobus-*like microorganisms.*

relatively slowly by comparison with the pyrite. However, it was frequently observed that when the pyrite and chalcopyrite were in contact, the trend normally observed was reversed, i.e. the $CuFeS_2$ reacted much faster than the FeS_2, which was essentially passivated as a result of its higher rest potential as compared with chalcopyrite in forming a galvanic couple. This is identical to the galvanic conversion described recently by Hiskey and Wadsworth (50) except that the conversion is accelerated, additionally, by the presence of bacteria which limit the formation of elemental sulfur through the reaction of Eq. [4]. Figure 11 illustrates this effect, and shows for comparison the differences in the surface reactions at a non-contacting FeS_2 phase. The chalcopyrite surface corrosion in Fig. 11 can also be compared with Fig. 8 (d).

The implications of this phenomena (as illustrated in Fig.11) in a low-grade waste heap or dump leaching operation would seem to have great significance because if a waste rock regime contained a large proportion of contacting sulfides, then the leaching rate would presumably be very large for the sulfide member having the lower rest potential. In the presence of bacteria, which is probably a characteristic of all leach dumps, the process would be enchanced as described above, and illustrated in Fig. 11. The galvanic interaction between chalcopyrite and pyrite during bacterial leaching is treated in more detail elsewhere (51).

IV. SUMMARY AND CONCLUSIONS

The morphologies of *T. thiooxidans*, *T. ferrooxidans*, and a *Sulfolobus*-like, thermophilic microorganisms attached to sulfide surfaces have been examined by scanning electron microscopy. The attachment of *Sulfolobus*-like microorganisms was more prolific than the *Thiobacilli*, and this was attributed to the difference in cell wall rigidity and the absence of a flagellum on the *Sulfolobus*-like microorganisms.

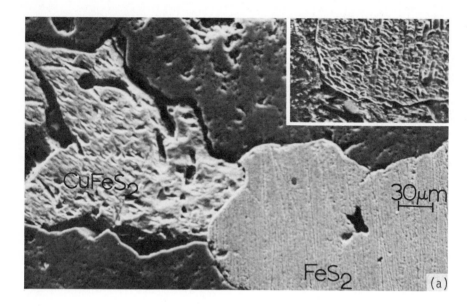

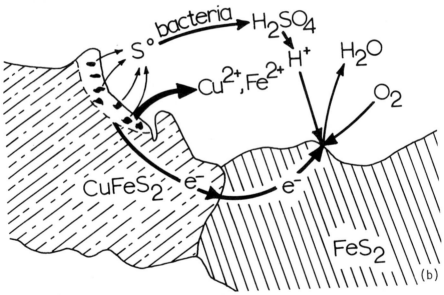

Fig. 11.   (a) Contacting pyrite and chalcopyrite phases showing selective galvanic corrosion of the anodically reacting chalcopyrite and the passivation of the pyrite after 8 weeks in medium inoculated with the Sulfolobus-like microorganism.   The insert shows a separate (non-contacting) pyrite surface similarly leached for only 6 weeks. (b) schematic representation of (a) depicting the galvanic conversion model for the $CuFeS_2/FeS_2$-acid system

In attempting to determine whether single microbes attach
selectively to crystal defects which intersect the surface,
chalcopyrite thin sections were prepared and observed in the
transmission electron microscope. While the technique has been
demonstrated to be a viable one, and dislocations have been observed, bacterial attachment to individual dislocations has not
been observed because for some reason it was not possible to get
the bacteria to attach to the electron transparent edge regions
of the chalcopyrite sections, possibly because of some edge
effects or surface tension effects. However, observations of
attached bacteria to the thicker regions of the specimen showed
them to be arranged in cell-like arrays resembling grain boundaries
or sub-grain boundaries. It may be that because of local variations in stoichiometry of the emergence sites of dislocations in
chalcopyrite, these may be favorable for attachment, but this
question remains open.

The attachment of bacteria was, however, observed to be
specific to sulfide phase regions on a polished waste rock surface
because these are regions offering an energy source. However,
while attachment was observed, and while it was unquestionably
specific to only the sulfide phase regions (FeS_2 and $CuFeS_2$), it
did not always occur. Nonetheless, bacterial catalysis was shown
to have been important in the leaching of the waste rock in flask
experiments, and correspondingly the surfaces of FeS_2 and $CuFeS_2$
phase regions were observed to have been reacted (corroded). The
corrosion observed was also varied, and the degree of corrosion
was observed to vary with crystallographic orientation.

When FeS_2 and $CuFeS_2$ phases were in intimate contact, the
normally rapidly corroding FeS_2 was observed to be passivated
while the $CuFeS_2$ was very vigorously corroded. This galvanic
conversion was enhanced in the presence of bacteria not only
because of the catalytic oxidation of the chalcopyrite, but also
because of the bacterial conversion of elemental sulfur formed

on the chalcopyrite to sulfuric acid. This observation can have significant applications in a variety of sulfide leaching processes.

There is no question that bacteria attach to mineral surfaces and that they attach to specific areas in a waste rock which can function as an energy source, namely sulfide phases in the present study. However attachment does not always occur, and bacterial catalysis can be very significant without direct attachment.

V. ACKNOWLEDGMENTS

This research was supported by the National Science Foundation (RANN) under grants AER-76-03758 and AER-76-03758-A01. The authors thank James A. and Corale L. Brierley for laboratory provisions and the provision of the high-temperature microorganism utilized in this investigation, as well as many fruitful discussions.

VI. REFERENCES

1. Taylor, J.H., and Whelan, P.F., *Trans. Inst. Min. Met.*, 52, 36(1943).

2. Razzell, W.E., and Trussell, P.C., *J. Bacteriol.*, 85, 595 (1963).

3. Rudolfs, W., *Soil. Sci.*, 14, 135 (1922).

4. Rudolfs, W., and Helbronner, A., *Soil Sci.*, 14, 459 (1922).

5. Colmer, A.R., and Hinkel, M.E., *Science*, 106, 253 (1947).

6. Bryner, L.C., Beck, J.V., Davis, D.B., and Wilson, D.G., *Ind. Engr. Chem.*, 46(12), 2587 (1954).

7. Bryner, L.C., and Anderson, R., *Ind. Engr. Chem.*, 49(10), 1721 (1957).

8. Malouf, E.E., and Prater, J.D., *J. Metals*, 353 (1961).

9. Winogradsky, S., *Botan. Ztg.*, 46, 261 (1888).

10. Dugan, P.R., and Lundgren, D.G., *J. Bacteriol.*, 89, 325 (1965).

11. Sinha, D.B., and Walden, C.C., *Can. J. Microbiol.*, 12, 1041 (1966).

12. Peck, H.D., *Bacteriol. Rev.*, 26, 67 (1962).

13. Brock, T.D., "Biology of Microorganisms", Prentice-Hall, Inc., Englewood Cliffs, New Jersey, 1970.

14. Torma, A.E., *Adv. Biochem. Engr.*, in Press (1977).

15. Tuovinen, O.H., and Kelly, D.P., *Int. Met. Rev.*, 19, 21 (1974).

16. Corrans, I.J., Harris, B., and Ralph, B.J., *J. South African Inst. Min. Met.*, March, 221 (1972).

17. Tuovinen, O.H., and Kelly, D.P., *Zeit. fur Allg. Mikrobiol.* 12 (4), 311 (1972).

18. Trudinger, P.A., *Minerals Sci. Engr.*, 3 (4), 13 (1971).

19. Silverman, M.P. and Ehrlich, H.L., *Adv. Appl. Microbiol.*, 6, 153 (1964).

20. Bruynesteyn, A., and Duncan, D.W., *Can. Met. Quart.*, 10(1), 57 (1970).

21. Pinches, A., Al-Jaid, F.O., and Williams, D.J.A., *Hydrometallurgy*, 2, 87 (1976).

22. Torma, A.E., in Proc. 3rd. Int. Biodegradation Symposium, (J.M. Sharpley and A.M. Kaplan, Eds.) p. 937 Applied Science Publishers, Ltd., London, 1976.

23. Sakaguchi, H., Silver, M., and Torma, A.E., *Biotechnol. Bioengr.*, XVIII (8), 1091 (1976).

24. Torma, A.E., and Itzkovitch, I.J., *Appl. and Environ. Microbiol.*, 32 (1), 102 (1976).

25. Bryner, L.C., and Jameson, A.K., *Appl. Microbiol.*, 6, 281 (1958).

26. Duncan, D.W., Walden, C.C., Trussell, P.C., and Lowe, E.A., *Trans. Soc. Min. Engrs.*, 238, 1 (1967).

27. Sutton, J.A., and Corrick, J.D., *Min. Engr.* 15, 37 (1963).

28. Sutton, J.A., and Corrick, J.P., U.S. Bureau of Mines Report Ins. 6423, p. 1, 1964.

29. Brock, T.D., Brock, K.M., Belly, R.T., and Weiss, R.C., *Arch Microbiol.*, 84, 54 (1972).

30. Brierley, C.L., and Brierley, J.A., *Can J. Microbiol.*, 19, 183 (1973).

31. de Rosa, M. Gambacorta, A., and Bulock, J.D., *J. Gen. Microbiol.*, 86, 156 (1975).

32. Silverman, M.P., *J. Bacteriol.*, 94, 1046 (1967).

33. McGoran, C.J.M., Duncan, D.W., and Walden, C.C., *Can. J. Microbiol.*, 15, 135 (1967).

34. Ehrlich, H.C., and Fox, S.I., *Biotechnol. Bioengr.*, 9, 471 (1967).

35. Duncan, D.W., and Drummond, A.D., *Can. J. Earth Sci.*, 10, 476 (1973).

36. Gormley, L.C., and Duncan, D.W., *Can. J. Microbiol.*, 20, 1453 (1974).

37. Brierley, C.L., Brierley, J.A., and Murr, L.E., *Res. Develop. Mag.;* 24 (8), 24 (1973).

38. Weiss, R.C., *J. Gen. Microbiol.*, 77, 501 (1973).

39. Baldensperger, J., Guarraia, L.J., and Humphreys, W.J., *Arch Microbiol.*, 99, 323 (1974).

40. Duguid, J.P., *J. Gen. Microbiol.*, 21, 271 (1959).

41. Hirsch, P., and Pankrantz, Sr. H., *Zeit. fur Allg. Mikrobiol.*, 10, 589 (1970).

42. Schaeffer, W.I., Holbert, P.E., and Umbreit, W.W., *J. Bacteriol.*, 85 137 (1963).

43. Meadows, P.S., *Arch fur Mikrobiol.* 75, 374 (1971).

44. Torma, A.E., Walden, C.C., and Branion, R.M.R., *Biotechnol. Bioengr.*, 12, 501 (1970).

45. Berry, V.K., and Murr, L.E., *Met. Trans.*, 6B, 488 (1975).

46. Murr, L.E., "Electron Optical Applications in Materials Science", McGraw-Hill Book Co., Inc., New York, 1970.

47. "Chemical Analysis of Metals",: Sampling and Analysis of Metal Bearing Ores", American Soc. for Testing and Materials, Philadelphia, Pennsylvania, 1969.

48. Collins, C.H., "Microbiolgical Methods", Plenum Press, New York, 1967.

49. Murr, L.E., and Berry, V.K., *Hydrometallurgy*, 2, 11 (1976).

50. Hiskey, J.B., and Wadsworth, M.E., *Met. Trans.*, 6B, 183 (1975).

51. Berry, V.K., Murr, L.E., and Hiskey, J.B., *Met. trans.*, 8B, in press.

GENETIC MECHANISMS IN

METAL-MICROBE INTERACTIONS

A.M. Chakrabarty

General Electric Company
Corporate Research & Development
Schenectady, New York, 12301 USA

The genetic bases of the interaction of microorganisms with metals have been reviewed. Many microorganisms develop resistance to toxic concentrations of inorganic or organic forms of heavy metals by reducing them to elemental metal. Genes specifying such reduction steps have been shown to be plasmid-borne in several instances. Plasmid-specified resistance to cadmium, on the other hand, is due to reduced uptake of this metal by the resistant cells. Microorganisms are also known to accumulate certain metal ions, which are rendered innocuous because of their binding with the intracellular protein. This ability has also been shown to be due to plasmid-borne genes in some instances. Finally, plasmids have also been implicated in enhancing methylation of metals such as mercury. Although nothing is known about the genetic basis of metal leaching by the Thiobacilli *group of bacteria, some areas where potential genetic improvements can be made with* Thiobacillus ferrooxidans *are suggested.*

I. INTRODUCTION

Microorganisms are ubiquitous in nature; so are various metal ions, which either occur in ores or natural rocks and minerals, or else are released to the environment via industrial and household uses of their compounds. Because metal ions, particularly the heavy metals, are toxic at high concentrations, microorganisms have evolved various biological systems to cope with their inter-actions with living cells. One such system has led to the devel-opment of resistant cells that can convert the reactive inorganic form of the metal to the less reactive elemental form. The metal can then be volatilized off (mercury, for example) or stored within the cells. Alternatively, resistance may arise from a change in the permeability properties of the cell membrane so that cellular uptake of the specific metal ion is greatly diminished. In addi-tion, several fungi and bacteria are known to methylate metal ions, which can then be easily volatilized off from the cells, or elaborate proteins which keep the metal in an innocuous form.

Microorganisms, particularly the chemolithotrophic *Thio-bacillus* group that can release soluble form of metals in an acidic medium from insoluble ores, which is of great industrial significance, are known. These microorganisms develop resistance to very high concentration of leached metals by mechanisms as yet unknown. An understanding of the genetic basis of leaching of metals by bacteria such as *Thiobacillus ferrooxidans*, as well as the development of resistance against the leached metals, could be of great value in the enhanced leaching and recovery of the metals for industrial use.

In this article, I would like to review some of the genetic and biochemical bases of the development of bacterial resistance and intracellular accumulation of metal ions. Additionally, I would like to point out some areas where genetic studies, particu-larly with *Thiobacillus ferrooxidans*, may contribute significantly towards enhanced leaching of metals such as copper or uranium.

II. METAL ION RESISTANCE, ACCUMULATION, AND METHYLATION

In general, microorganisms are sensitive to high concentra-tions of heavy metal salts. It has also been recognized, however,

that microorganisms can develop resistance against high, toxic
concentrations of many metal ions when adapted in presence of
increasingly higher concentrations. In addition, organisms can
also be isolated that exhibit high levels of resistance against
some metal ions in their naturally-occurring state. Such resistant bacteria often harbor plasmids that specify resistance
against the toxic metal (Table 1). Four alternative mechanisms
are used by the bacteria for achieving resistance:

(1) Reduction to free metal. This mode of resistance has
been widely studied both from genetic as well as enzymatic points
of view, particularly in regard to resistance against mercury.
Because of widespread industrial use of inorganic mercury salts
and the organomercurials, their consequent release into rivers
and lakes, and their ultimate capture by fish used as human food,
the environmental transformation of mercury has received much
attention in recent years. Various types of bacteria have been
demonstrated to play significant roles in the cycles of transformations from metallic to ionic to organomercurial states that
proceed in aquatic ecosystems and are responsible for the amounts
and nature of mercury compounds found in various parts of the
food chain (1,2).

It appears that most bacteria develop resistance against
mercury salts and organomercurials by reducing such species to
metallic mercury, which can then be volatilized off from the
cells. The genes specifying such enzymes have been shown to be
plasmid-borne and more than 200 of such plasmids in enteric
bacteria, *Staphylococcus aureus* and *Pseudomonas* have been characterized (3). Some plasmids determine resistance to mercuric ion
alone while others determine resistance to a range of mercurials
in addition to Hg^{2+}. Plasmid-bearing bacteria thus convert Hg^{2+}
to Hg^0, phenylmercuric acetate to Hg^0 + benzene and methylmercuric
chloride to Hg^0 plus methane. The enzyme systems specified by
such plasmids are invariably inducible, irrespective of whether

TABLE I   Plasmids Specifying
Resistance to Toxic Elements

Plasmid	Element Resisted[a]	References
PI 258	Hg, Cd, Bi, Pb, As, Sb	7
R factors	Ni, Co, Hg	33
pMG 101	Ag, Hg, Te	34
R 733	As	40
R 477	As, Hg, Te	18
R 3108	Hg	3
MER	Hg	4
pMG 6	Cr, Hg, Te	6
RMS 159	B, Hg, Te	6

[a] Only resistance to toxic elements has been shown.   Most of the plasmids also carry antibiotic resistance genes.

they attack inorganic mercury or the organomercurials, and generally consist of a reducing enzyme, a C-Hg splitting enzyme, and hydrolases specific for alkyl and aryl mercurials. The pertinent properties of such enzymes have recently been reviewed (4).

Microbial reductions of oxyanions of heavy metals, such as selenium and tellurium, have also been reported (5,6). In many cases, the reducing enzymes are known to be specified by plasmid-borne genes. Tellurite (or tellurate) reduction to metallic tellurium is specific for *Pseudomonas* plasmids belonging to P-2 incompatibility group. Plasmids belonging to other incompatibility groups can specify resistance to borates and chromates (6), but the specific chemical mechanisms invoked to produce the resistance are unknown.

(2) Reduced uptake of metal. In some cases, the resistance against metal ions is manifested by a reduced uptake of the ion. A typical example is resistance against Cd^{2+}, which in *Staphylococcus aureus* is known to be specified by a plasmid (7). The mechanism of Cd resistance results from a permeability change in the cells, so that resistant cells harboring the plasmid do not accumulate the toxic levels of cadmium normally accumulated by the

sensitive cells (8,9). Normally cadmium uptake by sensitive cells
is temperature-dependent, and inhibited by inhibition of energy
metabolism. There appears to be a highly specific relationship
between cadmium uptake and the manganese transport system of the
cell. With the cadmium sensitive cells, high Cd^{2+} rapidly
inhibits manganese transport, whereas in resistant cells, manganese transport system remains unaffected in the presence of high
concentrations of cadmium. The Cd^{2+}-resistance plasmid thus
appears to block an energy-dependent transport system in the
resistant cells (10).

   (3)   Intracellular concentration of metal. Resistance
against metal ions can also be manifested by an efficient intracellular accumulation of the ions, which are stored in apparently
innocuous forms. In some cases, this type of resistance has been
shown to be plasmid-specified. Thus, Kondo, et al. (8) have
demonstrated that more than 90% of Hg^{2+} ions present in a test
medium can be taken up by *S. aureus* cells harboring the plasmid
Pc-ase. The Pc-ase plasmid is known to afford resistance to
mercury, cadmium, arsenic, lead, and zinc ions, in addition to
penicillin. The mercury-accumulation process appears to be independent of temperature and presumably inducible. Similar accumulation of Hg^{2+} has been demonstrated for a resistant strain of
Enterobacter aerogenes, which was shown to concentrate 91% of the
total ^{203}Hg in the intracellular protein fraction in three hours
(11). Similar mercury accumulating and mercury-volatilizing
Pseudomonas strains have been described (12). The use of a
Pseudomonas culture, harboring aggregates of mercury resistance
and other plasmids, has been recommended for removal and recovery
of mercury from industrial wastes (13). The nature of the binding
of Hg^{2+} to the intracellular protein is not known, although
highly specific metal-binding metallothionen-like proteins, such
as cadmium-binding protein, have been described for mammalian
systems (14).

(4) Methylation of metals. The methylation of certain heavy metals by microorganisms is now well documented (2,15). Such methylated derivatives can be quite toxic; the initial studies of methylation of mercury were conducted because of the finding that essentially all the mercury in fish was the highly toxic methyl mercury. Although the anaerobic bacteria in lake or river sediments were initially believed to be the sole agents for conversion of inorganic mercury to $CH_3H_g^+$ and $(CH_3)_2Hg$, it is now well documented that many aerobic mercury-resistant bacteria can methylate Hg^{2+}. Thus *Pseudomonas* strains harboring the plasmids pMG1 and pMG2 can produce methyl mercury from Hg^{2+} (10). Similar methylations of other metals such as selenium (16), lead (17), arsenic, tellurium (18), tin and cadmium (19) by microorganisms present in the aquatic ecosystem are now known. Since most methylated metals are volatile, this represents a simple bacterial mechanism for developing resistance against inorganic metal ions. Transmethylation reactions, where methyl groups are transferred to mercury ions from biologically methylated tin compounds, are also known (20). While the genes specifying methylation of Hg^{2+} are known to be specified by plasmids pMG1 and pMG2 in some *Pseudomonas* strains (10), the disposition of methylation genes in other bacteria has not been investigated.

III. PROSPECTS FOR GENETIC IMPROVEMENT OF *THIOBACILLUS FERROOXIDANS*

Among microorganisms that are involved in the industrial production and recovery of various metals, *Thiobacillus ferrooxidans* is presumably the most widely used. It has been estimated that as much as 5% of the total world copper production is based on the solubilization and leaching of the metal from low-grade ores by *T. ferrooxidans* (21). This bacterium is capable of releasing more than 90% of copper from common sulfide ores such as chalcopyrite or chalcocite (22,23,24). In addition, *T. ferrooxidans* has also been shown to effect solubilization and release of such important metals as uranium, cobalt, nickel, zinc, and lead (25,26). Initial leaching of a low grade uranium ore containing 0.11% uranium with *T. ferrooxidans* for a period of 9 days has resulted in the extraction of 68% of the uranium, and the extent of extraction could be increased to 87% if the residue were reground and leached further (27). Optimization of the leaching process by adjusting the ore pulp density to 5% is reported to result in almost 100% extraction of uranium from a low grade ore (28).

Considerable work relating to the optimization of temperature, pH, aeration rate, ferrous iron concentration and pulp density suspension in the leaching of various metals from their ores by *T. ferrooxidans* is now in the literature (27,29). Scale-up of the laboratory and pilot plant studies to industrial scale operation particularly in the production of uranium, is now possible (28). However, the economic barriers to such scale-up, which would entail sizable expenditures for equipment and energy costs for grinding large quantities of low-grade ores to permit efficient microbial growth, are considerable. The inefficient but simple leaching of unprocessed tailing dumps is therefore favored on economic grounds.

In addition to optimizing the process for maximal leaching of metals, significant improvements can also be made in the performance of the bacteria. A study of the physiological parameters affecting leaching, as well as optimum nutritional and growth conditions (24) is essential for maximizing the recovery of the metal. In addition genetic improvements, leading to an enhancement of the rate and extent of metal solubilization, can and should be made. In this section, I would like to point out some areas where genetic improvements could be made to increase the effectiveness of *T. ferrooxidans* for metal solubilization.

A.   Oxidation of Fe^{2+} to Fe^{3+}

The chemolithotropic bacterium, *T. ferrooxidans*, is known to derive its energy from the oxidation of ferrous iron, metal sulfides and soluble sulfur compounds in an acidic medium. The oxidation processes can be described as follows:

$$2FeSO_4 + H_2SO_4 + \frac{1}{2}O_2 = Fe_2(SO_4)_3 + H_2O$$
$$4FeS_2 + 15O_2 + 2H_2O = 2Fe_2(SO_4)_3 + 2H_2SO_4$$
$$CuFeS_2 + 4O_2 = CuSO_4 + FeSO_4$$
$$MS + 2O_2 = MSO_4 \text{ (where M = Cu, Ni, Co, Zn, etc.)}$$

The oxidation of ferrous iron to ferric iron is an important reaction in the overall process, since ferric sulfate can react chemically with several ore minerals to oxidize them. Thus copper and uranium can react as follows:

$$CuFeS_2 + 2Fe_2(SO_4)_3 + 2H_2O + 3O_2 = CuSO_4$$
$$+ 5FeSO_4 + 2H_2SO_4$$
$$Cu_2S + 2Fe_2(SO_4)_3 = 2CuSO_4 + 4FeSO_4 + S$$
$$UO_2 + Fe_2(SO_4)_3 = UO_2SO_4 + 2FeSO_4$$

This reaction is particularly important for the extraction of uranium, since the tetravalent form of uranium present in the ore is insoluble, while the oxidized hexavalent form in the leach solution is soluble. The ferric ion, which serves as the oxidizing agent, is regenerated by the microorganisms via the reoxidation of the ferrous ion produced.

One of the primary rate-limiting steps in the leaching of metals from ores by *T. ferrooxidans* is this ferrous-ferric reoxidation. This is due to product inhibition of the oxidation step, i.e., ferric ion competitively inhibits the rate of ferrous ion oxidation. Kelly and Jones (this volume) have shown that ferric sulfate has no effect on the rate of CO_2 fixation and

affects the growth of *T. ferrooxidans* solely due to its inhibitory
effect on the ferrous ion oxidation step. Since production of
large quantities of ferric iron is necessary not only for a high
degree of uranium solubilization, but also to effect underground
leach mining of other valuable metals, it would be most desirable
to obtain a mutant of *T. ferrooxidans* where the oxidation step
would not be repressed by ferric sulfate. This would be reminis-
cent of the nitrogen-fixation derepressed mutants of *Klebsiella
pneumoniae*, where the fixation of atmospheric nitrogen is no longer subject to repression by the product NH_4^+ (30). Since *T. ferrooxidans* can grow rapidly in synthetic medium with ferrous sulfate
as an energy source, isolation of such mutants in presence of
high concentration of ferric sulfate but limiting concentrations
of ferrous sulfate in a chemostat should be feasible. Successful
isolation of such a mutant could not only lead to appreciable enhancement of the solubilization of uranium from ores such as
uraninite, but such mutants must also be used for direct dump
leaching of other metals such as copper, lead or zinc.

B. Toxic Effects of Metal Ions and UV-Sensitivity

   Thiobacillus ferrooxidans is known to develop resistance
against very high concentrations of the metals being leached.
Thus, resistance of *T. ferrooxidans* to copper ion concentrations
as high as 55 gm/liter (24) or to uranium concentrations of
12 gm U_3O_8/liter (31) is not unusual. However, there are other
contaminating metal ions that appear to inhibit growth even at
very low concentrations. Notable among those are silver, mercury,
and cadmium. Silver, even at concentrations as low as 1.0 ppm,
can appreciably inhibit the growth of *T. ferrooxidans* (32). Since
T. ferrooxidans might release soluble silver from its sulfide
(argentite) which is occasionally present with complex sulfide
ores, toxicity to silver is a major problem in the microbial
leaching. This problem, however, ought to be solvable by

introducing plasmids specifying resistance to metal ions such as silver, mercury, cadmium, etc. (3,33,34) to *T. ferrooxidans* derepressed mutants. This should markedly increase the resistance of *T. ferrooxidans* to the inhibitory effects of toxic metals. *T. ferrooxidans* is also known to be sensitive to UV, and plasmids are known that can greatly increase the resistance of the host cells towards UV-irradiation (35,36). For at least one plasmid, the resistance to UV has been shown to be due to a plasmid-specified DNA polymerase (36). Introduction of such plasmids should therefore protect *T. ferrooxidans* cells from the killing action of UV or γ-rays.

C. Nitrogen Nutrition

Another area in which genetic improvements for enhanced growth of *T. ferrooxidans* for dump leaching of metals might be made is in the self-generation of nitrogenous nutrients. Although the benefits of addition of nitrogen and phosphorus sources, as well as organic carbon compounds, to enhance metal dump leaching by *T. ferrooxidans* have not been thoroughly evaluated, it is likely that the exogenous supply of ammonia would help to enhance the growth of *T. ferrooxidans* and hence the consequent release of soluble metals. Enhanced growth and release of soluble metals during synergistic growth of *T. ferrooxidans* with the aerobic N_2-fixing organism *Beijerinckia lacticogenes* have been interpreted as due to the fixation and availability of assimilable nitrogenous compounds (37,38). Since it might be difficult for the N_2-fixing *Beijerinckia* species to grow synergistically with *T. ferrooxidans* at high acidity and metal ion concentrations during dump leaching, one convenient way to supply additional nitrogenous nutrients to *T. ferrooxidans* might be to genetically enable it to fix its own nitrogen. Since *T. ferrooxidans* is a gram-negative short rod-shaped bacterium just like *Pseudomonas*, it might be possible to introduce plasmids of P1-incompatibility groups such as RP4 into it. Plasmids like RP4 have a very wide host range and can be introduced into a large number of gram-negative bacteria (4). An RP4 plasmid having the *his nif* genes of *Klebsiella pneumoniae* which can also be transferred to a large number of bacterial genera has been constructed (39). One of the problems in nitrogen fixation by strictly aerobic bacteria is the lability of nitrogenase towards oxygen, and protective mechanisms, similar to those in *Azotobacter*, must be operative within such hosts. It is likely that RP4 plasmids containing the nitorgen fixation genes from *Klebsiella*, will be transmissible and stably maintained in *T. ferrooxidans*. Whether presence of such plasmids can lead to fixation of nitrogen by *T. ferrooxidans* and can result in the enhancement of growth and metal release will be very interesting areas of future research on *T. ferrooxidans* biology.

IV. REFERENCES

1. Jernelov, A., and Martin, A.L., *Ann. Rev. Microbiol.*, 29, 61 (1975).

2. Wood, J.M., *Science*, 183, 1049 (1974).

3. Schottel, J., Mandal, A., Clark, D., Silver, S., and Hedges, R.W., *Nature*, 251, 335 (1974).

4. Chakrabarty, A.M., *Ann. Rev. Genet.*, 10, 7 (1976).

5. Silverberg, B.A., Wong, P.T.S., and Chau, Y.K., *Arch. Microbiol.*, 107, 1 (1976).

6. Jacoby, G.A., in "*Pseudomonas aeruginosa:* Clinical Manifestations of Infection and Current Therapy", (R.G. Doggett, Ed.), Academic Press, New York (in press).

7. Novick, R.P., and Bouanchaud, D., *Ann. N.Y. Acad. Sci.*, 182, 279 (1971).

8. Kondo, I., Ishikawa, T., and Nakahara, H., *J. Bacteriol.*, 117, 1 (1974).

9. Chopra, I., *Antimicrob. Agents Chemother.*, 7, 8 (1975).

10. Silver, S., Schottel, J., and Weiss, A., in "Proceedings of the Third International Biodegradation Symposium", (J.M. Sharpley and A.M. Kaplan, Eds.), p. 899. Appl. Science Publishers, London, 1976.

11. Hamdy, M.K., and Noyes, O.R., *Appl. Microbiol.*, 30, 424 (1975).

12. Sayler, G.S., Nelson, J.D., and Colwell, R.R., *Appl. Microbiol.*, 30, 91 (1975).

13. Chakrabarty, A.M., Friello, D.A., and Mylroie, J.R., U.S. Patent 3,923,597 (1975).

14. Cherian, M.G., *Biochem. Biophys. Res. Commun.*, 61, 920 (1974).

15. Ridley, W.P., Dizikes, L.J., and Wood, J.M., *Science*, 197, 329 (1977).

16. Chau, Y.K., Wong, P.T.S., Silverberg, B.A., Luxon, P.L., and Bengert, G.A., *Science*, 192, 1130 (1976).

17. Wong, P.T.S., Chau, Y.K., and Luxon, P.L., *Nature*, 253, 263 (1975).

18. Summers, A.O., and Jacoby, G.A., *J. Bacteriol.*, 129, 276 (1977).

19. Huey, C.W., Brinckman, F.E., Grim, S.O., and Iverson, W.P., in "Proceedings of International Conference on Transport of Persistent Chemicals in Aquatic Ecosystems II", p. 73. Natl. Res. Council, Ottawa, 1974.

20. Brinckman, F.E., Iverson, W.P., and Blair, W., in "Proceedings of the Third International Biodegration Symposium", (J.M. Sharpley and A.M. Kaplan, Eds.), p. 919. Appl. Science Publishers, London, 1976.

21. Malouf, E.E., *Mining Eng.*, 23, 43 (1971).

22. Bruynesteyn, A., and Duncan, D.W., *Can. Met. Quart.*, 10, 57 (1971).

23. Silver, M., and Torma, A.E., *Can. J. Microbiol.*, 20, 141 (1974).

24. Sakaguchi, H., Torma, A.E., and Silver, M., *Appl. Environ. Microbiol.*, 31, 7 (1976).

25. Torma, A.E., *Rev. Can. Biol.*, 30, 209 (1971).

26. Tuovinen, O.H., and Kelly, D.P., *Int. Metall. Rev.*, 19, 21 (1974).

27. Guay, R., Silver, M., and Torma, A.E., *Biotechnol. Bioeng.*, 19, 727 (1977).

28. Guay, R., Silver, M., and Torma, A.E., *Eur. J. Appl. Microbiol.*, 3, 157 (1976).

29. Torma, A.E., Walden, C.C., Duncan, D.W., and Branion, R.M.R., *Biotechnol. Bioeng.*, 14, 777 (1972).

30. Andersen, K., Shanmugam, K.T., and Valentine, R.C., *Develop. Indust. Microbiol.*, 19 (in press).

31. Duncan, D.W., and Bruynesteyn, A., *Can. Min. Metall. Bull.*, 74, 116 (1971).

32. Hoffman, L.E., and Hendrix, J.L., *Biotechnol. Bioeng.*, 18, 1161 (1976).

33. Smith, D.H., *Science*, 156, 1114 (1967).

34. McHugh, G.L., Moellering, R.C., Hopkins, C.C., and Swartz, M.N., *Lancet*, i, 235 (1975).

35. Krishnapillai, V., *Mutation Res.*, 29, 363 (1975).

36. Lehrbach, P., Kung, A.H.C., Lee, B.T.O., and Jacoby, G.A., *J. Gen. Microbiol.*, 98, 167 (1977).

37. Tsuchiya, H.M., Trivedi, N.C., and Shuler, M.L., *Biotechnol. Bioeng.*, 16, 991 (1974).

38. Trivedi, N.C., and Tsuchiya, H.M., *Int. J. Min. Process*, 2, 1 (1975).

39. Dixon, R., Cannon, F., and Kondorosi, A., *Nature*, 260, 268 (1976).

40. Hedges, R.W., and Baumberg, S., *J. Bacteriol.*, 115, 459 (1973).

STRUCTURE-FUNCTION RELATIONSHIPS OF THIOBACILLUS

RELATIVE TO FERROUS IRON AND SULFIDE OXIDATIONS

Donald Lundgren

Syracuse University
Syracuse, New York, USA

and

Tatsuo Tano

Okayama University
Okayama, Japan

The acidophilic bacterium Thiobacillus ferrooxidans *used to develop a conceptual framework for explaining how the organism may oxidize reduced iron and sulfide. Structure-function relationships are shown using cell envelope and spheroplast-like preparations. The cell envelope is apprently involved in both iron and sulfide oxidations and the envelope somehow establishes microenvironments which enable the appropriate oxidases to function in an acid milieu.*

I. INTRODUCTION

Autotrophic bacteria restricted to inorganic oxidations for their energy are generally gram-negative bacteria. It has been the belief of the present investigators that the envelope of these bacteria is in some way involved in those reactions.

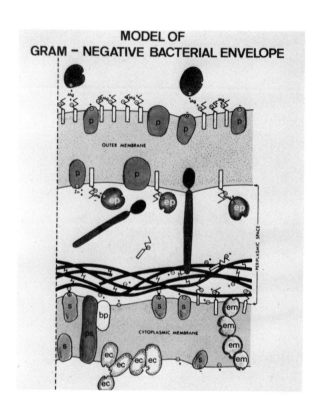

Fig. 1. *Modified drawing from Costerton et al (1) showing the layering of the gram-negative cell envelope. The drawing is used to underscore the complexity of structure and chemical components consisting of lipopolysaccharide, phospholipids, protein (enzyme) structures, cations, etc. The reader should consult the original article (1) for identification of specific labels.*

that allow microorganisms to use inorganic substrates such as
ferrous iron and reduced sulfur compounds for energy; the
energy in turn is used to support the cell's biosynthetic processes necessary for growth. The present conceptual framework
for the involvement of the envelope in chemoautotrophy reported
here comes from knowledge gleaned in published studies of
heterotrophic bacteria including *Escherichia, Salmonella, Pro-
teus* and *Pseudomonas* and from work with *Thiobacillus,* a chemo-
autotroph.

The complex cell envelope is schematically shown in Figure
1. An excellent overview of the importance of the cell envelope to general biological functions in heterotrophic bacteria
has been given (1). The presence of lipopolysaccharide in the
outer membrane layer establishes a zone outside the cytoplasmic
membrane and contains the movement of molecules into and out of
the space; this zone is referred to as the "periplasmic space".
The individual macromolecular and ion components of the cell
envelope layers influences the penetrability of both substrates
and end products of metabolism by binding or trapping molecules
and ions and thereby screens those molecules and ions that
reach the cytoplasmic membrane. Further, the composite of the
many chemical groups making up the structural architecture of
the cell envelope creates at times a variety of minienviron-
ments which undoubtedly are important for the proper function
of those enzyme catalysts promoting surface physiology and metabolism of the bacterium. An example of such an affect in a
heterotrophic bacterium is the lipopolysaccharide (LPS) influence on alkaline phosphatase, an enzyme found in association
with the cell envelope (1); here LPS provides a favorable
physicochemical environment for dimerization of subunits of
enzyme.

Relative to the subject of inorganic oxidations, envelope associated enzymes protected from a harmful environment such as low pH or high metal concentrations, could still have direct access to a non-soluble energy substrate such as FeS_2. Further, the sequestation of autotrophic processes in the cell envelope would serve to isolate toxic products of metabolism from the cytoplasm of the cell while allowing non-toxic products available for transport into the cell. As a consequence of this basic design of cell envelope architecture and the resulting function of the envelope in thiobacilli, a practical industrial application has occurred which couples the oxidation of an inorganic substrate to the leaching of a metal from a poor grade ore.

It is the purpose of this paper to review the structure-function relationships of the cell envelope of *Thiobacillus ferrooxidans*, a chemoautotrophic bacterium capable of oxidizing either reduced sulfur or iron. Some structure and function relationships for ferrous iron and sulfide oxidation are considered. It is the presence of both iron and sulfur in the form of pyrite (FeS_2) that makes biological leaching of metals economically feasible.

The subject will be introduced in the following manner: (a) The nature of the cell envelope of the gram-negative bacterium *Thiobacillus ferrooxidans*. (b) The involvement of the cell envelope in the oxidation of Fe^{++} and $S^=$.

II.   EXPERIMENTAL METHODS

The organism used was *Thiobacillus ferrooxidans*. Cultures were grown in the standard 9K medium (2) and when grown on sulfur, iron was replaced with colloidal sulfur at 5 gm/liter. Culture vessels used were 5-gallon carboys equipped with spargers and aeration was accomplished by passing humidified compressed air through the culture medium. The incubation temperature was 28 C and cultures were aerated for 96 hours with iron-grown cells and for 5-6 days with colloidal sulfur. All inoculations for iron-grown cells were made with fresh washed cell suspensions (15% w/v) and for sulfur-grown cells residual sulfur from a previously grown culture served as an inoculum.

Cell fractions, envelope preparations and spheroplasts were prepared using the French Pressure Cell (3) or a modified spheroplast procedure designed around that commonly used for *Escherichia coli* (4) and that procedure published for *T. ferrooxidans* (5). In the latter procedure the Tris-HCl buffer was replaced by phosphate buffer (pH 7.9) and the $CaCl_2$ was replaced by Mg^{++}. It was immediately apparent that *T. ferrooxidans* grown on either iron or sulfur did not respond to conventional spheroplasting or cell envelope separation procedures employed for heterotrophic bacteria and that many procedural adjustments had to be made.

Iron oxidation was assayed by measuring O_2 uptake with a Clark O_2 electrode.

Sulfide oxidation was followed by measuring O_2 utilization with the oxygen electrode using 0.12 μmoles of NaS_2 as the substrate in β-alanine sulfate buffer, pH 3.0. Electron microscopy was done following essentially the earlier procedures cited for *T. ferrooxidans* (3, 6).

Protein concentration was estimated colorimetrically (7). Prior to the assay, cells and cell envelopes were hydrolyzed for 30 minutes in 0.5N NaOH in a boiling water bath.

III. RESULTS AND DISCUSSION

A review of the structure and function relationships of the cell envelope of gram-negative bacteria has recently

appeared. Figure 1 is a drawing of the envelope of Gram-negative bacteria. Its structural complexity is apparent, but it
is this complexity that accounts for the wide range of growth
habitats of Gram-negative bacteria including the acid environment required by thiobacilli used in metal leaching. The
envelope consists of three major zones: (1) Cytoplasmic
membrane which constitutes the inner layer of the envelope
bordering on the cytoplasm. (2) The central zone comprising
both the rigid layer of peptidoglycan and the periplasmic
space. (3) The outer layer which contains the lipopolysaccharide and lipoprotein. Excellent discussions of each of
these zones in addition to that cited earlier have appeared
(8, 9, 10, 11).

Much of the fine structure details of *T. ferrooxidans* has
been reviewed (12). For general orientation, cells that have
been treated in three different ways to highlight their surface
features are shown in Figure 2. Figure 2a shows negatively
stained ferrobacilli; Figure 2b is a thin section of a part
of the cell envelope revealing the outer membrane (OM), peptidoglycan (PG) and cytoplasmic membrane (CM), the periplasmic
space resides between the outer and inner barriers of the cell.
Figure 2c is a frozen etched preparation of the same cell
revealing the aforementioned layering of the cell envelope.

Figure 3 is a sketch of the envelope showing layering of
a bacterium in juxtaposition to a crystal of pyrite (FeS_2)
modeled after an earlier published idea (13). How the outer
layer of the bacterium influences the solubilization and subsequent oxidation of the two substrates (Fe^{++} and $2S^=$) is
the important question relating to how these organisms function
in metal freeing from poor grade ores. Both Fe^{++} and $S^=$
substrates are capable of attack by the bacterium and both
the "iron oxidase" and "sulfide oxidase", two complex enzyme

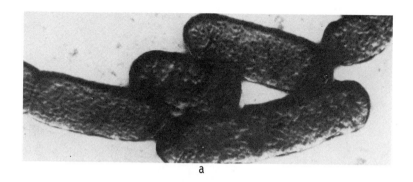

a

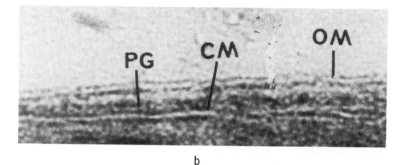

b

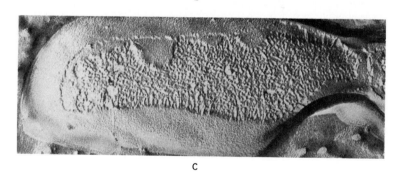

c

Fig. 2.   Electron micrographs of Thiobacillus ferrooxidans.
*a.   negative stained preparation.   b.   thin section of a chemi-
cally fixed cell.   c.   frozen etched preparation.*

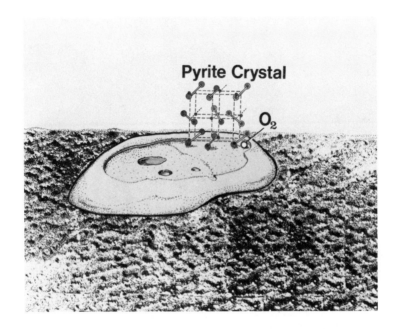

systems responsible for the respective oxidations, must co-exist and function depending upon what external environmental conditions are needed by the organism to support the appropriate biological reactions.

In the case of Fe^{++} oxidation the overall reaction is:

$$2Fe^{++} + \tfrac{1}{2}O_2 + 2H^+ \longrightarrow 2Fe^{+++} + H_2O$$

The two components of the overall reaction of iron oxidation are apparently spatially separated within the cell envelope (14). The energy associated reaction ($2e^- + \tfrac{1}{2}O_2 + 2H^+ \longrightarrow H_2O$) is probably located within the inner membrane whereas the ($2Fe^{++} \longrightarrow 2Fe^{+++} + 2e^-$) reaction is involved at the outer

membrane or perplasmic space level (14). How the two reactions
are coupled *in situ* is still unknown but the separation from a
practical standpoint is important for the product Fe^{+++} of this
reaction never has to enter the bacterium and be handled by a
special transport system.

It is apparent that within the cell envelope the organization of the necessary enzymes required to support the aforementioned oxidation reactions are protected from both the
external acid milieu and the high concentrations of metals.
Ferric iron is extremely insoluble at pH values above 4.0 and
large amounts of this metal accumulates following Fe^{++} ion
oxidation. These substrate "oxidases" are so structurally
organized in their respective layers that microenvironments
favorable to the functioning of enzymes are formed even though
the natural acid environment of the bacterium is harmful when
the enzymes are separated from the cell. This is shown with
cell envelopes, isolated from iron-grown thiobacilli, which
possess approximately 40% of the iron oxidizing ability of
intact cells. Figure 4 shows a thin section of enzymatically
active cell envelopes of *T. ferrooxidans* with an optimum pH
or iron oxidation of 2.5 (3). It is known that the optimum pH
of at least two partially purified enzyme components of the
"iron oxidase" complex (Fe^{++}-Cytochrome *c* reductase and
Cytochrome oxidase) are approximately pH 6.5 and 4.5 respectively (15, 16).

At this point in time relatively little is known about
the structure-function relationships of iron oxidation in *T.
ferrooxidans*. Suffice to say, an examination of Table 1
quickly reveals that compared to the cell envelopes of *Salmonella* and other heterotrophic bacteria little is also known
about the chemical composition of cell envelopes of *T. ferrooxidans*. Results of some physical and chemical studies of

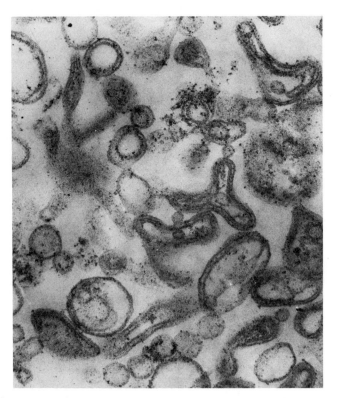

Fig. 4. Thin section of chemically fixed isolated cell envelopes from iron-grown Thiobacillus ferrooxidans.

T. ferrooxidans lipopolysaccharide have recently been published (17). Lipopolysaccharide isolated from iron-grown cells was stable to acid and alkali, possessed a ribbon-like morphology, contained no phosphorous and very little protein and possessed an unknown fatty acid component.

The oxidation of the sulfur component of pyrite raises comparable questions of structure-function relationships in the bacterium. The oxidation of $S^=$ is reported to follow those reactions given below (18, 19, 20);

TABLE I   Composition of Inner and Outer Membrane Fractions of
Salmonella typhimurium *and* Thiobacillus ferrooxidans

Salmonella typhimurium   Thiobacillus ferroxidans
                   Inner (Cytoplasmic) Membrane
  Phospholipids   *Phospholipids*
  Proteins, including: *Proteins, including:*
    Electron transfer   *Electron transfer*
      activities   *activities*
    Permease functions   *?*
    Phospholipids biosynthetic
      enzymes   *?*
    Lipopolysaccharide
      biosynthetic enzymes   *?*
                      Outer Membrane
  Lipopolysaccharide   *Lipopolysaccharide*
  Phospholipids   *Phospholipids*
  Proteins, including: *Proteins, including:*
    Phospholipase activity   *?*
    Murein lipoprotein   *Murein lipoprotein*
    Major polypeptides of
      known function   *?*
      A.   Iron binding and
             transport
      B.   Colicin binding
      C.   Vitamin B_{12} binding
      D.   Receptor F-pilus

$$SH^- \xrightarrow[\text{oxidase}]{\text{sulfide}} [S] + H^+ + 2e^-$$

$$2[S] \longrightarrow [S\text{-}S]$$

$$[S\text{-}S] + SH^- \longrightarrow {}^-S\text{-}S\text{-}SH$$

$$^-S\text{-}S\text{-}SH + X \longrightarrow X\text{-}S\text{-}S\text{-}SH$$

$$2[X\text{-}S\text{-}S\text{-}SH] \xrightarrow[\text{oxidase}]{\text{polysulfide}} X\text{-}S_6\text{-}X$$

$S^=$ oxidation proceeds in two stages. In the first stage, sulfide loses two electrons mediated by sulfide oxidase, and polymerization of the resulting sulfur atoms follows. The oxidation of a short-chain polysulfide to polymeric sulfur compounds occurs. These polysulfides are thought to be membrane bound. X in the scheme represents a hypothetical group linking the polysulfide chain to a membrane fraction of the cell envelope. Formation of this membrane bound polysulfide when studied using membrane fractions of *T. thiooxidans* required a cytoplasmic component (21). Polysulfide oxidase catalyzes the oxidation of the polysulfide.

Sulfide oxidation was studied using *T. ferrooxidans* grown on colloidal sulfur at pH 3.5; membrane preparations of these cells were prepared according to the procedures reported for iron-grown cells (3). However, with these membranes sulfide oxidation occurred only at pH 7.0, and not at pH 3.0. These preliminary results suggested that the structural organization of the membrane components involved in the oxidation was vital if cells were to mediate the first stage of sulfide oxidation in an acid milieu.

A much milder treatment of breaking intact cells than that of disruption by the French Press was used in an attempt to prepare cell-free membranes capable of supporting $S^=$ oxidation. The spheroplasting procedure earlier used for thiobacilli (5) was employed as a starting point in this study.

The spheroplasting treatment to sulfur-grown cells and the subsequent differential centrifugation of the treated cells on sucrose gradients led to the isolation of three types of membrane and cell envelope or leaky spheroplast preparations; Figure 5 shows thin sections of these preparations

identified as F-1, F-2, F-3 and intact bacterium. Sulfide
oxidation at a low pH was observed in the F-3 preparation and

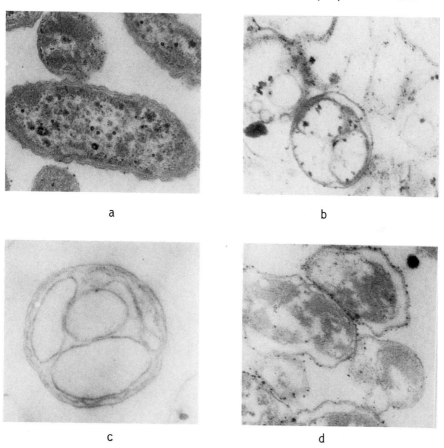

Fig. 5. Electron micrographs showing thin sections of
sulfur-grown Thiobacillus ferroxidans treated for spheroplast
formation:
(a) Intact cell - Thin section of an intact sulfur-grown cell.
(b) F-1 fraction - Membrane collected at a 40 ∿ 46% sucrose
     gradient in 0.01M phosphate buffer (pH 7.0) containing
     10 mM MgSO$_4$.
(c) F-2 fraction - Envelope membranes collected at a 46 ∿ 51%
     sucrose gradient as above.
(d) F-3 fraction - Envelope membranes and leaky spheroplasts
     collected at a 52 ∿ 54% sucrose gradient.

this fraction also possessed both NADH oxidase and succinate dehydrogenase activities. The fraction contained spheroplast-like structures which had lost their rigidity but were still structurally organized so that the respective cell envelope layers appeared to be intact. The authors conclude that enough structure is intact to establish the necessary micro-environments for the enzymes to permit the oxidative reactions within the natural acid environment of the bacterium.

Figure 6 shows the pH activity curve for sulfide oxidation of the F-3 preparation; neither the F-1 nor F-2 preparations

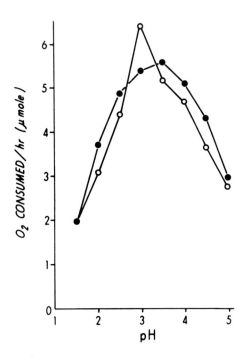

Fig. 6. pH activity curves for sulfide oxidation by intact Thiobacillus ferrooxidans *and the F-3 fraction. Reaction mixture contained for intact cells (0——0) β-alanine-H₂SO₄ buffer; cells, 60 mg/dry wt.; Na₂S, 0.12 μmoles; total volume, 2 ml. For the F-3 fraction the reaction mixture contained spheroplasts and membranes, 2.74 mg protein; β-alanine-H₂SO₄ containing 1 mM MgSO₄ and 10% sucrose; Na₂S, 0.12 μmole. Temperature was 28 C.*

possessed sulfide oxidase activity. Further, F-1 and F-2 preparations did not possess NADH oxidase nor succinate dehydrogenase activities.

From these preliminary results, it is the authors' feeling that the two membrane systems, that is the lipopolysaccharide-containing outer membrane and the inner cytoplasmic membrane constitute the necessary structural organization needed to support sulfide oxidation in an acid milieu. Hopefully, a successful approach in isolating active membrane fractions will enable us to answer this question.

IV. REFERENCES

1. Costerton, J.W., Ingram, J.M., and Cheng, K.J., *Bacteriol. Rev.*, Vol. 38, No. 1, 87 (1974).

2. Silverman, M., and Lundgren, D., *J. Bacteriol.*, 77, 642 (1959).

3. Bodo, C., and Lungren, D.G., *Can. J. Microbiol.*, 20, 1647 (1974).

4. Osborn, M. J., Gander, J.E., Parisi, E., and Carson J., *J. Biol. Chem.*, 247, 3962 (1972).

5. Howard, A., and Lundgren, D.G., *Can. J. Biochem.*, 48, 12, 1302 (1970).

6. Remsen, C., and Lundgren, D.G., *J. Bacteriol.*, 92, 6, 1765 (1966).

7. Lowry, O., Rosebrough, M., Farr, A., and Randall, R., *J. Biol. Chem.*, 193, 265(1951).

8. Bayer, M.E., in "Mode of Action of Antibiotics on Microbial Walls and Membranes", (M.R.J. Salton and A. Tomasz, Eds.), Vol. 235, p. 6. N.Y. Acad. Sci., 1974.

9. Osborn, M.J., Rick, P.D., Lehmann, V., Rupprecht, E., and Singh, M., in "Mode of Action of Antibiotics on Microbial Walls and Membranes", (M.R.J. Salton and A. Tomasz, Eds.), Vol. 235, p. 52. N.Y. Acad. Sci., 1974.

10. Leive, L., in "Mode of Action of Antibiotics on Microbial Walls and Membranes", (M.R.J. Salton and A. Tomasz, Eds.), Vol. 235, p. 109, N.Y. Acad. Sci., 1974.

11. Salton, M.R.J., and Owen, P., in "Annual Review of Microbiology", (M.P. Starr, J.L. Ingraham, and S. Raffell, Eds.), Vol. 30, p. 451. Annu. Rev. Inc., California, 1976.

12. Lundgren, D.J., Vestal, J.R., and Tabita, F.R., in "Water Pollution Microbiology", (R. Mitchell, Ed.), p. 69. John Wiley & Sons, Inc., 1972.

13. Dugan, P.R., "Biochemical Ecology of Water Pollution", p. 124. Plenum Press, New York, 1972.

14. Ingledew, W.J., Cox, J.C., and Helling, P.J., *Proc. Soc. Gen. Microbiol.*, Vol. 4, Part 2, 74 (1977).

15. Din, G.A., Suzuki, I., and Lees, H., *Can. J. Biochem.*, 45, 1523 (1967).

16. Blaylock, B.A., and Nason, A., *J. Biol. Chem.*, 20, 3453 (1963).

17. Hirt, W.E., and Vestal, J.R., *J. Bacteriol.*, 123, 642 (1975).

18. Moriarty, D.J.W., and Nicholas, D.J.D., *Biochem. Biophys. Acta*, 184, 114 (1969).

19. Aminuddin, M., and Nicholas, D.J.D., *Biochim. Biophys. Acta*, 325, 81 (1973).

20. Aleem, M.I.H., *Plant and Soil*, 587 (1975).

21. Tano, T., Oka, M., and Lundgren, D.G., *Abstr. Annu. Meeting Amer. Soc. Microbiol.*, 15 (1977).

LEACHING OF MINERALS USING BACTERIA OTHER THAN THIOBACILLI

Norman W. Le Roux
Don S. Wakerley
and
Vivian F. Perry

Warren Spring Laboratory
Stevenage
Herts SG1 2BX
England

The aim was to investigate the technical feasibility of leaching metals using bacteria other than thiobacilli. Bacteria were isolated from soil and grown on organic media so that metabolic products were produced which were capable of leaching nickel and copper from a copper/nickel concentrate. The bacteria were not examined in detail but most of those used in the tests were rod-shaped cells of various sizes. On solid media, they grew on malt plus peptone agar plates at pH 5.0 as smooth, cream-coloured colonies. Some indication of metal selectivity and adaptation to metal inhibition was observed. The bacteria produced acids from the organic substrates and in one case 99% nickel and 42% copper were leached. Comparative tests using sulphuric acid for leaching were less efficient than the microbial tests. Nickel solubilization, which was greater than copper, was favoured by low pH and was directly related to the development of acidity. Metal leaching was also favoured by elevated temperature and agitation. The amount of nickel solubilized increased but the percentage amounts of ore solubilized decreased with increase in ore pulp density. At 25% pulp density the concentration of nickel in solution was about 10g per litre.

I.   INTRODUCTION

The ability of thiobacilli to assist in the recovery of metals from ores is now well known and their contribution to the solibilization of metal sulphides, while the mechanism is not completely clear, is accepted and well documented e.g. (1). They are the only microbes to be used for metal recovery on a commercial scale.

In general, microbes are ubiquitous on this planet and are capable of producing many varied metabolities under different environmental conditions. It has been shown that the mobilization of metals can result from microbial activity (2). The extent and magnitude of these 'natural' phenomena are not well documented for at least two reasons. Various rate limiting factors lead to very slow reaction kinetics, the amounts of metal involved are very small and the time span long when considered against a man's working life. However, the amounts and effects can be large when measured in geological time. What are the reactions taking place in the natural environment? What are the rate limiting factors? Can these factors be optimized so that the ability of microbes to mobilize metals can be harnessed for the benefit of man? These are far-reaching questions and cannot yet be answered.

In the preliminary studies reported here, further evidence was sought to reinforce these ideas. The technical feasibility for metal solubilization by microbes other than thiobacilli was studied on a laboratory scale. It is emphasized that conditions were not optimized and economic viability was not sought. The authors are well aware of the many problems in bridging the gap between technical feasibility and economic/commercial viability.

II.   EXPERIMENTAL

A.   Metal Source

Two copper/nickel sulphide concentrates were used as a source of metals. Sample A contained 3.8% copper and 11.5% nickel, and

Sample B, 4.5% copper and 7.1% nickel. The copper and nickel
were present as chalcopyrite and pentlandite. Sample B was only
used for the higher pulp density tests.

B. Analyses

Copper, nickel, and iron in the filtered supernatant liquors
and in the residues, which were dried at room temperature, were
measured by atomic absorption spectrophotometry.

C. Inocula

Two solids were used as sources for microbes. Soil P was a
Black Norfolk Peat and Soil C a Hertfordshire Clay adjacent to an
organic compost heap. No particular significance should be
attached to these soil samples. They were chosen for their con-
venience and the potential presence of a large variety of microbes.

D. Media

Two common bacterial media were used, Malt Extract (Oxoid
Ltd), rich in carbohydrates and Bacteriological Peptone (BDH Ltd),
rich in proteins. These were prepared individually as 1.66g Malt
Extract/l and 0.34g Peptone/l and were also used mixed.

For tests with *Thiobacillus ferrooxidans* the inorganic salts
medium of Silverman and Lundgren was used (3).

E. Isolation of Microbes

Bacteria were isolated on agar plates prepared by adding 1%
Ionagar (BDH Ltd) and an acid indicator (Bromecresol Green or
BDH 4.5) to mixed Malt Extract/Peptone medium. After sterilization by autoclaving (121°C for 20 minutes) the medium was poured
into petri dishes, cooled, streaked with samples from culture
flasks and incubated at 30°C.

F. Microbial Testwork

A series of tests were made with the media and inocula described and the variables pH, temperature, agitation, and pulp
density studied.

Known weights (about 0.4g) of copper/nickel concentrate were
added to 100 ml of medium in 250 ml conical flasks. The flasks
were shaken in an orbital incubator (A.E. Gallenkamp Ltd) at 200
rpm and 30°C. Solution analyses for copper and nickel and pH
measurements were done on the supernatant liquors every 3 to 5
days. The shaken tests were allowed to stand for 4h before
samples were withdrawn for analysis. The initial tests which
were done at controlled pH values, were readjusted when necessary
by the addition of NaOH or H_2SO_4 when the solutions were analysed.
Tests which showed the highest rates of leaching were selected for
making subcultures which were made with a 1% inoculum. Sterile
control tests were carried out by adding a biocide (5 ml of 1%
"Panacide", BDH Ltd).

III. RESULTS AND DISCUSSION

A. Initial Tests

An initial series of tests were done to determine the rates
of leaching of nickel and copper from Sample A in a mixed malt +
peptone medium adjusted to pH 5.0 or 7.0. Pairs of flasks were
either continuously shaken to give good aeration or kept stationary to give poor aeration. Either peat or clay soil (1g) was
used as an inoculum. It was intended that all tests would be
done at the controlled temperature of 30°C. However, after the
first day the production of foul-smelling fermentation products
from the stationary tests was so obvious that these flasks had to
be removed to a fume hood at room temperature (about 20°C) for the
remainder of the test period.

In the first 50 days three tests showed good solubilization
of nickel (Fig. 1). The highest rate of nickel solubilization
occurred in the aerated flask containing the more acid, clay soil
culture. About 220 ppm Ni (50%) was solubilized in 50 days.
Lower rates of nickel extraction were obtained using 'non-aerated'
(stationary) cultures but this might have been due to the fact
that the tests were carried out at room temperature (20°C) rather
than 30°C.

In the test with clay soil at pH 7.0, nickel which had been
solubilized was precipitated. In all these initial tests the
amount of copper in solution was less than 3 ppm.

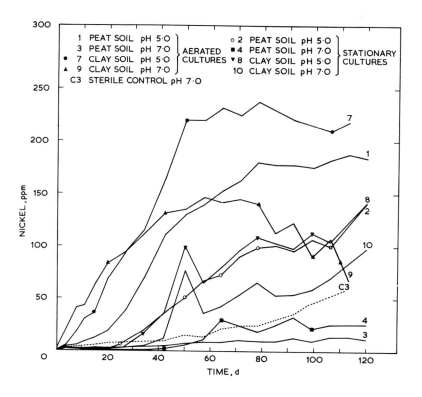

*Fig. 1.   Leaching of nickel using malt + peptone medium at
pH 5.0 and 7.0 with peat or clay soil as inoculum.*

In the tests at pH 5.0 and to a lesser extent in those at
pH 7.0, acid was produced in the flasks. The solutions originally
at pH 5.0 became acid, pH 3.5 to 4.0 every 3 to 5 days despite the
fact that the solution pH was restored to 5.0 each time pH measurements were made. This acid production was associated with
nickel solubilization and maximum nickel extraction occurred when
acid production was most rapid.

B. Effect of Subculturing

In an attempt to increase the beneficial microbial activity
sucessive subcultures were made from the clay soil culture at pH
5.0. Some increases in the rate of nickel solubilization and

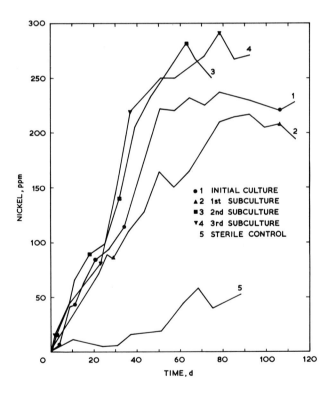

*Fig. 2. Leaching of nickel with successive clay soil sub-
cultures on malt + peptone medium at pH 5.0.*

yield were achieved (Fig. 2). Residue analyses showed that with
the second and third subcultures over 80% nickel and about 20%
copper had been extracted (Table 1).

C. Tests Without pH Control

The initial tests had suggested that nickel solubilization
was favoured by low pH. Moreover, cultures originally at pH 5.0
and to a lesser extent at pH 7.0 produced a considerable amount
of acid and needed repeated pH adjustment. To examine the effect
of dispensing with pH adjustment a subculture from the aerated
clay soil culture at pH 5.0 was made into malt + peptone medium
initially adjusted to pH 5.0 but thereafter no further pH adjustments were made. The rate of nickel dissolution obtained (Fig.
3a) was similar to those obtained in the initial aerobic tests at
pH 5.0. Also, like the initial tests copper in solution was very
low. This was unlikely to be due to a pH effect as the pH started
at 5.0 and then decreased during the course of the experiments.
Residue analysis, which also included the measurement of nickel
and copper which had been solubilized and then precipitated
(Table 1) showed that 99% nickel and 42% copper had been extracted
in 66 days, which was considerably more than had been obtained
previously. The pH of the culture was also determined and good
correlation was shown between the rate of nickel solubilization
and the development of acidity (Fig. 3a).

Repeated subcultures of this test were made and increases in
the rates of nickel leaching and acid production were obtained
(Figs. 3b-3f). After 5 subcultures the time for maximum nickel
extraction was reduced from 73 to about 30 days (Fig. 3f). The
amounts of nickel and copper extracted, as determined by residue
analyses remained between 81-98% and 25-42%, respectively
(Table 1). In all these subcultures, solution analyses showed
that after some time nickel was being precipitated. It was
noticed that, simultaneously with the onset of nickel precipitation,

TABLE I Residue and Solution Analyses of Bacterial Leaching Tests on Malt and Peptone Media

Test Medium	Wt of ore (mg)	Wt of residue (mg)	Nickel					Copper				
			In Ore (ppm)	In Residue (ppm)	Solubilized* (ppm)	Solubilized* (%)	In Solution (ppm)	In Ore (ppm)	In Residue (ppm)	Solubilized* (ppm)	Solubilized* (%)	In Solution (ppm)
Initial Tests at pH 5.0												
M+P 2nd subculture	461	382	530	101	429	81	283	176	130	46	26	0.6
M+P 3rd subculture	393	373	450	70	380	84	270	150	123	27	18	0.3
Tests without pH Control												
M+P	351	236	403	6	397	99	250	134	78	56	42	16
M+P 1st subculture	439	393	505	15	490	97	335	168	125	43	26	8
M+P 2nd subculture	467	426	536	22	514	96	308	155	115	40	26	11
M+P 3rd subculture	425	393	488	86	402	82	290	162	118	44	27	11
M+P 4th subculture	419	416	482	87	395	82	315	161	121	41	25	10
M+P 5th subculture	452	353	518	17	501	97	473	172	127	45	26	21
M+P 30°C then 40°C	428	366	493	35	458	93	320	164	117	47	29	9
M+P 40°C	399	403	460	14	446	97	433	153	131	22	14	14

* Calculated from residue analyses.

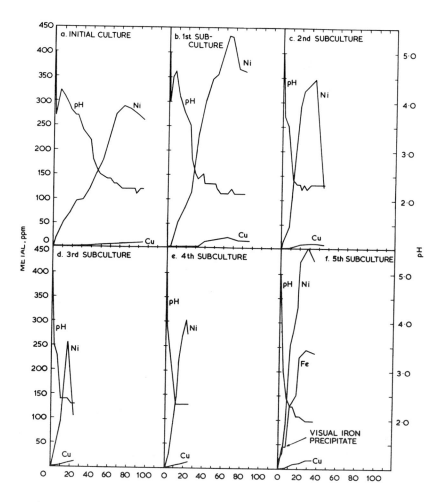

Fig. 3. (3a-3f) Bacterial leaching of nickel and copper using malt + peptone medium without pH control.

iron precipitation could be visually observed. The time at which iron precipitation was first observed is indicated in Fig. 3f and also in some later figures. Because of this observation the behaviour of iron in solution was determined in the fifth culture. The maximum iron in solution was 247 ppm (17%) and the leaching curve for iron had a similar pattern to that for nickel (Fig. 3f).

A test was also made to determine if an inoculum taken from a culture which was actively leaching nickel after 7 days was more effective than inocula taken when nickel solubilization and acid production were diminishing. Contrary to expectation the latter were found to be the most active inocula although the numbers of bacteria per unit volume were not counted.

D.   Effect of Temperature and Agitation

Tests were done in shaken and stationary flasks at both 30°C and room temperature (circa 20°C) using malt + peptone medium at pH 5.0. The clay soil culture at pH 5.0 from the initial tests was used as inoculum for each flask.

The rates of leaching of nickel were initially more rapid in both the shaken tests. After 15 days the rate of leaching in the stationary test at 30°C increased to that obtained in the shaken test at 30°C and a similar quantity of nickel was finally extracted in both tests (Fig. 4).

The rate of leaching in the stationary test at room temperature was very slow although the rate did increase after day 36. In the shaken test at room temperature leaching was initially reasonably rapid but after about day 15 the rate decreased sharply. This was probably the result of rough shaking with loss of solution on to the wool plug in the neck of the flask. For this reason the test must be regarded as unrepresentative. Although the results indicate that the leaching of nickel is favoured at 30°C and by shaking it is not possible to assess which factor is most important without further testing.

E.   Tests at 40°C Without pH Control

The earlier leaching tests had shown that better leaching took place at 30°C than at room temperature. Thus, tests were carried out to determine the rate of leaching at 40°C in malt + peptone medium at an initial pH of 5.0 but with no subsequent

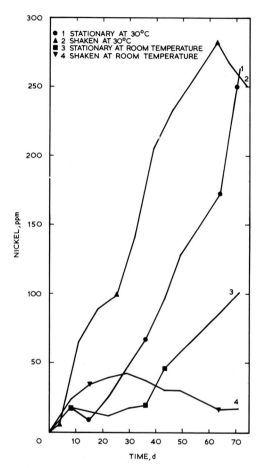

Fig. 4. Effect of aeration and temperature on the bacterial leaching of nickel using malt + peptone medium at pH 5.0.

pH control. In one of the tests the culture was grown at 30°C for 10 days and then transferred to 40°C while in a parallel test, the culture was maintained at 40°C throughout the test period.

There was an increase in the rate of leaching of nickel when the culture was transferred from 30 to 40°C after 10 days but in the test at 40°C, nickel extraction was erratic and at no time did the rate exceed that obtained at 30°C (Fig. 5). A pH curve is

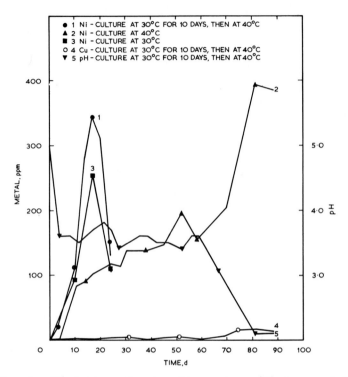

Fig. 5.   Variation of extraction rates with temperature for nickel and copper on malt + peptone medium.

also plotted and shows a correlation between the rate of nickel extraction and rate of acid production.

Residue analysis (Table 1) showed that copper solubilization was only about half that obtained in comparable tests at 30°C, whereas nickel solubilization was unaffected.

In the test at 40°C, microscopic examination showed that the bacteria became very elongated after only a day at 40°C suggesting that microbial growth was being adversely affected.

F.   Effect of Pulp Density

Tests were made with copper/nickel concentrate at pulp densities of 1.2, 4.0, 10.0, 15.0, and 25.0% using cultures grown on malt + peptone medium. The test at 1.2% pulp density was done with Concentrate A but because of availability Concentrate B was used for tests at higher pulp density. A sterile control test was set up using 4% pulp density. The rates of nickel, copper, and iron solubilization and pH changes are shown in Figs. 6a to 6f.

A comparison of the rates of nickel solubilization obtained in these tests and also in a previous test using 0.4% pulp density (Fig. 3f) shows a steady increase with increase in pulp density (Fig. 6g).

The calculated average nickel extraction rates at different pulp densities for a 26 day period of maximum solubilization, showed the efficiency of nickel extraction decreased with increasing pulp density (Table 2). It should be noted that Table 2 shows results averaged only for a good 26 day period and not for the completed test.

At the end of the tests nickel in solution reached quite high concentrations. For example, at 25% pulp density the concentration of nickel was almost 10g/l. The decrease in acidity and the rate of nickel solubilization after 60 days in this test suggested that the malt + peptone medium was being exhausted and that with fresh medium high rates of nickel solubilization would

TABLE II   *Effect of Pulp Density on the Efficiency of Nickel Solubilization*

Pulp Density	Ni Solubilized /100 ml Slurry/ 26 days $\frac{}{g}$	Ni Solubilized / 26 days %
0.4	0.052	100.0
1.2	0.065	48.1
4.0	0.174	61.0
10.0	0.179	25.2
15.0	0.200	18.7
25.0	0.322	17.7

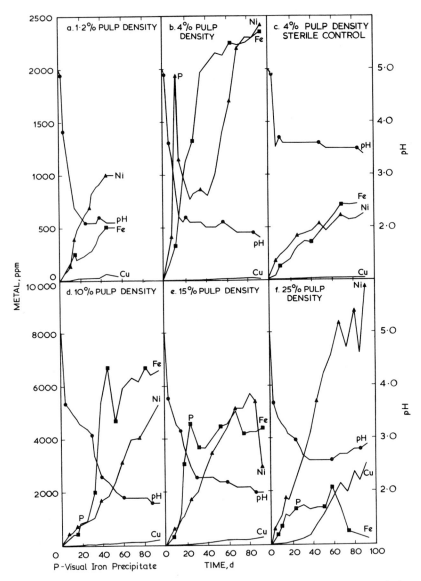

Fig. 6. (6a-6f) *Effect of pulp density on the bacterial leaching of nickel, copper and iron (malt + peptone medium).*

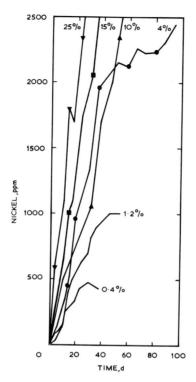

Fig. 6g. Effect of pulp density on the bacterial leaching of nickel (malt + peptone medium).

be resumed. Residue analyses showed that about 80% of the nickel had been solubilized at 10 and 15% pulp density and about 32% at 25% pulp density. These figures are only approximate as difficulty was experienced in drying the residues at room temperature due to the gelatinous nature of the residue.

The rates of copper extraction remained low at all pulp densities except for the unexplained high result at 25%. In this test appreciable quantities of copper were solubilized after 40 days (Fig. 6f).

In the sterile test using 4% pulp density much less acid was produced and the rate of nickel solubilization was less than a quarter of that in the comparable inoculated tests (Fig. 6c).

A check on the residue from the test with concentrate at 25% pulp density showed that of the 68% nickel left in the leached residue, 10% was soluble in 20% HCl (after correcting for nickel solubilized by 20% HCl from unleached ore). This nickel had been solubilized and reprecipitated in an acid soluble form. Thus, the total nickel leached in this test was about 42%. It is possible that in some of the other tests, nickel which was reported in the residue had in fact been solubilized and then reprecipitated. This could also have happened to copper. This should be kept in mind when considering the solubilized nickel and copper values in Tables 1, 2, and 3 which were calculated from residue analyses. They are probably minimal values.

G. Leaching Tests With Sulphuric Acid

In the microbial tests the leaching of nickel was apparently associated with the production of acid. Therefore, tests were carried out to determine the amount of nickel which could be solubilized from this ore by an acid leach, using similar conditions at 30°C. The tests were set up using copper/nickel concentrate and sulphuric acid at pH 2.0 and 3.0 with subsequent pH control, and at 3.0 and 5.0 without pH control. The highest rate of nickel solubilization was obtained in the test at pH 2.0 but there was little difference in the rates in the other tests (Fig. 7a). Even at pH 2.0 the rates were lower than those attained in the best microbial tests. The pH curves for these tests are shown in Fig. 7b. The amounts of copper solubilized in the sulphuric acid leaching tests were also less than those typically obtained in the microbial tests (Table 3).

In order to simulate the microbial tests more closely two additional tests were set up. One used a subculture from a

TABLE III   Residue and Solution Analyses of Sulphuric Acid and T. ferrooxidans Leaching Tests

Test Medium	Wt of ore	Wt of residue	Nickel					Copper				
			In Ore	In Residue	Solubilized*	Solubilized*	In Solution	In Ore	In Residue	Solubilized*	Solubilized*	In Solution
	mg	mg	ppm	ppm	ppm	%	ppm	ppm	ppm	ppm	%	ppm
pH 2.0 controlled	440	305	505	18	487	97	330	168	147	22	13	16
pH 3.0 controlled	424	282	488	44	444	91	–	158	107	51	32	–
pH 3.0 no pH control	409	288	470	87	383	82	220	156	153	3	2	5.3
pH 5.0 no pH control	396	342	455	12	343	75	373	151	113	39	26	21
T. ferrooxidans	403	833	462	3	460	99	330	153	77	76	50	25

* Calculated from residue analyses.

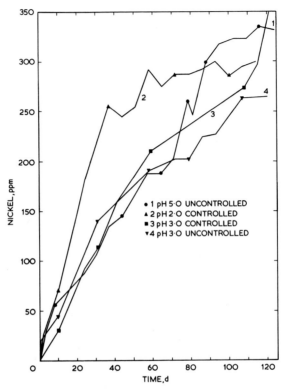

Fig. 7a.   Leaching of nickel with sulphuric acid.

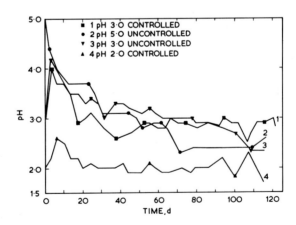

Fig. 7b.   pH of sulphuric acid leaching tests (the pH curves of the controlled tests were obtained from measurements taken immediately before pH adjustments were made).

previous microbial test in malt + peptone medium and the other
had sulphuric acid alone with the pH adjusted every 3-4 days to
the value attained by the microbial culture. The nickel leaching
curves show that leaching was more rapid in the microbial test
(Fig. 8).

Although the pH values in the sulphuric acid test lagged
behind those in the microbial culture the difference seemen insufficient to account for the difference in leaching rates. The
pH curve of the sulphuric acid test was plotted from measurements
taken immediately before an adjustment was made to bring the pH
value to that of the microbial test. Thus, the pH values were
measured when there was a maximum difference between them. In

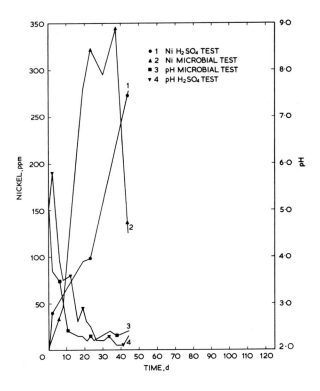

*Fig. 8. Nickel leaching in a microbial test (malt + peptone
medium) v. in a sulphuric acid leaching test, at similar pH values.*

with the pH curves obtained in a previous similar microbial test
with concentrate present and in the sulphuric acid leaching test
at pH 5.0 is shown in Fig. 9. This indicates that production of
acid is due to both the microbes and concentrate.

H.   Leaching with *Thiobacillus ferrooxidans*

The previous tests had shown that the solubilization of
nickel and copper was, at least in part, dependent on the production of acid from an organic medium by the microbes. The iron and
sulphur oxidizing bacterium *T. ferrooxidans* is the organism normally considered for the acid leaching of sulphide minerals and a
comparative test was therefore made to assess the rates of solubilization that could be achieved with a strain of *T. ferrooxidans*
growing on Concentrate A.

The concentrate (0.4g) in 100 ml of an inorganic salts medium
containing 2g Fe^{++}/l at pH 2.0 was inoculated with an active
culture of *T. ferrooxidans*. Some 65% of the nickel was leached in
10 days. Thereafter there were fluctuations in the nickel concentration until about 95% was solubilized in 50 days (Fig. 10).

The fluctuations in the nickel leaching curve might have
been associated with the changes in the form of iron precipitates.
The initial rate of leaching of nickel was faster than in the most
active of the tests using malt + peptone media and soil cultures.
The total amounts of nickel and copper solubilized, based on
residue analysis were 99 and 50%, respectively (Table 3).

I.   Microbial Studies

During the tests with malt and peptone media the supernatant
solutions were frequently examined for microbial growth. Apart
from a few protozoa and fungi seen in the initial tests, only
bacteria were observed and the majority of these were rod-shaped
cells of various sizes. The largest number of bacteria were seen

practice the average pH of the sulphuric acid test would have been closer to that of the microbial test than is indicated in Fig. 8.

The indications from this were that the acid produced in the microbial leaching tests was more efficient than sulphuric acid or that there were additional factors produced by the microbes which influenced leaching.

In the sulphuric acid leaching tests at pH 3.0 and 5.0 without pH control the pH curves showed a marked fall (Fig. 7b). This suggested that acid could be released or produced from the copper/ nickel concentrate and may have contributed to the production of acid in the microbial cultures. To determine the relative contributions of the concentrate and the microbes to the production of acid a microbial test was also set up to follow the pH changes in malt + peptone medium, initially at pH 5.0, in the absence of concentrate. A comparison of the pH curve obtained in this test

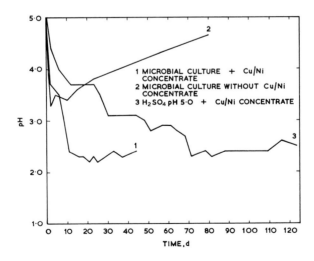

Fig. 9. pH variation in leaching tests initially at pH 5.0; 1) microbial test in malt + peptone medium with Cu/Ni concentrate; 2) microbial test in malt + peptone medium without Cu/Ni concentrate; 3) H_2SO_4 leaching test with Cu/Ni concentrate.

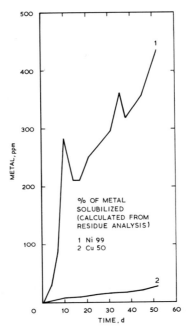

Fig. 10. Leaching of nickel and copper using T. ferrooxidans
at pH 2.0.

within 24h of inoculating a test. Thereafter the numbers observed
in solution decreased.

Samples from some of the early tests were inoculated on to
agar plates at pH 5.0. Numerous, very small, smooth, cream-
coloured colonies consisting of motile rods (1.0-2.0 x 0.5 micro-
metres) grew on all plates after 48h. Tests for acid production
by these bacteria were negative. In an attempt to isolate acid-
producing organisms, samples from the tests on malt + peptone
without pH control were plated on agar at pH 5.0, 3.0, and 2.6.
The concentration of malt + peptone was also increased in an
effort to increase the size of the colonies. Bacteria similar in
size and colonial appearance to those isolated from the initial
tests were obtained but only at pH 5.0. At pH 3.0 and 2.6 only a
few yeast colonies grew after 30 days.

The isolated bacteria were similar in appearance to $T.$ *ferro-oxidans* but less motile. As $T.$ *ferrooxidans* is known to grow on organic substrates (4) it was possible that the bacteria were strains of this organism. Limited attempts were made to grow the bacteria on ferrous iron and mixtures of ferrous iron and malt + peptone media but no growth was obtained. Although the tests were of short duration it seemed unlikely that the bacteria were $T.$ *ferrooxidans*. Oxidation of reduced sulphur compounds was not tested.

In retrospect it is possible that concentrates were not wholly suitable for tests such as these. They are not 'natural' metal sources having been concentrated by man and would give high concentrations of soluble metal ions in a comparatively short time. It might have been more satisfactory to use rocks containing low concentrations of metal for these early studies.

Although these studies were directed at a preliminary examination of the technical feasibility for metal solubilization by microbes it is interesting to reflect on these other forms of microbial leaching. In some cases there could be interest in dilute solutions of metals. To illustrate this statement a comparison can be made between, for example, recovery of gold or uranium and zinc or iron from solution. In the case of the two former metals it is economic in some cases to recover them from solutions containing 1-3 ppm of gold and 10-40 ppm of uranium. It is not difficult to find 'waste' solutions having zinc or iron concentrations much higher than these i.e. 1-2 gl^{-1}. These solutions do not at present arouse much interest except perhaps as potential pollution problems. An important factor is obviously the volume of solution involved and hence the total amount of metal present. Uranium occurs in copper dump leach liquors in low concentrations, 2-15 ppm. The volume of liquid in circuit can be high, 250 m^3 min^{-1} (50,000 galls min^{-1}). The amount of uranium is large enough to tempt an economic recovery. As metals such as zinc and iron

are not recovered at present from solutions containing a few grams
per litre, this should not preclude the possibility of recovery in
the future of metals from 'dilute' solutions. Perhaps microbial
leaching leading to dilute solutions of metals could be the first
stage in an economically viable process. Therefore innovation
and the technology to concentrate and recover these metals must
advance to complement microbial leaching research. The concentration step we are seeking for the future might also be microbial,
but that is another story.

IV. CONCLUSIONS

The results have indicated that bacteria other than thio
bacilli can be used for leaching metals. Although the mechanism
appeared to be acid production there was evidence that the acid or
acids produced were more effective than equivalent amounts of
sulphuric acid.

The rates of leaching obtained with the soil isolated were
much slower than when a strain of $T.$ $ferrooxidans$ was used. This
latter strain had been adapted over long periods of time to tolerate high concentrations of metals. The soil isolates after 5
subcultures reduced the leaching time by a factor of 3.5 and it is
possible that further adaptation would have taken place thus increasing their activity and tolerance to metals.

V. ACKNOWLEDGMENT

Warren Spring Laboratory are pleased to acknowledge the
financial support and interest of Consolidated Gold Fields Ltd,
London, in this work and for their permission to publish it.

VI. REFERENCES

1. Tuovenen, O.H., and Kelly, D.P., *Z. Allg. Mikrobiol.*, 12, 311-346 (1972).

2. Pareš, Y., *Rev. Ind. Miner.*, 50, 408-414 (1968).

3. Silverman, M.P., and Lundgren, D.G., *J. Bacteriol*, 77, 642 (1959).

4. Tabita, R., and Lundgren, D.G., *J. Bacteriol*, 108, 322 (1971).

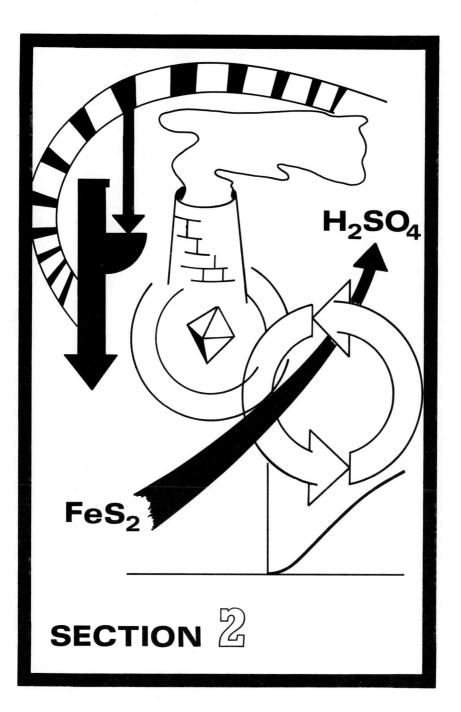

H₂SO₄

FeS₂

SECTION 2

II WASTE TREATMENT AND ENVIRONMENTAL CONSIDERATIONS

In many countries where enormous quantities of waste products are accumulating, a need has been created to develop effective methods to deal with metal extraction problems associated with these various wastes, and with related environmental considerations.

The microbiological activity involved in the recovery of metal values from mining and processing waste materials is very complex. The reaction mechanisms involved in the associated metal dissolution processes are also not completely known. In this regard, the topics presented in this section contribute toward a solution to these problems and to a better understanding of the leaching phenomena involved.

The first chapter of this section deals with the microbiological leaching of waste materials such as fly ash, slags, and jarosite-type leach residues. The aim of this research is to extract toxic metal ions (such as Cd, As, etc.) from the waste materials and as a consequence to reduce the possibility of contaminating the environment by these industrial wastes. The second chapter is an attempt to provide a microbiological process for the recovery of sulfur in its elemental form from sulfate and sulfuric acid-containing aqueous wastes which are to be released in the environment. The final chapter of this section describes a microbiological leaching process for the removal of inorganic sulfur contained in coal. This process may be considered as a practical alternative to the chemical coal desulfurization process. The effect of the parameters influencing the attendant oxidation process is also discussed.

METAL RECOVERY AND ENVIRONMENTAL PROTECTION
BY BACTERIAL LEACHING OF INORGANIC WASTE
MATERIALS

Hans Georg Ebner

University of Dortmund
Dortmund, Federal Republic of Germany

Industrial waste materials contain valuable metals in high total amounts but in low concentrations. Sulfidic dust concentrates and oxidic fly ash from a copper process, slag from a lead smelting process and "Jarosite" from a zinc electrolysis were leached with different thiobacillus species. The leaching efficiency was specific to materials, treatment and inoculated bacteria species. The maximum output for zinc was around 95%, for copper around 70%. We were concerned about the high concentrations of 32 mg Cd/l, 5 mg Pb/l and 30 mg As/l in the leaching liquors because these values exceed the allowed toxic levels in drinking water.

I. INTRODUCTION

The foreseeable shortage of raw materials is one of the major problems faced by the world today. But solely declaiming the depletion of oil as an energy source means forgetting about the exhaustion of mineral ore deposits which are necessary for the production of metals. While fossile energy sources are exchangeable against each other this will - in most cases - not become

195

TABLE I   *Reserves of the World's Energy and Mineral Resources*

Nickel	46 yrs.	Lignite	270 yrs.
Aluminum	44 yrs.	Coal	190 yrs.
Zinc	43 yrs.	Gas	54 yrs.
Lead	29 yrs.	Oil	34 yrs.
Copper	25 yrs.		
Mercury	20 yrs.		
Silver	12 yrs.		
Gold	10 yrs.		

feasible for metals. Recent studies (1, 2, 3) about the estimated lifetime of resources show that - even under the aspects of increasing haulage - the resources of fossile materials will last many decades longer than those of metallic ores (Table I).

The depletion of high-grade ore deposits corresponds with an exponential growth rate in the demand for metals. Between 1975 and the year 2000 the requirement for zinc is estimated to rise from 6 to 12 million tons, for copper from 8 to 24 million tons and for aluminum from 13 to 53 million tons (2, 3).

Another problem of our century is in solving industrial pollution. Forced by strictly growing governmental regulations industry has to pretreat a lot more waste materials before depositing them as in former years. These data correspond pretty well with the predictive model of Massachusetts Institute of Technology which concludes that without the development of new and more economical technologies the industrial production of our world will break down within the first decades of the coming century owing to depleted resources, a growing population and an increasing pollution (4).

Many inorganic industrial wastes still contain valuable metals. Although their metal concentrations are mostly very low their total amount itself is often high because of the large masses which arise during industrial processes. A development of a cheap and simple process to pretreat these residues before their storage on deponies can be regarded at least under two beneficial aspects: toxic substances will be removed from the wastes which lowers the cost for their disposal and at the same time valuable metals can be gained from the waste materials.

To meet the needs of raw-material supply and to diminish environmental pollution we investigated the possibilities of recycling inorganic industrial waste by bacterial leaching. Included in the investigations were especially those residues which contain remarkable concentrations of valuable metals or which have to be stored on special deposits because of their high contents of toxic elements or which accumulate in large quantities during industrial processes.

II. MATERIALS AND METHODS

The waste materials were supplied by different German industrial companies. All materials possessed a very small grain size below 0.1 mm except the slag from a lead smelting process which had to be crushed before the bacterial leaching treatment.

Jarosite, J-1, was a residue which amasses during the production of zinc. After roasting of the zinc ore the originated ore concentrate is leached chemically with sulfuric acid. The zinc pregnant liquor is pumped to the recovery plant whereas the iron rich residue called "Jarosite" has to be stored on special deposits. KS-1 and KS-2 were sulfidic dust concentrates from copper processes in which the ores are chemically leached with hydrochloric acid. The oxidic fly ash, KO-3, is filtered from the fumes of a pyrite roasting process. BS-1 was a basic slag derived from a lead smelting process.

Different *Thiobacillus thiooxidans* and *Thiobacillus ferrooxidans* strains had been isolated during former experiments from several locations of European and American countries (mines, rivers,

vulcanoes, sea water). They were grown either on normal or diluted LEATHEN's respectively STARKEY's medium at incubation temperatures of 30° C or higher.

Leaching experiments were performed in 300 and 500 ml Erlenmeyer-flasks as static and/or stirred cultures. To create
applied conditions for further in-situ processes the pH values of
the culture media were neither lowered at the beginning nor maintained during the experiments with additional acid supply. The
pH was measured electrometrically with INGOLD-pH-electrodes and
a KNICK-pH-meter (Type 700).

The elements in solution were analysed by atomic absorption
spectroscopy with a PERKIN-ELMER AAS 400. Before their determination the leaching liquors had to pass a filtration unit and the
remaining solids on the filter were washed twice with distilled
water. The measured element concentrations of the liquids multi-
plied by their volumes gave the total amount of the leached metals
in solution. Correlating these results with the original metal
contents the solids had before their treatment expresses the
leaching efficiency in per cent.

For the performance of the arsenic analyses we gratefully
acknowledge the cooperation of the Institut für Spektrochemie,
Dortmund.

III. RESULTS AND DISCUSSION

The chemical analyses of the waste materials proved interesting amounts of several valuable metals (Table II). The dust
concentrates KS-1 and KS-2 from the copper processes contained
18.3% and 4.4% copper as well as 10.1% and even 29.5% zinc. The
fly ash, KO-3, from the pyrite roasting however possessed only
0.9% copper and 2.5% zinc. Jarosite, J-1, held 0.38% copper and
still 9.6% zinc. Similar concentrations showed the basic lead
slag with 0.19% copper and 9.2% zinc. The total iron amount in

TABLE II   *Chemical Analyses of Industrial Waste*
Materials (%)

	KS-1 dust conc. sulfidic	KS-2 dust conc. sulfidic	KO-3 fly ash oxidic	J-1 jarosite	BS-1 lead slag
Cu	18.3	4.4	0.9	0.38	0.19
Zn	10.1	29.5	2.5	9.6	0.2
Fe_{tot}	22.9	25	37.4	27.2	29.2
S_{tot}	37.2	33.5	trace	9	1.23
Pb	6.1	1.6	19.4	8.6	1.56
As	0.4	2.2	4.1	0.21	0.015
Cd				0.15	0.005

in all materials varied between "a trace" up to 37.2%. Remarkable were also the relative high concentrations of hazardous
components in the residues like lead, arsenic and cadmium which
create the problems for the deposition of these waste materials.

Our leaching experiments in Erlenmeyer-flasks with several
strains of *Thiobacillus thiooxidans* and *ferrooxidans* demonstrated
that the complexity of the materials in mineral and ion content
requires special treatments for each residue. About 10 out of 30
different thiobacillus strains could not even survive concentrations of 1% w/v of the waste in solution. From the remaining 20
strains meanwhile 3 of them are adapted to higher tolerances
against the complex materials. Without any difficulty we were
able to suspend 15% of KS-1 or KO-3, 10% of the lead slag, but
only 5% of the sulfidic dust concentrate KS-2 or the Jarosite
J-1 in the media. In special experiments the bacteria could
survive concentrations up to 20% of the waste in solution.

To test the dependency of different strains in different
waste suspensions we examined the growth of *Thiobacillus thiooxi-*
dans strains T-E 1 and T-E 2 and of *Thiobacillus ferrooxidans*
F-E 2 in several concentrations of KS-1, KS-2 and KO-3 (Table
III). During these experiments *Thiobacillus ferrooxidans*, strain

TABLE III Maximal Endured Waste Concentrations by Different
T. thiooxidans *and* T. ferrooxidans *Strains (% w/v).*

	KS-1	KS-2	KO-3
Concentration T. thiooxidans strain	10%	0.75%	15%
	T-E 2	*T-E 1* *T-E 2*	*T-E 1*
Concentration T. ferrooxidans *strain*	15% *F-E 2*	5% *F-E 2*	15% *F-E 2*

F-E 2, withstood the highest concentrations in the test. Never-theless we selected *Thiobacillus thiooxidans* strain T-E 2, from our collection for the leaching experiments with the fly ash KO-3. Although this strain could not endure the same high waste concen-trations like the strain F-E 2 it had a better capability to leach elements from this fly ash.

During the experiments our particular interest became mainly concentrated on the feasibility of leaching the lead slag BS-1 and the Jarosite J-1. Leaching the Jarosite with *Thiobacillus ferro-oxidans* was not quite successful. Although the bacteria survived more than 30 days in the solutions with 1% and 5% Jarosite the pH values did not decrease, but remained close to those of the sterile control (5). In the experiments with the basic lead BS-1 the pH values of the inoculated solutions and the sterile control even rose to pH 7 with 1% of lead slag and to pH 8.3 with 5% of lead slag in solution (5). An inoculation of *Thiobacillus thio-oxidans* gave far better results. With 5% of Jarosite in solution the pH dropped to 0.9 after 20 days (5).

In further investigations we tried to explore at which day(s) during the progress of the leaching processes the metal concentra-tions became their maxima. At the beginning of the tests a static culture time of seven days was given to let the bacteria attach to the particles (Fig. 1). At the end of this time the pH began

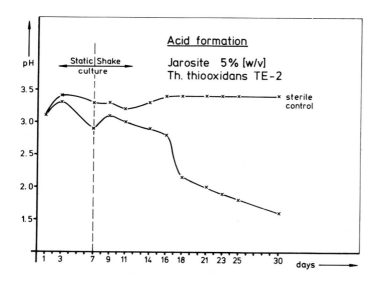

Fig. 1.   pH-curves of Jarosite suspensions inoculated with
T. thiooxidans *and sterile control.*

already to fall slightly, but after starting to shake the culture
flasks the pH almost immediately went up again. After two more
days of shaking the pH anewed to decrease and finally fell sharply
between the 16th and 18th day. We will have to prove now whether
the long time of seven days for an attachment of the bacteria to
the particles can be reduced or may even not be necessary.

Following the leaching rates over 23 days of shaking - respectively 30 days in total - around 36% of the cadmium content
from the Jarosite had been leached (Fig. 2). In the sterile
control, plotted as black bars in the figure, considerable
amount of cadmium was leached as well. The leaching rate of
zinc rose within the first seven days remained almost constant
until the 18th day after which it climbed again (Fig. 3). Correlated with the decrease of the pH-values in the solution there had
also been a sharp decline of the pH-curve on the same 18th day of

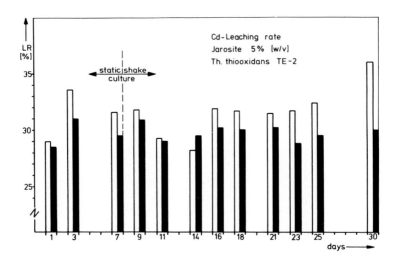

Fig. 2.   Leaching rates of cadmium from Jarosite with and without bacteria (black bars) in per cent from the total original content.

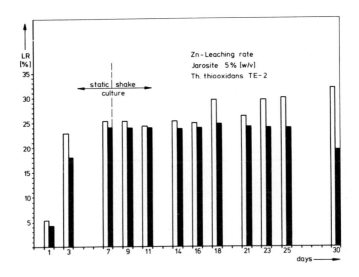

Fig. 3.   Leaching rates of zinc from Jarosite with and without bacteria (black bars) in per cent from the total original content.

these experiments. It is worth noting that the dissolved zinc
concentrations in the sterile control declimbed after the 18th
day.   We suppose that a process of precipitation or rearrangement
of the zinc ions takes place. Regarding the copper output we
observed similar tendencies as during the zinc leaching (Fig. 4).
However in these experiments the bacteria did not dissolve more
than 7% of the total copper content which is unsufficient for
bacterial leaching processes insitu. More experimental work to
clarify these problems will be necessary.

During further approaches it became obvious that the leaching
efficiency was specific to materials, treatment and inoculated
bacteria. Leaching the concentrates KS-1 and KS-2 with *Thiobacillus ferrooxidans* for example was much more effective than with
Thiobacillus thiooxidans. Leaching KS-1 with a mixed culture of
Thiobacillus ferrooxidans strains F-E 2 and F-E 21 gave better
results than leaching this concentrate with one of these strains

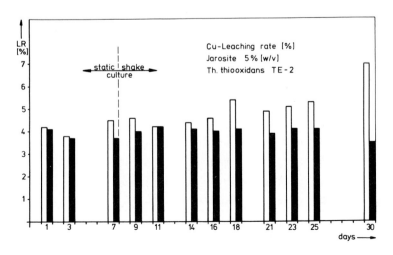

*Fig. 4.   Leaching rates of copper from Jarosite with and without
bacteria (black bars) in per cent from the total original content.*

separately. On the other hand *Thiobacillus thiooxidans* created
the maximal efficiency when leaching the fly ash KO-3, Jarosite
and the lead slag.

The results (that were obtained in this study) with optimized
growth conditions are encouraging enough for further experiments
(Table IV). The average relative yields of copper from all residues lie in the region of 35 to 40% and for zinc above 50%. The
maximal yield of zinc which was 94% was leached from the concentrate KS-2. The maximal output of copper came from the Jarosite
with 69%. From the fly ash KO-3 the bacteria had been able to
bring 81% of the arsenic content into solution.

Not all of our examined residues are already stored on special
deposits. As we could isolate thiobacillus from a number of
deposits we are concerned about the high concentrations of toxic
elements which can be leached by the bacteria from the waste
materials (Table V). The concentrations of lead, cadmium and
arsenic leached from these wastes are severalfold higher than the
permitted values in potable water in Germany. Uncontrolled deposits contaminated by bacteria therefore can cause severe ground
water intoxication.

From the results it can be seen that inorganic industrial
residue materials contain valuable amounts of metals which are too
good for depositing or wasting especially at a time when natural
resources are getting scarce and the demand of developing industries is growing exponentially. With bacterial leaching it should
be possible to create economically feasible processes with which
metals from inorganic wastes can be brought into solution and then
gained by conventional recovery. Our experiments also proved that
on industrial deposits - in many cases - bacteria are already
growing and thus are leaching hazardous elements from the wastes.
We know that we just scratched the envelope of the parcel with
industrial waste materials but we are sure that it contains a lot
more gainable elements and many other interesting residues or

Maximal Leaching Efficiency from different Wastes (%)

	KS-1			KS-2			KO-3			J-1		BS-1	
	Cu	Zn	As	Cu	Zn	As	Cu	Zn	As	Cu	Zn	Cu	Zn
Bact. Leaching	12	38	3	35	94	14	42	85	81	69	52	33	50
Ster. Control	0.1	16	-	0.4	10	0.2	10	16	-	5	10	0	0
Th. Strain	F-E2 F-E21	F-E2	F-E2	F-E2	F-E2	F-E2	T-E2	T-E2	T-E2	T-E2	T-E2	T-E2	T-E2

F = Thiobacillus ferrooxidans

T = Thiobacillus thiooxidans

TABLE IV Maximal Leaching Efficiency from Different Industrial Wastes by Inoculation with Different Thiobacillus trains vs. Sterile Controls (%).

TABLE V   Concentrations of Hazardous Elements in Microbiologically Leached Industrial Wastes and Maximal Allowed Concentrations in Drinking Water in Germany (mg/l).

	Cd	*Pb*	*As*
KS-1	*8.7*	*2.0*	*8.4*
KS-2	*5.1*	*2.5*	*18.0*
KO-3	*0.7*	*2.3*	*30.1*
J-1	*31.8*	*5.0*	*–*
BS-1	*0.7*	*3.1*	*–*
allowed in drinking water	*0.006*	*0.04*	*0.04*

tailings. It is to hope that our work will be encouraging enough to lead more scientists to the field of microbiological recycling.

IV. ACKNOWLEDGMENT

Thanks are due to Mrs. Brigitte Steinfeld for her excellent and skillful technical assistance.

V. REFERENCES

1. Bundesanstalt für Geowissenschaften und Rohstoffe: "Die künftige Entwicklung der Energienachfrage und deren Deckung", Abschnitt III, Das Angebot von Energierohstoffen, Hannover, 1976.

2. Deutsche Forschungsgemeinschaft: "Denkschrift zur Lage der Lagerstättenforschung", Teil I, Mineralische Roshstoffe, H. Boldt Verlag KG, Boppard, 1975.

3. von Engelhardt, "Raubbau an den Erzvorräten", *Bild der Wissenschaft*, 11, 78 (1976).

4. Meadows, D., "The Limits of Growth", Universe Books, New York, 1972.

5. Ebner, H.G., in "Conference Bacterial Leaching 1977", (W. Schwartz, Ed.), GBF Monograph Series, No. 4, Verlag Chemie, Veinheim - New York, 1977.

SULFATE DECOMPOSITION: A MICROBIOLOGICAL PROCESS

Douglas J. Cork
and
Michael A. Cusanovich

University of Arizona
Tucson, Arizona, U.S.A.

The disposal and treatment of sulfur, a major process stream constituent of industry and mining, is rapidly expanding into one of its most formidable challenges. Significant potential exists for the treatment of industry's major effluents including excess sulfuric acid, gypsum, coal de-sulfurization by-products, acid mine waters, and general metallurgical effluents.

A variety of microorganisms are capable of metabolizing sulfate to elemental sulfur. In nature, these processes are generally slow and yield products toxic to the environment (e.g. hydrogen sulfide). We are reporting here on studies designed to quantitatively convert sulfate to elemental sulfur utilizing **Desulfovibrio desulfuricans** *for sulfate conversion to sulfide and photosynthetic bacteria for the sulfide's subsequent conversion to sulfur.*

We have developed a purge system, using an inert carrier gas, to continuously remove sulfide from actively growing cultures of **Desulfovibrio** *and supply this substance to either* **Chlorobium** *thiosulfatophilum or* **Chromatium vinosum**. *This technique has allowed us to achieve quantitative conversion of sulfate (initial concentrations of 30 to 40 mm) to elemental sulfur. Conditions for optimizing the process have been studied and the approach has been successfully applied to solvent extraction raffinates.*

Previous studies on static cultures at various sulfate and sulfate concentrations or on mixed cultures of **Desulfovibrio**

207

and photosynthetic bacteria yielded discouraging results. These approaches did not quantitatively convert sulfate to sulfur and bacterial growth was poor. The studies described here provides elemental sulfur.

I. INTRODUCTION

A variety of metallurgical effluents including solvent extraction raffinates contain high concentrations of sulfate ion. These effluents present serious disposal problems in that the sulfate must be converted to a stable, non-leachable form. We herein propose a solution to this problem which utilizes the natural biological activities of *Desulfovibrio desulfuricans, Chlorobium thiosulfatophilum,* and/or *Chromatium vinosum,* obligate anaerobes of the aquatic sulfur cycle (1,2). *Desulfovibrio* reduces sulfate to sulfide, and the green and purple sulfur bacteria *Chlorobium* and *Chromatium* photosynthetically oxidize sulfide to elemental sulfur. *Chlorobium thiosulfatophilum* excretes the colloidal sulfur, while *Chromatium vinosum* stores sulfur intracellularly. In principle, a solvent extraction raffinate would provide sulfate to *Desulfovibrio,* thereby initiating the sulfur cycle. The sulfide thus produced can be made available to the green and purple sulfur bacteria being quantitatively oxidized to elemental sulfur. These photosynthetic bacteria grow on autotrophic media and fix CO_2.

It is the purpose of the studies described herein to demonstrate that a mutualism may be established between *Desulfovibrio* and either *Chromatium* or *Chlorobium,* so as to quantitatively convert sulfate to elemental sulfur. Further, we have found that the addition of a purge system utilizing an inert carrier gas leads to a more efficient conversion of sulfate to sulfide, and then sulfide to sulfur, than can be obtained from static cultures of each of the bacteria. Finally data will be presented which demonstrate that this purge system can be adapted to solvent extraction raffinate.

II. MATERIALS AND METHODS

A. Organisms

Experiments were carried out with *Desulfovibrio desulfuricans* ATCC7757 *Chromatium vinosum* (strain D), and *Chlorobium thiosulfatophilum* (strain PM).

B. Media and Static Growth Conditions

Desulfovibrio was grown in a modified media of Starkey (3):

Compound	gm/liter
$NaSO_4$	3.0
$MgSO_4$	1.5
NH_4Cl	1.0
$K_2HPO_4 \cdot 3H_2O$	0.65
$CaCl_2$	0.08
$Fe(NH_4)_2(SO_4)_2 \cdot 6H_2O$	0.01
Lactic Acid (85%)	5.0 ml
Yeast Extract	0.2

The pH was adjusted to 7.0 with NaOH and growth was carried out in completely filled 1 liter prescription bottles which were immersed in a 30°C water bath.

Strictly autotrophic media was used for the photosynthetic bacteria. The media of Bartsch (4) was modified to contain:

Compound	gm/liter
$NaCl$	10.0
NH_4Cl	1.2
$K_2HPO_4 \cdot 3H_2O$	1.3
$MgCl_2$	0.55
$CaCl_2$	0.12
Larson's Trace Elements (5)	2.0 ml
$NaHCO_3$	2.0

Solutions containing the appropriate amount of sulfide were autoclaved separately and added to the bulk media. The pH was adjusted to 7.0 with HCl. Growth was routinely carried out in completely filled one liter prescription bottles which were immersed in a 30°C water bath. The center of each bottle containing photosynthetic bacteria was placed five centimeters from two forty watt General Electric showcase lamps. These photosynthetic cultures were illuminated continuously.

C.   Gas Purged Bacterial Growth Conditions

A gas purged system of bacterial growth is particularly amenable to the following sulfur metabolism by *Desulfovibrio* and either *Chlorobium* or *Chromatium*:

$$SO_4^= \xrightarrow{\quad \textit{Desulfovibrio} \quad} H_2S \xrightarrow[\textit{Chromatium}]{\textit{Chlorobium} \text{ or}} S^\circ$$

This sulfur pathway reveals that the common intermediate is volatile hydrogen sulfide, which may be transferred from the *Desulfovibrio* culture to the photosynthetic culture by an inert carrier gas (75% Ar, 25% CO_2), which was also used to maintain a constant pH ($\sim$7).

The purge system is shown schematically in Figure 1. *Desulfovibrio* and the photosynthetic organisms are grown in separate one liter prescription bottles, each culture occupying 80% of the bottle volume. The bottles containing separate cultures are connected by latex tubing. Gas dispersion tubes deliver the inert carrier gas to *Desulfovibrio*, and the inert carrier gas plus the volatile sulfide to the photosynthetic culture bottle. Lighting and temperature conditions are identical with those described for static cultures. Constant gas pressure was maintained while media was withdrawn for assay.

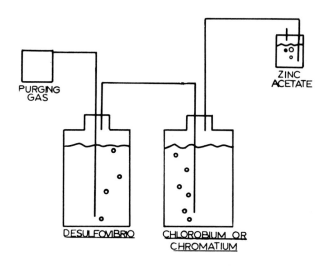

Fig. 1. The gas purged mutualistic system utilizing Desulfovibrio *and either* Chlorobium *or* Chromatium.

D. Chemical Analysis of Bacteria

Sulfide was determined according to the method of Truper and Schlegel (6). Sulfate was determined by a $BaCl_2$ - gelatin turbidimetric method (7). Thiosulfate was analyzed by the method of Sorbo, et al. (8). Elemental sulfur, *Chromatium* bacteriochlorophyll and protein were determined as described by Schmidt and Kamen (9). *Chlorobium* chlorophyll was extracted in absolute methanol and quantitatively determined by use of the absorption coefficients reported by Stanier and Smith (10).

After assaying for sulfide, elemental sulfur, and bacterio-chlorophyll, the cells were centrifuged at 15,000 rpm. The supernatant was analyzed for sulfate and the pellet was washed in 4% trichloroacetic acid and then resuspended three times in 7:2 acetone-methanol and recentrifuged in a clinical centrifuge before proceeding with the Biuret protein assay as described by Schmidt and Kamen (9). For all assays, absorbances were recorded on a Cary 118 Spectrophotometer.

III. RESULTS

As has been determined for a variety of phototrophic bacteria (10), static cultures of both *Chlorobium thiosulfatophilum* and *Chromatium vinosum* are inhibited by high concentrations of sulfide (greater than 4 and 9 mm $Na_2S \cdot 9H_2O$, respectively). Figure 2 presents data on bacteriochlorophyll, protein, elemental sulfur, sulfate, thiosulfate, and sulfide production or utilization for static cultures of *Chlorobium* and *Chromatium* over a twenty-four hour period. As seen in Figure 2, sulfide is rapidly oxidized in both cases: 60% and 31% of the sulfide is oxidized to elemental sulfur by *Chlorobium* and *Chromatium*, respectively. However, at least 40% of the sulfide is oxidized to thiosulfate and sulfate in each case.

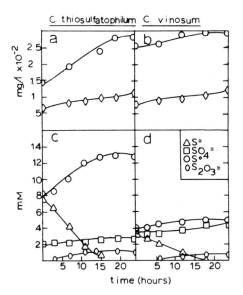

Fig. 2. Growth parameters for static cultures of Chlorobium *and* Chromatium. a *and* b: *Protein (o) and bacteriochlorophyll* (◊) *versus time for* Chlorobium *and* Chromatium, *respectively.* c *and* d: *sulfide (Δ), sulfur, (o), sulfate (□) and thiosulfate (0) versus time for* Chlorobium *and* Chromatium, *respectively. The ordinate for bacteriochlorophyll concentration is* $x\ 10^{-1}$.

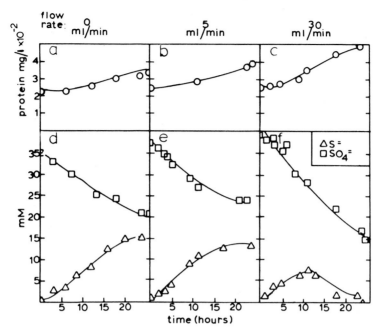

Fig. 3.   The effect of gas purge rates on growth parameters of Desulfovibrio. *a, b, and c: Protein (o) versus time, d, e, and f: Sulfate (□) and sulfide (△) versus time.*

Typical results with a static culture of *Desulfovibrio* inoculated in the presence of 35 mM sulfate are presented in Figure 3a and 3d. Although bacterial growth is slow, substantial reduction of sulfate to sulfide takes place in twenty-four hours.

A.   Sulfate Reduction by Purged Cultures of *Desulfovibrio*

The purging of cultures of *Desulfovibrio* with an inert carrier gas (see Materials and Methods for the experimental procedure) results in a profound effect on the reduction of sulfate. Figures 3b and 3e present results at a gas purge rate of 5 ml/min and Figures 3c and 3f at 30 ml/min. For a twenty-four hour period, a 5 ml/min purge yields results almost identical to those found in static cultures (Figures 3a and 3d).

However, at a purge rate of 30 ml/min, a substantial increase in the amount of sulfate reduced and protein synthesized is noted at twenty-four hours. Further, sulfide production does not exceed 8 mM and appears to drop off to a low steady-state rate of production (2-3 mM) after 10-12 hours. These results suggest that high concentrations of sulfide were responsible at least in part for the limited reduction of sulfate observed at a purge rate of 5 ml/min and in static cultures. Thus high purge rates keep the amount of sulfide in the *Desulfovibrio* culture at a low level and stimulate both cell growth and sulfate reduction.

B.   Effect of Purging on Mutualistic Cultures

Figure 4 presents data on the effects of providing the effluent gas of a purged culture (30 ml/min) of *Desulfovibrio* to a culture of *Chlorobium thiosulfatophilum*. Over a twenty-four hour period, approximately 90% of the sulfate reduced by *Desulfovibrio* is converted to elemental sulfur by *Chlorobium* (Figures 4a and 4b). Further, unlike static cultures (Figure 2), no thiosulfate is produced and less than 4mM sulfate is formed. Compared to the static culture of *Chlorobium*, approximately four times as much protein and nine times as much bacteriochlorophyll is synthesized in the purged system in twenty four hours (compare Figure 2a to Figure 4a). Finally, about 4.5 times as much elemental sulfur is formed in 24 hours in the purged system as compared to a static culture.

Studies on elemental sulfur production by *Chromatium* in the same purging system are presented in Figure 5. For these experiments a purge rate of 5 ml/min was used as we observed an inhibition of the growth and extent of sulfide oxidation by *Chromatium* at higher purge rates. As observed with *Chlorobium*, no thiosulfate was formed when sulfide was provided by the carrier gas. However, in the first twenty four hours, little or no elemental sulfur was produced by *Chromatium*, and the total

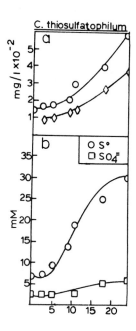

Fig. 4. *Growth parameters for purged cultures of* Chlorobium.
a: *Protein (o) and bacteriochlorophyll (◊) versus time.* b:
Sulfur (o) and sulfate (□) versus time. The ordinate for
bacteriochlorophyll is x 10^{-1}. The gas purge rate is 30
ml/min.

amount of elemental sulfur was substantially less than formed by
Chlorobium. Further at later times (> 40 hours) large amounts
of sulfate are formed. Thus, in terms of net production of
elemental sulfur from sulfate, *Chlorobium* is far superior to
Chromatium.

C. Raffinate Studies

Twenty-five percent (162 mM sulfate) and 50% (324 mM sulfate)
hydrometallurgical solvent extraction raffinates (composition
given in Table 1) were supplemented with lactic acid (47 or 94 mM),
yeast extract (1 gm/liter) K_2HPO_4 · $3H_2O$ (0.65 gm/liter and
$MgCl_2$ (1 gm/liter). Undiluted raffinate was unable to support
Desulfovibrio growth.

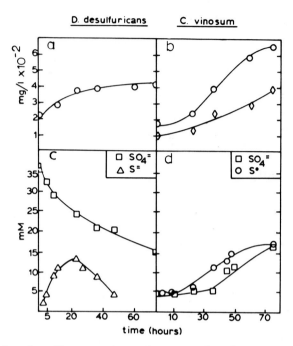

Fig. 5. Growth parameters for purged cultures of Chromatium.
a: Protein (o) versus time for Desulfovibrio. b: Protein (o)
and bacteriochlorophyll (◊) versus time for Chromatium.
c: sulfate (□) and sulfide (△) versus time for Chromatium. The
ordinate for bacteriochlorophyll concentration is x 10^{-1}. The
gas purge rate is 5 ml/min.

Figure 6 presents typical results for sulfide production by
Desulfovibrio obtained with 25% raffinate (47 and 94 mm lactic
acid, Figure 6a and 6b, respectively), and 50% raffinate (94 mM
lactic acid, Figure 6c). For these experiments, the Desulfovibrio
culture was purged at a rate of 30 ml/min into a zinc acetate
trap for quantitative determination of sulfide (6).

The cultures in 25% raffinate yielded a net amount of
sulfide of 18-22 moles/24 hours irrespective of the initial
lactic acid concentration. These levels of sulfide production
compare favorably with that obtained with Desulfovibrio grown on a
defined media (Figure 3f) and indicate that diluted raffinate
contains no compounds detrimental to Desulfovibrio growth.

TABLE I   *Chemical Analysis of Solvent Extraction Raffinate*

Constituent	Concentration (mM)[a]
$SO_4^=$	646
NH_4^+	1170
$PO_4^=$	<.001
Mg^{++}	1.42
Fe^{++}	.54
Ca^{++}	6.50
Na^+	17
K^+	.80
NO_3^-	.03
As	.12
Ni^{++}	.004
Co^{++}	.05
Zn^{++}	.009
Cu^{++}	<.001

[a] *Chemical analysis was done by the University Analytical Center, Department of Chemistry, University of Arizona.*

The experiments done in the presence of 50% raffinate (Figure 6c) indicate a substantial decrease in the rate of production of sulfide in twenty-four hours and suggest that the concentration of some substance deleterious to the growth of *Desulfovibrio* is sufficiently high to inhibit growth.

IV. DISCUSSION

In Table 2 we have summarized the results of our studies in order to provide a comparison of static and purged cultures of *Chromatium* and *Chlorobium* in terms of the various parameters

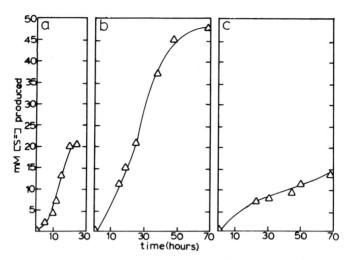

Fig. 6. Sulfide production for purged cultures of Desulfovibrio *grown on diluted raffinate. a: 25% raffinate, 47 mM lactic acid. b: 25% raffinate, 94 mM lactic acid. c: 50% raffinate, 94 mM lactic acid. The gas purge rate is 30 ml/min. See text for concentrations of other constituents added to the raffinate.*

measured. Because sulfate reduction was becoming limiting at 24 hours in static cultures of *Desulfovibrio*, we used the data from this time point for all cultures in these comparisons. Presented in Table 2 are the amount of elemental sulfur formed ($\Delta[S°]$) in millimoles/liter, the ratio of elemental sulfur formed to sulfate and thiosulfate formed ($\Delta[S°]/(\Delta[SO_4^=] + [S_2O_3^=])$), the ratio of elemental sulfur formed by the photosynthetic bacteria to sulfate reduced by *Desulfovibrio* in mutualistic cultures ($\Delta[S°]/\Delta[SO_4^=]_R$), the amount of protein synthesized ($\Delta[protein]$) in mg/liter, and the amount of bacteriochlorophyll synthesized ($\Delta[Bschl]$) in mg/liter.

For optimum conversion of sulfate to elemental sulfur, the ratio $\Delta[S°]/\Delta[SO_4^=]_R$ should approach 1.0 and the ratio $\Delta[S°]/(\Delta[SO_4^=] + \Delta[S_2O_3^=])$ should approach infinity. Although we have not achieved the ideal state to date, the data presented

TABLE II   *Comparison of static and purged cultures after 24 hours of growth*

System	Purged Chlorobium	Static Chlorobium	Purged Chromatium	Static Chromatium
$\Delta[S^\circ]^a$	21.5	4.6	2.3	0.9
$\Delta[S^\circ]/\Delta[SO_4^= + S_2O_3^=]^b$	6.1	1.4	1.6	0.4
$\Delta[S^\circ]/\Delta[SO_4^=]_R^c$	0.9	—	<0.0	—
$\Delta[protein]^d$	450	135	80	50
$\Delta[Bchl]^e$	27	3.2	3.1	3.2

[a] *the increase in elemental sulfur concentration (mM) in 24 hours.*

[b] *the ratio of the increase in elemental sulfur concentration to the combined increase of sulfate and thiosulfate concentration in 24 hours.*

[c] *the ratio of the increase in elemental sulfur concentration to the total concentration of sulfate reduced by* Desulfovibrio *in 24 hours.*

[d] *the increase in protein concentration (mg/l) in 24 hours.*

[e] *the increase in bacteriochlorophyll concentration (mg/l) in 24 hours.*

in Table 2 clearly demonstrates that the purge system utilizing *Desulfovibrio* and *Chlorobium* approaches this condition. *Chlorobium* is able to convert approximately 90% of the sulfate reduced by *Desulfovibrio* to elemental sulfur in twenty-four hours with the net yield of approximately 20 mmoles of elemental sulfur per liter. An overall process conversion for a two step single stage reduction, single stage oxidation system is 55%. Further, the elemental sulfur to oxidized sulfur ratio is substantially improved in the purged system. *Chlorobium* offers the added advantage that the elemental sulfur produced is excreted extracellularly and can be easily isolated by differential centrifugation.

In terms of conversion of sulfide to elemental sulfur *Chromatium* is marginal at best. In static cultures, *Chromatium* can only tolerate low levels of sulfide and large amounts of sulfate are produced. In the purged system, *Chromatium* produces elemental sulfur at reasonably high levels; however, this process is much slower than observed with *Chlorobium* (50-70 hours are required for optimum levels), and the elemental sulfur to sulfate ratio never exceeds 2.0.

The studies reported here demonstrate that *Desulfovibrio* can survive on diluted solvent extraction raffinate and can efficiently convert sulfate to sulfide in the concentration range 25-50% raffinate. These observations reaffirm the technical feasibility of using *Desulfovibrio* for industrial applications.

Overall the studies reported here establish that a gas-purged mutualistic system of *Desulfovibrio* and *Chlorobium* can be used for the efficient conversion of sulfate to elemental sulfur. Presently,studies are underway utilizing continuous culture methods to determine optimum conditions for elemental sulfur production in terms of pH, temperature, and substrates. It is anticipated that these studies will yield an even more efficient process applicable to an industrial scale.

Having established the technical feasibility of using microorganisms for the conversion of sulfate to elemental sulfur, it is of some interest to address the economics. The primary "energy currency" of the system is lactic acid, the electron and carbon source utilized by *Desulfovibrio* for sulfate reduction and cell growth. In part the energy requirements can be minimized by recycling *Chlorobium* i.e., feeding it to *Desulfovibrio* after separating it from the elemental sulfur. However, a substantial amount of fixed carbon is still required to drive sulfate reduction. At the present time, studies are underway testing alternate carbon sources including raw sewage. Considering the many options available for further optimization of the microbiological conversion of sulfate to elemental sulfur, it is our view that a viable industrial process can be developed. We feel the approach described here potentially offers an economical and environmentally sound industrial procedure for the disposal of sulfate containing effluents in a non-leachable form.

V.   ACKNOWLEDGEMENTS

Support for this research was kindly provided by Anaconda Copper Company. We are particularly indebted to Dr. T. McNulty and J. Spisak of Anaconda Copper for much encouragement and many fruitful discussions.

VI.   REFERENCES

1. Roy, A.B., and Trudinger, P.A., in "The Biochemistry of Inorganic Compounds of Sulfur", p. 3, Cambridge University Press, New York, 1970.

2. Postgate, J.R., in "Inorganic Sulfur Chemistry", (G. Nickless, Ed.) p. 259. Elsevier Publishing Company, New York, 1968.

3. Starkey, R.L. *Arch Mikrobiol.*, 9,268 (1938).

4. Bartsch, R.G., in "Bacterial Photosynthesis" (H. Gest, A. San Pietro, L.P. Vernon, Eds.), p. 501. Antioch Press, Yellow Springs, 1963.

5. Larsen, H., *J. Bacteriol.*, 64, 187 (1952).

6. Truper, H.G., and Schlegel, H.G., *Antonie van Leeuwenhoek J. Microbiol. Serol.*, 30, 225 (1964).

7. Tabatabai, M.A., *Sulfur Inst. J.*, 10, 11 (1974).

8. Sorbo, B. *Biochim. Biophys. Acta.*, 27, 324 (1958).

9. Schmidt, G.L., and Kamen, M.D., *Arch. Mikrobiol.*, 73 (1970).

10. Stanier, R.Y., and Smith, J.H. C., *Biochem. Biophys. Acta*, 41, 478 (1960).

MICROBIOLOGICAL DESULFURIZATION

OF COAL

Patrick R. Dugan
William A. Apel

The Ohio State University
Columbus, Ohio 43210

Mixed enrichment cultures of acidophilic microorganisms are effective in the removal of pyritic sulfur from 20% slurries of a commercial grade of pulverized coal in the laboratory. Neither T. ferrooxidans *nor* T. thiooxidans *alone were effective in sulfur removal from the coal. Microbial pyrite oxidation was most effective when the initial pH of the coal slurries was in the range 2.5 to 2.0 and when the slurries were supplemented with* $(NH_4)_2SO_4$. *Under these conditions approximately 97% of the pyritic sulfur was removed within 5 days from coal which had an initial pyritic sulfur content of 3.1%. This resulted in a beneficiation from an average 4.6% total sulfur to about 1.5% total sulfur.*

I. INTRODUCTION

A. The High Sulfur Coal Problem

It is anticipated that coal production in the U.S. will more than double by the year 1990 and the largest source of this coal will continue to be the Appalachian fields (33,43,45). All coal contains sulfur in varying amounts and much of the coal found in the Appalachian fields contains a relatively high total sulfur content of 3.0 to 5.5% (84), much of which is in the form of pyritic minerals deposited during the period of coal formation.

There are three forms of sulfur in coal: (i) organic sulfur in which the sulfur is either covalently bonded to carbon in the general forms -R-S-S-R-, and -R-S-R- or bound as a sulfate in the general form R-O-SO$_3$ (15,34), (ii) pyritic sulfur in the form of iron pyrite and marcasite, both of which have the same chemical composition, FeS$_2$, but differ in physical structure, and (iii) sulfate. The total sulfur content of U.S. coal types ranges from about 0.25% to nearly 7.0% (81). The percentage of sulfate in non-weathered (non-oxidized) coal is nearly always less than 0.1% and is therefore a minor consideration in the present discussion. Appalachian coal usually contains from 0.5% to 2.0% organic sulfur plus 0.5% to 4.0% pyritic sulfur, although there are deposits in Kentucky, Virginia, West Virginia and elsewhere that contain considerably less than 1% total sulfur. As a gross generality, the content of pyritic sulfur usually exceeds the content of organic sulfur in coals which contain more than 1% total sulfur.

The Clean Air Act Amendments of 1970 provided for the establishment of national ambient air quality standards which set a limit on SO$_2$ of 0.03 ppm (80 μg/m^3) as an annual arithmetic mean concentration with an allowable 24 hour maximum of 0.14 ppm (365 μg/m^3), not to be exceeded more than once each year. When translated to source of performance standards, power plant emissions are restricted in some locations to 1.2 lbs per million BTU. Depending upon how calculations are made, this could restrict the combustion of coal at large power plants without scrubbers to that coal which contains somewhere between 0.5% and 1.5% total sulfur.

Combustion of coal results in conversion of most of the sulfur present in the coal to SO$_2$ and ultimately to SO$_4^{-2}$. In

general, the greater the sulfur content of the coal the greater
is the stack emission of SO_2 and SO_4^{-2} into the atmosphere. In
order to maintain satisfactory sulfur oxide levels in the atmosphere in regions where large scale coal combustion occurs, either
coal with a total sulfur content of less than 1.5% (some regulatory authorities argue less than 1.0%) must be used or provisions must be made to scrub sulfur oxides out of the stack gas
before entry into the atmosphere. Coal of less than 1% total
sulfur is generally considered to be "low sulfur coal" and coal
with greater than about 1.5% to 2.0% total sulfur is considered
by many regulatory authorities to be "high" sulfur coal. It is
estimated that approximately 33% of our available coal cannot
presently be used without stack scrubbers because the sulfur content is too high to meet air quality standards.

Sixty-two percent of the low-sulfur reserves in the U.S. are
found West of the Mississippi whereas 90% of the present coal use
for electric power generation is East of the Mississippi (26).
Increased costs associated both with production and shipping (approximately $.006 per ton mile) of Western coal to Eastern mar-
kets provides an incentive to remove sulfur from high sulfur
Eastern coal.

In this report the phrases: pyrite oxidation, sulfur oxidation, solubilization of pyritic minerals and sulfate release are
all tantamount to removal of pyritic sulfur from coal via a solubilization and leaching process. Sulfate production can be due
to oxidation and solubilization of both pyritic and organically
bound sulfur.

B. The Oxidation of Pyrite

Exposure of the chemically reduced compound FeS_2 to oxygen
and water results in oxidation of the FeS_2 to ferric sulfate by
a complex series of chemical reactions which are summarized in
the following reactions:

(I) $Fe^{+2} \longrightarrow Fe^{+3}$ + electron

(II) $2 S^{-2} + 3O_2 + 2H_2O \longrightarrow 2(SO_4^{-2})$ + 16 electrons + $4H^+$

(III)   Sum: $FeS_2 + 3O_2 + 2H_2O \longrightarrow 2H_2SO_4 + Fe^{+3}$

The oxidized iron (Fe^{+3}) formed, subsequently reacts with water to produce ferric hydroxide and more acid according to the following equation:

(IV) $Fe^{+3} + 3H_2O \longrightarrow Fe(OH)_3 + 3H^+$(14,16,17,23,25,35,36,64)

The initial report by Colmer and Hinkle in 1947 (16) on the isolation of *T. ferrooxidans* from acidic coal mine drainage generated considerable interest, both in the basic biology and ecology of the organism and in the potential for exploitation of the organism to oxidize pyritic minerals.

Examination of the acidophilic iron and sulfur oxidizing bacteria relative to the oxidation of pyritic minerals found in coal appears to have been originally reported both by Leathen, et al.(35) and by Temple and Delchamps (68). In these early reports the motivating interest for the research was in understanding the formation of sulfuric acid from pyrite (i.e. the production of acidic coal mine drainage). Application of the acidophilic iron and sulfur oxidizing bacteria for the removal of pyrites in coal was studied by Zarubina, et al.(83) and by Ashmead (3) who demonstrated that the natural microbial flora of acidic mine waters increased the rate of oxidation of pyrite in a 4% pyritic sulfur coal sample compared to oxidation in the absence of added bacteria. Silverman, et al.(59) subsequently reported that *Ferrobacillus* (presently *T. ferrooxidans*) accelerated the oxidation of samples of pyrite and coarsely crystalline marcasite extracted from coal (greater than 60% pyrite content) but that the cells were inactive on coarsely crystalline iron pyrite. Oxidation rates in the presence of *T. ferrooxidans* were increased by reducing the pyrite particle size. *T. thiooxidans* cells were inactive on all the pyritic samples they examined. These same investigators subsequently reported (50,60) that T. *ferrooxidans* catalysed removal of appreciable amounts of pyrite from coal within 3 to 4 days and that particle size reduction as well as pretreatment with 2 N HCl increased the susceptibility of most coal samples to bacterial desulfurization. They also reported that addition of $Fe_2(SO_4)_3$ increased pyrite removal from acid treated coal. There was and still is disagreement as **to the** ability of T. *thiooxidans* to oxidize pyritic materials; (37,38, 68,69). However, there appears to be universal agreement that T. *ferrooxidans* oxidizes pyritic minerals by catalyzing reaction III. During the period between 1947 and 1965 attention was also directed toward bacterial leaching of other sulfide minerals (57, 65) and interest in the bacterial leaching of sulfur from coal subsequently diminished until stimulated by the energy publicity of the early 1970's.

The only commercial scale data on bacterial removal of pyrite from coal appears to be that of Capes, et al.(13) who blended various percentages of weathered coal as their inoculum of bacteria with run of mine (r.o.m.) coal. They were able to show that addition of 10% weathered coal resulted in a reduction of the sulfur content of r.o.m. coal from 6.1% to 2.7% during subsequent agglomeration of pyrite from pulverized coal, using a flotation technique. These results were reported to be much superior to those obtained using chemical depressants. It is important to note that relatively little sulfate was formed during the bacterial processing, indicating that pyrite rejection during agglomeration in their system cannot be ascribed to simple leaching but involves preferential wetting of coal due to the rendering of pyrite hydrophobic by bacterial action that aids in pyrite rejection during agglomeration. It must be pointed out however, that use of weathered coal as inoculum may have been fortuitous as will be discussed subsequently. At any rate it

seems unreasonable to ascribe the total benefit obtained exclusively to *T. ferrooxidans* but rather to the mixed microbial
population.

Reports that iron oxidizing bacteria are more active than
the exclusively sulfur oxidizing bacteria relative to rates of
pyrite oxidation (35,60,80) in view of prior observations that
ferric sulfate could oxidize pyrite (69) led to speculation that
the primary **role** of bacteria in pyrite oxidation was the production of ferric ions and that the ferric ions thus produced, oxidized more pyrite with concomitant regeneration of ferrous ions
according to Equation V. The recycled iron could again be
oxidized by the bacteria and the cycle would continue (35,59,62).

$$(V) \quad 14\ Fe^{+3} + FeS_2 + 8\ H_2O \longrightarrow 15\ Fe^{+2} + 2(S)\ \underset{or\ O_2}{\overset{Thiobacilli}{\longrightarrow}}$$
$$\overset{\longmapsto T.\ ferrooxidans \longrightarrow}{} \quad 2SO_4^{-2} + 16\ H^{+} \longleftarrow$$

This microbiological process is analogous to a chemically
catalysed reaction process. That is, living bacteria are oxidation catalysts promoting the oxidation of insoluble metallic sulfide to soluble sulfate, which is then removed by leaching. The
bacteria utilize the pyrite nutritionally and grow in the system.

Singer and Stumm (63) have concluded that under acidic conditions below pH 4.0, the rate of pyrite oxidation by ferric ion
is considerably greater than the rate of ferrous ion oxidation
in the absence of bacteria. Hence, the bacteria must catalyze
the oxidation of Fe^{+2} to Fe^{+3} ion in order to supply the Fe^{+3}
to oxidize the pyrite. They, therefore, concluded that the bacterially catalyzed reaction controls the rate of pyrite oxidation under acidic conditions. Lau, et al.(34) also came to essentially the same conclusion in that the bacteria were essential
to maintenance of the high Fe^{+3} to Fe^{+2} ion ratio in solution
which is necessary to chemically oxidize pyrite. The mechanism
of sulfur oxidation by *T. thiooxidans* would be somewhat different
in that sulfur is essentially insoluble, therefore requiring

direct contact of bacterium to substrate (71,80).

C. The Microorganisms of Highly Acidic Environments

Thiobacillus ferrooxidans is classified on the basis of its ability to utilize the oxidation of ferrous iron as its sole energy source and CO_2 as its sole carbon source. It is placed in the genus *Thiobacillus* because it can also oxidize reduced sulfur as a sole energy source. *T. ferrooxidans* has been reported to adapt from autotrophic growth on Fe^{+2} to heterotrophic growth on glucose as a carbon and energy source after at least two transfers in media containing both Fe^{+2} and glucose (46,54,55,66). Adapted cells contain a complete tricarboxylic acid cycle when grown heterotrophically, but unadapted cells lack alpha-keto glutarate dehydrogenase and reduced nicotinamide adenine dinucleotide oxidase when grown on Fe^{+2}, and the presence of Fe^{+2} is reported to repress glucose dissimilation (66). With reference to alternate substrates for chemolithotrophic growth, the iron oxidizing enzymes are reported to be inducible (40). It has been proposed that the facultative heterotrophic bacteria derived from *T. ferrooxidans* cultures, which grow on either organic or reduced sulfur, but not on Fe^{+2}, be designated as a new species: *T. acidophilus* (27). Differences in (i) enzymes (75), (ii) mole % guanine plus cytosine, 55-57.1% for iron grown and 62.9-63.2% for glucose grown (22,27,49), and (iii) lack of the iron oxidizing enzyme in glucose grown cells, indicate that the heterotrophic isolates are different from the autotrophic isolates and that they are selected from *T. ferrooxidans* cultures (54,55,67). Furthermore, the glucose oxidizing acidophiles examined do not react with fluorescein isothiocyanate labeled rabbit anti-*T. ferrooxidans* IgG, which is specific for *T. ferrooxidans* (2).

It appears that *T. ferrooxidans* strains are chemolithotrophic whereas the organic oxidizers which retain the ability to oxidize reduced sulfur compounds are either mixotrophic or facultatively heterotrophic. Also there are many different strains of *T. ferrooxidans* with different metabolic capabilities (8,31) present in the environment. There is also divergence of opinion relative to the use of organic substrates by chemolithotrophic bacteria (30,42,44,47,48,79).

Walsh et al. (82) questioned how freshly mined spoils initially achieve sufficient acidity to be optimal for the acidophilic *Thiobacilli*. These investigators isolated a different microorganism; *Metallogenium*, which produces some acid via the oxidation of Fe^{+2} to Fe^{+3} when accompanied by subsequent hydrolysis. The acid then creates an environment suitable for the succession of the acidophilic *Thiobacilli*. Although this could be an ecological explanation, it need not be a prerequisite to establishing *T. ferrooxidans* growth since chemical oxidation of pyrite would accomplish a lowering of pH to about 4.0 over a longer time span. Also many non-iron oxidizing acidophilic *Thiobacillus* isolates actually grow better when the pH is above 3.5 to 4.0 thereby lowering their own environmental pH sufficiently. Leaching of sandstone will also produce acid sufficient to lower the pH down to 4.0.

A microorganism capable of oxidizing iron and sulfur as well as growth at thermophilic temperatures (45^{0}-70^{0}C) and low pH was isolated by Brierley from a thermal region in Yellowstone Park (9). The organism designated "ferrolobus" was later characterized more thoroughly (10). A similar but different organism, *Sulfolobus acidocalderius* was isolated and characterized by Brock, et al. (12) and was reported to have an optimum temperature of 70^{0}-75^{0}C with a range of 55^{0}-80^{0}C and a pH optimum of 2 to 3. Both organisms oxidize reduced iron and sulfur and grow on simple organic compounds and yeast extract (10,11,12) and both have been shown to oxidize metal sulfide ores with potential for commercial metal extraction processes (11) where high temperatures

(45°-80°C) can be maintained.   The organisms do not appear to grow much below 45°C.

D. Field Studies

Field studies of the activity of bacteria in coal refuse showed a strong correlation between uptake of $^{14}CO_2$ and most probable numbers (MPN) of iron oxidizing bacteria but not with the acid tolerant heterotrophic microorganisms which were also present in the refuse.   Maximal $^{14}CO_2$ uptake occurred in 2 to 3 year old coal refuse with only slight incorporation in fresh refuse or material 40 years old.   Maximal uptake was always found in samples taken from the surface above 8 to 10 cm depth, at temperatures between 20° and 30° C and at moisture content of between 23 and 35%, (5) all of which agree with data previously published on the basis of laboratory investigations (20,34,56, 61,70,76).

Apel and Dugan developed a fluorescent antibody technique which allowed them to follow increasing numbers of *Thiobacillus*

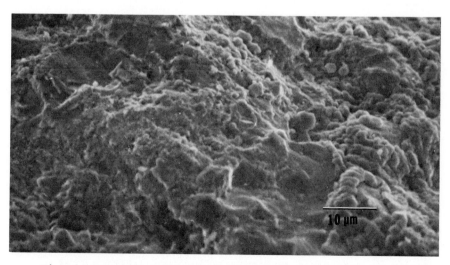

Fig. 1. Scanning electron micrograph (EM) of a coal refuse sample showing surface porosity and what appear to be microorganisms on the surface.

ferrooxidans in coal refuse (2). *T. ferrooxidans* cells were detected in the surface washings of the refuse, but none were detected in refuse which had been prewashed and then pulverized, indicating that *T. ferrooxidans* cells were on the coal refuse
surface but not in the internal pores of the refuse materials.
Fig. 1 is a scanning electron micrograph showing the porosity
of coal refuse. Typical bacterial growth curves with time were
observed and the exponential phase of these curves corresponded
to increases in titratable acidity in the refuse samples, while
over the same period acid production in sterilized refuse controls was considerably less (2). The correlation between the
acid production rate and the exponential growth phase of *T.*
ferrooxidans in coal refuse suggests that bacteria are a dominant
stimulus leading to acid formation from pyrite in coal refuse (a
non-marketable coal with a high sulfur and mineral content) and
previous work on refuse is directly applicable to higher grades
of coal.

The acidophilic *Thiobacilli* are known to produce autotoxic
metabolic by-products which inhibit iron and sulfur oxidation by
the cells when present in sufficient concentration (7,24,52,78,
79). Lower molecular weight organic acids, particularly alpha-
keto acids, some of which are intermediates of the cells metabolic pathways inhibit metabolism of iron and sulfur by *Thio-*
bacilli by causing the cell membrane to become leaky and ultimately disrupt (79) which may also allow intolerable amounts of
H^+ to enter the cell (1). Acid tolerant heterotrophs are also
present in the environments of the acidophilic *Thiobacilli* (5,
20,21,22,41,76,77) and may therefore be an indirect aid in
sulfur oxidation by virtue of their removal of the autotoxic organic by-products (78,79). It is known that certain organic substances retard iron and sulfur oxidation by *Thiobacilli* (19,24)
and also to retard the rate of metallurgical leaching from
sulfides (28,51,72,73,74). The biochemical reactions which

solubilize organically bound sulfur are different from those of pyrite oxidation. The splitting of SO_4^{-2} from $R-O-SO_3$ is due to sulfhydrolase enzymes, whereas the oxidation of $-R-S-S-R-$ and $-R-S-R-$ would be caused by desulfhydrase enzymes. This topic has not been adequately studied relative to coal beneficiation.

An indigenous population of acid tolerant heterotrophic bacteria is responsible for production of slime streamers in highly acid mine water (pH 2.8). This group of bacteria has not been well characterized but at least one isolate has a pH optimum near neutrality (21,22). This suggests that the organisms colonize in the stream creating a localized microcosm that is different from the surrounding, highly acid environment.

Bohlool and Brock have reported the growth of a thermophilic mycoplasma, *Thermoplasma acidophilum*, in refuse piles (6). Yeasts and filamentous fungi have frequently been isolated from acidic mine water (5,18,20,22,29,32). The source of organic nutrient required to support the rather extensive amount of heterotrophic growth in mine drainage is not evident. One possible source would be from metabolic by-products produced by the acidophilic autotrophs. Schnaitman and Lundgren (52) reported that organic acids accumulate from autotrophic growth of *T. ferrooxidans*. Several species of algae, especially *Euglena* and *Ulothrix*, have also been observed in rather large numbers in acid drainage at pH 3.0 in the presence of sunlight (18,20,29,32), which also may either enrich the stream with organic substances or remove by-products that are toxic to the iron and sulfur oxidizers. Other sources of heterotrophic nutrient would be from coal components such as phenolics and the organic content of other sedimentary deposits.

Microbiological leaching of sulfur from coal has many features that are similar to metallurgical bacterial leaching processes and some that are quite different. Both processes involve the oxidation of sulfide minerals by a variety of

acidophilic microorganisms, including *Thiobacillus* species; ultimately to soluble SO_4^{-2}. In the case of iron pyrite, bacteria
also oxidize Fe^{+2} to Fe^{+3}. In contrast to metallurgical leaching
where oxidation of sulfide mineral releases a commercially useful
metal ion from the ore pile that is recovered from the ensuing
acidic leachate, it is the coal leach substrate which is the
desired end product; whereas, the acidic solution of solubilized
mineral ions presently represents an undesirable waste for which
a commercial use is being sought. Again, by comparison, although
a coal sulfur content in excess of about 1.5% is considered to be
high, this represents a small concentration of nutritional iron
and/or sulfur for the bacteria which depend upon it for growth.
Finally, it must be emphasized that the sulfur content of coal
is usually reported on the basis of analytic averages from large
samples and there is tremendous variability within short distances in coal seams or within short distances in coal piles
unless precautions are taken to obtain representative homogeneous
samples. Pyrite crystals may be many centimeters in size or in
some coal samples lenses often less than 10 micrometers in size
remain intact, embedded within the coal (53) and exposure of
these microscopic pyrite lenses to bacterial action requires extensive pulverization of the coal. The *Thiobacilli* which cata-
lyse the oxidation have an approximate diameter of 1 micrometer.
The purpose of this paper is to describe some of the parameters
which influence removal of sulfur from coal slurries by microbial
metabolism in an effort to demonstrate that the process has
potential for commercial exploitation.

II. MATERIALS AND METHODS

A. Inocula

T. ferrooxidans was grown under forced aeration at ambient
temperature ($23\pm 1^{\circ}C$) in "9K" medium (58). After 5 days incubation, the cells were harvested, washed, suspended in pH 3 H_2SO_4

solution, and stored as previously described (2). Cell suspensions were used as inoculum in 1 week or less after harvesting.

T. thiooxidans was grown on ATCC medium #450, harvested, and stored using the same method described for *T. ferrooxidans*.

Natural coal enrichment cultures were obtained by either inoculating a 10% slurry (pH 3) of pulverized non-sterile, high sulfur coal with acid mine drainage and incubating at ambient temperature for 4 weeks on a gyratory shaker (200 rpm), or simply incubating a 10% slurry (pH 3) of non-sterile, uninoculated, pulverized high sulfur coal on a gyratory shaker (200 rpm) for 4 weeks at ambient temperature. The former coal enrichment was designated coal enrichment #3, while the latter was designated coal enrichment #4.

Pyrite enrichment cultures were obtained by inoculating slurries of pyrite adjusted to pH 3 with H_2SO_4 with acid mine drainage and incubating under the conditions described for the coal enrichments. All acid coal mine drainage samples were from an acid stream (pH 3.5) located approximately 10 miles east of McArthur, Ohio. After the initial 4 week incubation period, all enrichment cultures were maintained by transferring to fresh slurries every 2 weeks.

B. Effects of Initial pH

Coal slurries (20% coal in H_2O) were prepared with a commercially pulverized coal (Table 1) designated as coal sample #1 which was courteously provided by the Columbus and Southern Ohio Electric Co. The slurries were adjusted to a variety of pH with 10 N H_2SO_4 and 12.5 ml aliquots were dispensed in 250 ml Erlenmeyer flasks and sterilized by autoclaving at 20 p.s.i. for 25 min. After cooling to ambient temperature, duplicate flasks containing slurries at each of the above pH were inoculated with 5% inoculum from the coal enrichment culture #3, while 2

additional flasks from each of the above pH slurries remained uninoculated and served as sterile controls. Both inoculated, and
sterile flasks were placed on a gyratory shaker (200 rpm) and
incubated at ambient temperature. At 2-3 day intervals during the
incubation period, 1 ml aliquots were aseptically removed from
each of the flasks and used to determine pH and sulfate concentration. The pH of the samples was determined utilizing a
Corning model 12 expanded scale pH meter equipped with a Corning
475060 pH electrode. Sulfate concentrations were determined as
outlined below.

C. Quantitative Determination of Sulfate

The procedure used for determination of sulfate was a modification of a barium sulfate turbidimetric procedure in which
samples were collected, diluted in H_2O (if necessary), and
treated with excess $BaCl_2$. The turbidity of each sample was
determined in a Shimadsu model 14PS-50L spectrophotometer set at
a wave length of 450 nm. Percent transmission of each sample
was recorded and compared to a standard sulfate curve prepared
in the range of 0 to 600 ppm of sulfate.

D. Effects of Bacterial Inoculum on Desulfurization

Twenty percent coal-water slurries at pH 2.5 were prepared
as described above and were inoculated with either 5% by volume
coal enrichment culture #3, 5% pyrite enrichment culture, 5%
T. ferrooxidans suspension (10^7 cells/ml), 5% *T. thiooxidans* suspension (10^7 cells/ml), or 5% suspension containing half *T. ferrooxidans* and half *T. thiooxidans* (10^7 cells/ml). All flasks
were prepared in duplicate, including abiotic controls, with pH
and sulfate concentrations in each flask determined at 2-3 day
intervals as previously described.

E.   Supplemental Nutrients

In initial studies on the effects of supplemental nutrients
on the desulfurization of coal, sub 200 mesh coal (Table 1) was
employed in 20% slurries (initial pH approximately 3.5) set up as
shaker cultures as described above. "9K" basal salts (58) were
added to one set of slurries while another set of slurries
remained unsupplemented. Two flasks of supplemented and 2 flasks
of unsupplemented coal slurry were inoculated individually with
one of the following: 5% pyrite enrichment culture, 5% coal enrichment culture #3, 5% $T.$ $ferrooxidans$ (10^7 cells/ml), and 5%
acid mine drainage obtained from the previously described source.
Again, pH and sulfate concentrations were determined at 2-3 day
intervals.

The effects of additional supplemental nutrients on commercially blended coal (coal sample 1) also was examined using 20%
slurries in 250 ml shaker cultures as previously described. Rep-
licates of slurries containing: no supplement, ATCC medium #450
trace minerals, 9K basal salts, 0.05% K_2HPO_4, and 0.3% $(NH_4)_2SO_4$
were inoculated with coal enrichment #3 and incubated on a gyratory shaker at ambient temperature. In addition, inoculated,
unsupplemented slurries were incubated at ambient temperature on
a gyratory shaker under a 5% CO_2 - 95% air atmosphere. In all
cases comparisons between inoculated slurries were made to comparable uninoculated controls. All values reported are averages
of duplicate samples.

F.   Scanning Electron Microscopy

Coal specimens for scanning electron microscopy (SEM) were
attached to Dag 154 (Achison Colloids Co., Port Huron, Mi. 48060)
coated aluminum specimen stubs with double sided tape, after
which they were sputter coated with 100 $\overset{o}{A}$ of gold. Specimens
were examined at various magnifications (20° angle) with an

Hitachi model S-500 scanning electron microscope. Photomicrographs were taken utilizing Polaroid 4" x 5" Land Film (type 55).

III. RESULTS AND DISCUSSION

Coal sample #1, a power generating blend of pulverized coal assumed to be a representative sample of commercial coal, was separated into 3 mesh size fractions and analysed for total, organic and pyritic sulfur as shown in Table 1.

Table I   Values Showing % Sulfur Content
of Various Fractions of the Pulverized Coal

Mesh size composition	Sulfur Content as % of Fraction			
	Total	*Pyritic*	*SO_4^{-2}*	*Organic*
10% 50 to 100 (.297 to .149 mm)	-	-	-	-
45% 100 to 200 (.149 to .074 mm)	*4.2*	*2.9*	*0.1*	*1.2*
45% sub 200 (sub .074 mm)	*5.4*	*4.2*	*0.2*	*1.0*
Combined Coal Mixture	*4.6*	*3.1*	*0.1*	*1.4*

Scanning electron micrographs of the coal sample and *T. ferooxidans* are shown in Figs. 2 and 3.

Fig. 2.   A scanning EM of the pulverized coal sample showing particles and what appear to be microbes on the surface. — = 5 *um*

Fig. 3.   Scanning EM of T. ferrooxidans *for size comparison to Fig. 2.* — = 0.5 *um*

Fractions of the coal shaken in Erlenmeyer flasks as 20% slurries in distilled H_2O (wt. to vol.) in the presence of various types of inocula were examined for production of soluble SO_4^{-2} and changes in pH. Figure 4 presents data from the 100 to 200 mesh fraction versus time in days. It can be seen that different enrichment culture inocula varied with respect to rates of SO_4^{-2} released from the coal sample. The culture of *T. ferrooxidans* used in the experiment had virtually no effect on SO_4^{-2} release when compared to release from the sterile (uninoculated) control. It is also noted that the cultures with the most active SO_4^{-2} release also had the most dramatic drop in pH over the same time span. The coal sample inoculated with coal enrichment #3 (Fig. 4) contained 1.8% total sulfur and 0.1% pyritic sulfur at the end of the experiment.

Data similar to that in Fig. 4 for smaller particle size coal (sub 200 mesh) is presented in Fig. 5. In every case there was more SO_4^{-2} released and at higher rates in the sub 200 mesh coal than from the 100 to 200 mesh coal (Fig. 4), which is consistent with previous reports (35,39). Again, the *T. ferrooxidans* strain used had little effect on either SO_4^{-2} release or acid production and the pH drop in other samples had a positive correlation with SO_4^{-2} release. In all cases (Fig. 4 and Fig. 5) a 6 to 7 day lag in production of SO_4^{-2} was observed. The sample inoculated with coal enrichment #3 contained 2% total sulfur and 0.5% pyritic sulfur at the end of the experiment, whereas the control still contained 5.2% sulfur.

When considering the greater total amount of SO_4^{-2} released from sub 200 mesh coal (Figs. 5 and 6) compared to 100-200 mesh (Fig. 4) it must be emphasized that the total sulfur content is higher in the smaller particle fraction (see Table 1) and differences in results are attributable both to particle size and original pyrite content.

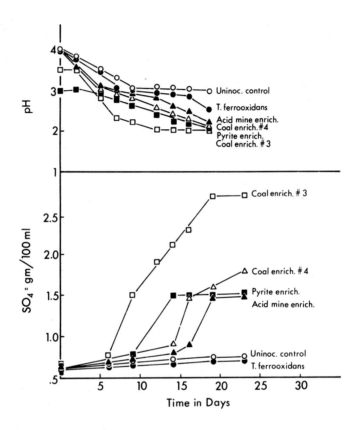

Fig. 4. Curves showing SO_4^{-2} release from 100-200 mesh coal and pH change vs time when in the presence of various types of inoculum.

In an attempt to reduce the lag period, a series of 20% slurries of sub 200 mesh coal were supplemented with the "9K" basal salts (58), inoculated with coal enrichment #3 and monitored for pH and SO_4^{-2}. Results are presented in Fig. 6 and clearly show larger amounts of SO_4^{-2} release than non-nutrient supplemented flasks (Figs. 4 and 5). However, the lag period was not significantly reduced.

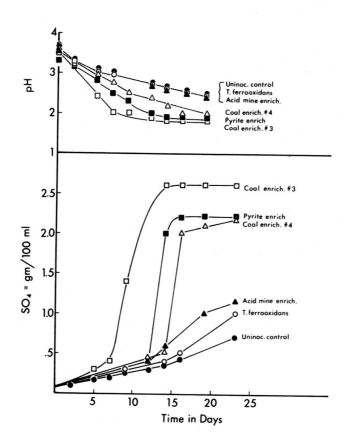

Fig. 5.   Curves showing SO_4^{-2} release from sub 200 mesh coal and pH change vs time when in the presence of various types of inoculum.

When pH data in Figs. 4, 5 and 6 are examined collectively, rapid release of SO_4^{-2} is observed after the pH drops below 2.5 and the highest rates of release are seen in the pH range 2.5 to 2.0, suggesting that the pyrite oxidizers are most active in this pH range and that slurries should be initially adjusted to about 2.5. Previous unpublished experiments showed an

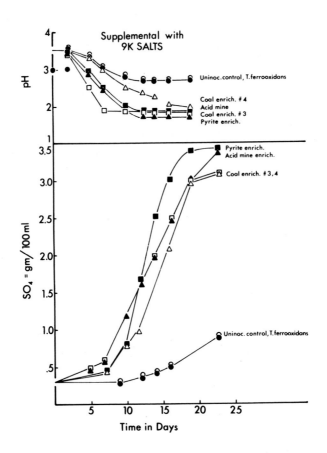

Fig. 6. Curves showing SO_4^{-2} release and pH change from sub 200 mesh coal when supplemented with "9K" salts solution vs time when in the presence of various types of inoculum.

optimum pH for iron oxidation by our strain of *T. ferrooxidans* in the range 2.3 to 2.8 with drastic reduction of iron oxidation below 2.0 and above 3.5, all of which is consistent with our present data. Sulfate release versus time from a series of pH adjusted 20% slurries of coal Sample #1 are shown in Fig. 7. The lag period was significantly reduced in the pH range 2.5 to 2.0.

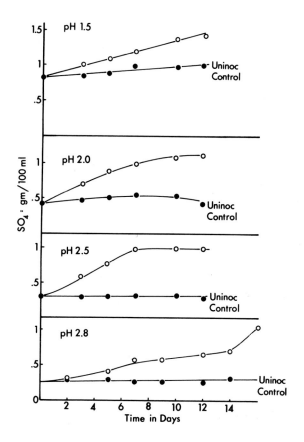

Fig. 7. Curves showing decrease in lag period for SO_4^{-2} release vs time as influenced by initial pH of a 20% slurry of the commercial coal blend.

Addition of supplemental nutrients was then examined using a 20% slurry of the commercial coal #1 when inoculated with coal enrichment #3 at an initial pH of 2.5. These data are shown in Fig. 8. In these experiments there was no observable increase in rate or amount of SO_4^{-2} release by addition of either trace minerals or by 0.05% K_2HPO_4. The slightly decreased SO_4^{-2} release in the presence of "9K" salts is unexplained. Addition of 5% CO_2 as a gas in the head space of a controlled growth chamber resulted initially in an increased lag possibly as a result of buffering the pH, followed by a slight positive effect

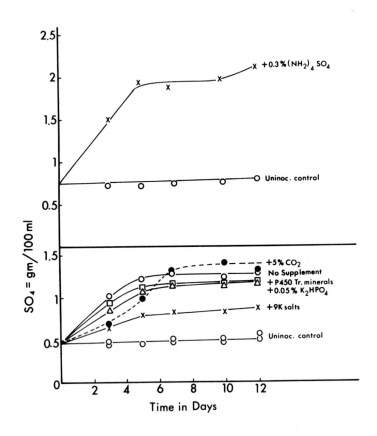

Fig. 8. Curves showing SO_4^{-2} release vs time from 20% slurry of coal blend, initially adjusted to pH 2.5 when various nutrient supplements were added.

on SO_4^{-2} release. Addition of 0.3% $(NH_4)_2SO_4$ had a beneficial influence on both the rate and amount of SO_4^{-2} released. Experiments are continuing in our efforts to optimize the rate and amount of SO_4^{-2} release prior to expanded volume (scale up) experiments. Finally, we have also used *T. thiooxidans* as an inoculum without being able to show significant SO_4^{-2} release. However, when *T. thiooxidans* and *T. ferrooxidans* are inoculated together there was a significant increase in SO_4^{-2} release over the sterile controls.

The above experiments on a single pulverized coal sample (or fractions thereof) show that SO_4^{-2} ions were released at greater rates in the presence of mixed enrichments of acidophiles than by either of the species of *Thiobacillus* used in the experiments. Since the *T. ferrooxidans* was a very active iron oxidizing strain it is possible either that oxidation of pyrite involves more than production of Fe^{+3} as an oxidant, or that another organism is responsible for initiating the attack on pyrite surfaces to provide the initial release of Fe^{+3} necessary to carry out the oxidation, or that neither bacterium alone could oxidize pyrite all the way to SO_4^{-2} because we did not assay for intermediate sulfur compounds. There is no *a priori* reason to attribute pyrite oxidation exclusively to either a direct or indirect mechanism. Undoubtedly both occur in the continually changing coal slurry environment and several different species and strains contribute to pyrite oxidation. It is our view that some organisms aid the iron and sulfur oxidizing autotrophs by removing otherwise autotoxic by-products of metabolism. Although there was little SO_4^{-2} released by *T. ferrooxidans*, the cells remained active Fe^{+2} oxidizers and SO_4^{-2} analysis may not account for activity of this organism in this system - a point which would be consistent with the observation of Capes, et al. (13) although they were not using pure cultures of *T. ferrooxidans*.

It can also be concluded that greater rates of SO_4^{-2} release as well as reduction of lag periods are seen when initial pH is adjusted to the range 2.5 to 2.0. Addition of $(NH_4)_2SO_4$ as a supplemental nutrient resulted in greater oxidation of pyrite from the coal samples.

Although the data confirms previous reports of more rapid pyrite oxidation as particle size is reduced; the smaller particulate fractions also contain a higher percentage of sulfur and would be expected to release more total SO_4^{-2}, necessitating a distinction between rate and amount released.

It can be concluded that microorganisms effectively remove pyritic sulfur from coal via oxidation and solubilization in laboratory scale experiments thereby demonstrating the potential for commercial scale operations. The organic sulfur content of coal was not solubilized in these experiments and efforts are now being directed toward the possibility of microbial removal of the organic sulfur fractions of coal and to examine the effect of other variables on coal desulfurization.

IV. ACKNOWLEDGMENT

This work was supported in part by The Ohio State University Development Fund.

V. REFERENCES

1. Apel, W.A., and P.R. Dugan, This symposium (1977).

2. Apel, W.A., P.R. Dugan, J.A. Filppi, and M.S. Rheins, *Appl. Environ. Microbiol.*, 32, 159 (1976).

3. Ashmead, D., *Colliery Guardian*, 190, 694-698 (1955).

4. Beck, J., *J. Bacteriol.*, 79, 502 (1960).

5. Belly, R.T., and T.D. Brock, *J. Bacteriol.*, 117, 726 (1974).

6. Bohlool, B.B., and T.D. Brock, *Appl. Microbiol.*, 28, 11 (1974).

7. Borichewski, R.M., *J. Bacteriol.*, 93, 597-599 (1967).

8. Bounds, H.C., and A.R. Colmer, *Can. J. Microbiol.*, 18, 735 (1972).

9. Brierley,J.A., Ph.D. Thesis, Montana State Univ., 104 pp. (Diss. Abstr., 67, 1342 (1966)).

10. Brierley,C.L., and J.A. Brierley,*Can. J. Microbiol.*, 19, 183 (1973).

11. Brierley,C.L., *Develop. Industrial Microbiol.*, 273 (1977).

12. Brock, T.D., K.M. Brock, R.T. Belly, and R.L. Weiss, *Arch. Mikrobiol.*, 84, 54 (1972).

13. Capes, C.E., A.E. McJLHinney, A.F. Sirianni, and I.E. Puddington, *Canadian Mining and Metallurgical Bull.*, 88-91 (1973).

14. Carpenter, E.V., and L.K. Henderson, Res. Bull. No. 10. Engineer. Exper. Sta., Univ. of West Virginia (1933).

15. Casagrande, D., and K. Diefert, *Science*, 195, 675-676 (1977).

16. Colmer, A.R., and M.E. Hinkle, *Science*, 106, 253 (1947).

17. Colmer, A.R., K.L. Temple, and M.E. Hinkle, *J. Bacteriol.*, 59, 317 (1949).

18. Cook, W.B., *Proc. Indust. Wast. Conf.*, 21, 258 (1966).

19. Dugan, P.R., and D.G. Lundgren, *Develop. in Industrial Microbiology*, 5, 250 (1964).

20. Dugan, P.R., and C.I. Randles, Rept. Ohio State Univ. Water Resources Center, Columbus, Ohio, 123 pp. (1968).

21. Dugan, P.R., C.B. Macmillan, and R.M. Pfister, *J. Bacteriol.*, 101, 973 (1970).

22. Dugan, P.R., C.B. Macmillan, and R.M. Pfister, *J. Bacteriol.*, 101, 982 (1970).

23. Dugan, P.R., "Biochemical Ecology of Water Pollution", pp. 123-137, Plenum Publ. Co., N.Y., N.Y. (1972).

24. Dugan, P.R., *Ohio J. Sci.*, 75, 266 (1975).

25. Duncan, D.W., J. Landesman, and C.C. Walden, *Can. J. Microbiol.*, 13, 397 (1967).

26. Gary, J.H., R.M. Baldwin, C.Y. Bao, M. Kirchner, and J.D. Golden, Research and Develop. Rept. No. 77 prepared for U.S. Office of Coal Research, U.S. Dept. Interior, 99 pp. (1973).

27. Guay, R., and M. Silver, *Can. J. Microbiol.*, 21, 281 (1975).

28. Itzkovitch, I.J., and A.E. Torma, *IRCS Med. Sci.*, 4, 155 (1976).

29. Joseph, J.M., *Ohio J. Sci*, 53, 123 (1953).

30. Kelly, D.P., *Ann. Rev. Microbiol.*, 25, 177 (1971).

31. Kelly, D.P., and O.H. Tuovinen, *Int. J. Syst. Bacteriol.*, 12, 170 (1972).

32. Lackey, J.B., Public Health Reports 53, 1499 (1938).

33. Land Reborn, Rept. by: Board on Unreclaimed Strip Mined Lands and Ohio Dept. Natural Resources. Columbus, 91 pp. (1974).

34. Lau, C.M., K.S. Shumate, and E.E. Smith, Proceed 3rd Symp. on Coal Mine Drainage. Mellon Inst., Pittsburgh, Pa., p. 114 (1970).

35. Leathan, W.W., S.A. Braley, and L.D. McIntyre, *Appl. Microbiol.*, 1, 61-64. *Appl. Microbiol.*, 1, 65-68 (1953).

36. Lees, H., S.C. Kwok, and I. Suzuki, *Can. J. Microbiol.*, 15, 43 (1969).

37. Lorenz, W.C., and E.C. Tarpley, U.S. Bureau of Mines Rept. No. 6247, U.S. Dept. Interior, 13 pp. (1963).

38. Lorenz, W.C., and R.W. Stephan, Rept. Bureau of Mines, U.S. Dept. Interior, Pittsburgh, Pa. (1967).

39. Lundgren, D.C., J.R. Vestal, and F.R. Tabita, In Microbial iron Metabolism. (Ed.) J.B. Neilands, Ch. 18, Academic Press, N.Y., N.Y. (1974).

40. Margalith, P., M. Silver, and D.G. Lundgren, *J. Bacteriol.*, 92, 1706 (1966).

41. McCoy, B. and P.R. Dugan, In 2nd Symp. on Coal Mine Drainage Research. Mellon Inst., Pittsburgh, Pa., pp. 64-79 (1968).

42. Myers, P.S., and W.N. Millan, *Appl. Microbiol.*, 30, 884-886 (1975).

43. Osborn, E.F., *Science*, 183, 477 (1974).

44. Peck, H.D., *Ann. Rev. Microbiol.*, 22, 489 (1968).

45. Project Independence - Coal. FEA Task Force Rept., U.S. Supt. Documents No. 041-018-00015-1 (1974).

46. Remsen, C.C., and D.G. Lundgren, *Bacteriol. Proc.*, 33 (1963).

47. Rittenberg, S.C., *Advan. Microbiol. Physiol.*, 3, 159 (1969).

48. Rittenburg, S.C., *Antonie van Leeuwenhoek*, 38, 457 (1962).

49. Rodgers, D.R., M.Sc. Thesis submitted to The Ohio State Univ. (1974).

50. Rogoff, M.H., and I. Wender, *Nature*, 192, 378-379 (1961).

51. Sakaguchi, H., A.E. Torma, and M. Silver, *Appl. and Environ. Microbiol.*, 31, 7-10 (1976).

52. Schnaitman, C., and D.G. Lundgren, *Can. J. Microbiol.*, 11, 23 (1965).

53. Schopf, J.M., U.S. Geological Survey Bull. 1111-B (1960).

54. Shafia, F., R.F. Wilkinson, *J. Bacteriol.*, 97, 256 (1969).

55. Shafia, F., K.R. Brinson, M.W. Heinzman, and J.M. Brady, *J. Bacteriol.*, 111, 56 (1972).

56. Shumate, K.S., E.E. Smith, P.R. Dugan, R.A. Brandt, and C.I. Randles, Rept. U.S. Environ. Protection Agency, Water Poll. Control Ser. DAST-42, 14010 FPR. Washington, D.C. (1971).

57. Silver, M., and A.E. Torma, *Can. J. Microbiol.*, 20, 141-147 (1974).

58. Silverman, M.P., and D.G. Lundgren, *J. Bacteriol.*, 78, 326-331 (1959).

59. Silverman, M.P., M.H. Rogoff, and I. Wender, *Appl. Microbiol.* 9, 491-496 (1961).

60. Silverman, M.P., M.H. Rogoff, and I. Wender, *Fuel*, 42, 113 (1963).

61. Silverman, M.P., and H.L. Ehrlich, *Adv. in Appl. Microbiol.*, 4, 153 (1964).

62. Silverman, M.P., *J. Bacteriol.*, 94, 1046 (1967).

63. Singer, P.C., and W. Stumm, *Science*, 167, 1121 (1970).

64. Sokolova, G.A., and G.I. Karavaiko, Physiology and Geochemical Activity of *Thiobacilli*. Trans. from Russian. Israel Program for Scientific Transl., Jerusalem. Through U.S. Dept. Commerce, Springfield, Va. 293 pp. (1968).

65. Sutton, J.A., and J.D. Corrick, U.S. Bu. Mines Rept. 5839 (1961).

66. Tabita, R., and D.G. Lundgren, *J. Bacteriol.*, 108, 328 (1971).

67. Tabita, R., and D.G. Lundgren, *J. Bacteriol.*, 108, 334 (1971).

68. Temple, K.L., and E.W. Delchamps, *Appl. Microbiol.*, 1, 255-258 (1953).

69. Temple, K.L., and W.A. Kohler, Res. Bull. No. 25, Engineering Experiment Station, Univ. of West Virginia, Morgantown (1954).

70. Truax-Traer Coal Co., Rept. U.S. Environ. Protection Agency, Water Poll. Control Research Series 14010DDH, 148 pp. (1971).

71. Trudinger, P.A., Rev. *Pure and Applied Chem.*, 17, 1 (1967).

72. Torma, A.E., and G.G. Gabra, *IRCS Med. Science*, 3, 228 (1975).

73. Torma, A.E., G.G. Gabra, R. Guay, and M. Silver, *Hydrometallurgy*, 1, 301-309 (1976).

74. Torma, A.E., and R. Guay, *Naturalistic Can.*, 103, 133-138 (1976).

75. Tuovinen, O.H., B.C. Kelley, and O.J. Donald Nicholas, *Can. J. Microbiol.*, 22, 109 (1976).

76. Tuttle, J.H., C.I. Randles, and P.R. Dugan, *J. Bacteriol.*, 95, 1495 (1968).

77. Tuttle, J.H., P.R. Dugan, C.B. Macmillan, and C.I. Randles, *J. Bacteriol.*, 97, 594 (1969).

78. Tuttle, J.H., and P.R. Dugan, *Can. J. Microbiol.*, 22, 719 (1976).

79. Tuttle, J.H., P.R. Dugan, and W.A. Apel, *Appl. and Environ. Microbiol.*, 33, 459-469 (1977).

80. Vogler, K.G., and W.W. Umbreit, *Soil Sci.*, 51, 331 (1941).

81. Walker, F.E., and F.E. Hartner, U.S. Bureau of Mines Information Circular No. 8301, 51 pp. (1966).

82. Walsh, F., and R. Mitchell, *Environ. Science and Technol.*, 6, 809 (1972).

83. Zarubina, Z.M., N.N. Lyalikova and Ye. I. Shmuk, *Izvest. Akad. Navk. S.S.S.R., Otdel. Tekh. Nauk. Metallurgiya i Toplivo*, 1, 117-119 (1959).

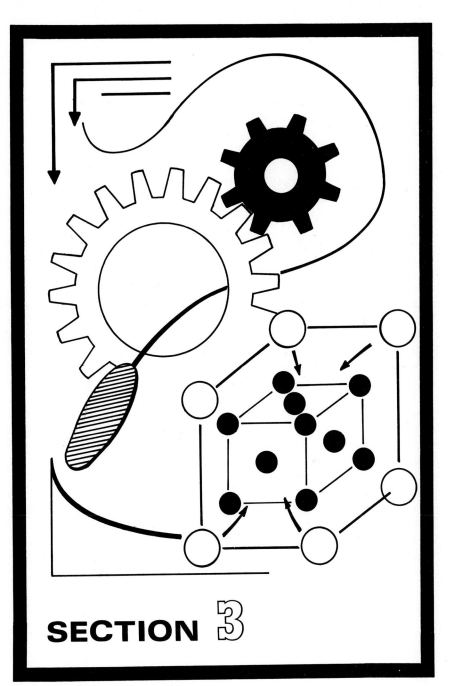

SECTION 3

III BIOEXTRACTIVE APPLICATIONS AND OPTIMIZATION

The fact that the majority of the papers composing this volume are contained in this section is somewhat consistent with the relationship between the theoretical or basic research areas and the applications. In a practical sense, this is of course a healthy feature.

This section begins with a treatment of microfloral observations in an industrial dump leaching operation and is clearly a bridge between the basic research areas in Section I and the more applied areas which follow. The mechanism of bacterial leaching applied to industrial dumps and conditions for optimizing the conditions for leaching are presented. Many of the chapters of this section deal with specific applications of bacterial leaching, the optimization of the leaching conditions, and process optimization based upon economic considerations. Many of the test results reported relate to laboratory and small pilot-plant experiments. However, large-scale data is presented beginning with the very first chapter of this section and especially in the concluding chapter. The final chapter also makes an attempt at pointing out the advantages of large-scale testing compared with laboratory testing.

The objectives of this book and the symposium upon which it was based, are strongly amplified in the chapters composing this section. The applications of fundamental concepts and basic research as outlined in the chapters composing Section I are very clearly exemplified in this section, and the synergistic aspects of metallurgy and microbiology in the extraction of metal values by leaching are certainly apparent throughout.

OBSERVATIONS ON THE MICROFLORA IN AN INDUSTRIAL COPPER

DUMP LEACHING OPERATION

Stoyan N. Groudev,
Fratio N. Genchev
and
Stefan S. Gaidarjiev

Higher Institute of Mining and Geology
Sofia - Darvenitza, Bulgaria

The paper presents many years of observations on the micro-flora in a large dump in Vlaikov vrah, Bulgaria, containing about twenty million tons of low-grade copper sulfide and oxide ores. The observations cover two periods: from 1968 to 1972, when the dump was an area of natural leaching processes, and since 1972, when industry began using it for recovering copper by means of bacterial leaching. The microflora of the dump is quite varied and the paper gives the specific part played in the leaching processes by some of its main representatives. Cases of microbial mutualism in leaching, as well as cases of inhibition of the ferrobacilli by some other bacteria, moulds and protozoa are described. The ferrobacilli in the dump are a very mixed population and some rather special wild strains have been isolated: thermophilic, facultative autotrophic and strains, possessing systems for substrate phosphorylation during the oxidation of sulfides and sulfur, and capable of growth at low rH_2. The microflora of each separate component of the complex production process is described: in the dump itself, the circulating solutions, the regeneration pond and the cementation unit. The paper presents the results of the attempts to stimulate the useful microflora by a suitable change of some environmental factors and by adding surface active *agents. In 1974 and in 1975, a laboratory selected mutant strain having a high level of oxidizing activity and a well-distinguishable marker, was introduced into part of the dump after being grown in fermentors, built next to the dump. The marker is a mutation, caused by a post-replication repair mechanism following UV-irradiation, and consists in the possession by the strain of a specific direct metabolic pathway from S^{2-} and S^0 to sulphate via polythionates. The fate of this selected strain and its influence on the leaching are shown. The possibility for and the usefulness of introducing laboratory selected strains into leaching operations are discussed.*

I.   INTRODUCTION

The ore piles in the proximity of the Vlaikov vrah mine, Bulgaria, contain about twenty million tons of waste low-grade copper sulfide and oxide ores, with an average copper content of about 0.05%. Some parts of the dumps have a higher copper content-about 0.15%. The principal copper-containing mineral is chalcopyrite, followed by covellite and chalcocite. Pyrite is well-represented, too and quartz and feldspars are the basic minerals of the host rock.

The observations conducted in that area during 1968 showed that after rainfall, acid drainage waters with a high content of copper and iron ions (over 0.5 and 2 g/ℓ, respectively) and of bacteria from the species *Thiobacillus ferrooxidans* and *T. thiooxidans* flow out of the dumps. That finding served as an argument for the construction of an industrial plant for bacterial leaching, which was commissioned in 1972.

Since 1968 there have been regular checks of the microflora in the dumps. After 1972 these checks also covered the separate stages of the industrial plant and many attempts were made to stimulate the bacterial leaching processes. Some of the more interesting results and conclusions from these experiments are shown in the present paper.

II. PROCEDURES, RESULTS, AND DISCUSSION

The isolated microorganisms were identified by means of
the guides of Bergey, Gilman and Lodder (1,2,3). The ferrobacilli
in the solution were estimated by direct counting under a phase-
contrast microscope, using a bacterial counting chamber, and by
the most probable number method (MPN); the cells, attached to the
solid surface were estimated by means of determining bacterial
nitrogen. The number of *T. thiooxidans* was found by using solution plating on the Waksman solid thiosulphate medium.

Regardless of the selective ecological conditions, the
microflora of the industrial plant is characterized by considerable
variety of its species of the (*T. ferrooxidans, T. thiooxidans,
T. thioparus* and *T. denitrificans*), various sulphur bacteria
(colourless, green and purple), iron bacteria (*Gallionella* and
Leptothrix), sulphate-reducing and other heterotrophic bacteria
(*Desulfovibrio, Bacillus, Enterobacter (Aerobactor), Pseudomonas,
Caulobacter* etc.), moulds and yeasts (*Cladosporium, Penicillium,
Trichosporon, Rhodotorula* etc.), microscopic algae (*Ulothrix*) and
protozoa (*Amoeba, Euglena, Eutrepia*) were found. The green algae
Hormidium fluitans, whose role in bacterial leaching was recently
shown by Madgwick and Ralph (4), grow luxuriantly in the acid
dump effluents.

After the beginning of the industrial recovering of copper
the composition of the dump microflora suffered only negligible
changes, the most important of which was the drop in the amount
of heterotrophic microorganisms.

The ferrobacilli in the dumps are a very mixed population,
but some more interesting isolates deserve special mention.

Thermophilic ferrobacilli, capable of growth and oxidation of
ferrous iron and sulphur at 60°C, were isolated from some rich-in-
pyrite parts of the dump. The temperature in these parts exceeds
sometimes 80°C, a finding noted also by Beck (5) for dumps in the

United States. Thermophils from the genus *Sulfolobus* (6) and the *Sulfolobus*-like microbe (7) have not been isolated. The thermophilic ferrobacilli leached pyrite faster than the mesophilic under laboratory conditions. Unfortunately, their activity on chalcopyrite, even after a preliminary adaption, was not higher (leaching rates, comparable with these, noted by Brierley and Murr (8) with the *Sulfolobus*-like microbe, were observed).

A considerable part (about 40%) of the isolated ferrobacilli were facultative autotrophs and were capable of using various organic compounds as a source of energy. The facultative autotrophic nature of these isolates was established beyond doubt by means of the appropriate genetic tests. On the other hand, the heterotrophic companion of the *T. ferrooxidans* (9) was found in the dump, too. The facultative autotrophic ferrobacilli dissimilated the glucose by the Entner-Doudoroff pathway, and, in part, by the pentose-phosphaste pathway, and this corroborates the findings of Tabita and Lundgren (10) concerning the heterotrophic metabolism of the *T. ferrooxidans*. An interesting fact must be mentioned: no correlation was found to exist between the number of the ferrobacilli and the quantity of organic matter in the acid dump effluents. The facultative autotrophic ferrobacilli were isolated mostly from the dump sections with a lower rH_2. It seems that under normal conditions the organic matter in the acid drainage waters is used exclusively by the heterotrophic microorganisms, belonging to other taxonomic genera.

Ferrobacilli incapable of sulphur-oxidizing activity were isolated. They are very similar to the species *Ferrobacillus ferrooxidans*, described by Leathen,et al. (11). Under laboratory conditions we have so far failed to induce sulphur-oxidizing activity in these isolates. The DNA base composition of these isolates was 57.3 $\pm$ 0.09 % GC, which is very close to that found by Guay,et al. (12) for *T. ferrooxidans*, grown on iron. It is interesting to note that the isolates, which possess a sulphur-

oxidizing activity and which oxidize most actively the chalcopyrite, had a % GC always over 58 - a finding, which coincides
also with the data obtained by the above-mentioned authors. These
results indicate that the ability of the ferrobacilli to adapt to
various substrates can be observed also under natural conditions.
It is difficult to say what is the genetic nature of that adaptation.

Some ferrobacilli, possessing systems for substrate phosphorylation during the oxidation of sulphides and sulphur have been
isolated from the depth of the dump (about 30 m below the surface).
These strains grow well at rH_2 < 20 and retain their viability
even at rH_2 = 14-16. Their molar economic coefficient (Y_m) was
always greater than that of the strains possessing systems only
for oxidative phosphorylation. There are reasons to accept that
ferrobacilli with substrate phosphorylation use ferric iron as
electronic acceptor during the oxidation of sulphur compounds,
i.e. act in accordance with the mechanism, suggested by Brock and
Gustafson (13).

Among the various isolates, the "typical" strain from Vlaikov
vrah dump is characterized by the data, shown in Table I.

Two terminal types of isolates can be differential: one,
which includes isolates, possessing a pronounced ferrous iron-
oxidizing activity and a negligible (arising only after induction)
sulphur-oxidizing activity, and the other, including isolates
with pronounced sulphur-oxidizing and moderate ferrous iron-oxidizing activity. The isolates from the first type were found mainly
in the dump solutions, and those from the second type mainly on
the ore surface. The second type isolates oxidized the chalcopyrite much more quickly.

The different points of the industrial plant had different
ferrobacilli counts (Table II).

TABLE I. *Physiological Characterization of* Thiobacillus ferrooxidans, *Isolated from Vlaikov vrah Dump*

Parameter	Substrate	
	Fe^{2+}	S^O
Maximum $Qo_2(N)$ *value,* $\mu\ell/mg.h$	20 840	680
Minimum doubling time h	7.5	16
Maximum specific growth rate, h^{-1}	0.092	0.043
Economic coefficient (Y),		
g dry wt/g iron oxidized	0.006	0.112
Molar economic coefficient (Y_m),		
g dry wt/g atom iron oxidized	0.315	3.650
Power coefficient (Y_{ATP}),		
g dry wt/mole ATP	0.630	?
Effectiveness of CO_2 *fixation,*		
moles $CO_2/100$ *moles* O_2	2.5	22.5
Effectiveness of free energy		
utilization, %	22	15
Optimum pH	2.3	2.8
Optimum temperature, $^O C$	35	37
Tolerance to heavy metal ions,		
inhibitory level, g/ℓ		
Cu	> 12	> 5
Zn	> 15	>10
Ni	> 10	> 5
Co	> 10	> 3

TABLE II. *The Number of Ferrobacilli in Various Points in the Industrial Plant during the Different Months*

Point	May	July	September	October
I	$9.10^3 - 4.10^5$	$2.10^4 - 6.10^5$	$1.10^4 - 3.10^5$	$2.10^4 - 2.10^5$
II	$7.10^3 - 9.10^3$	$9.10^3 - 9.10^4$	$7.10^3 - 1.10^4$	$9.10^3 - 1.10^4$
III	$3.10^4 - 5.10^8$	$2.10^5 - 4.10^8$	$8.10^4 - 1.10^5$	$7.10^4 - 8.10^4$
IV	$6.10^7 - 4.10^8$	$7.10^7 - 2.10^8$	$8.10^7 - 2.10^8$	$5.10^7 - 9.10^7$
V	$8.10^7 - 4.10^8$	$5.10^7 - 5.10^8$	$5.10^7 - 5.10^8$	$6.10^7 - 4.10^8$

I - *pregnant solution (dump effluent)*
II - *barren solution (precipitation plant effluent)*
III - *solution taken from the collection pond II*
IV - *solution taken from the regeneration pond*
V - *ore taken from a dump/a depth of 50 cm/*
   Number of ferrobacilli: in the liquid samples are indicated cells/ml; in the solid samples - cells/g.

The basic "reservoir" of ferrobacilli in the plant is the leaching material itself, i.e. the ore in the dumps. Both the absolute and relative numbers of ferrobacilli in the different levels of the dump are not exactly known, but it seems that the greatest number of them are located in the upper part - down to 50 cm from the surface. Ferrobacilli have been found also farther down the dump (down to as far as 30 m below the surface), but only in and around the gaps in the ore mass. Many of the ferrobacilli, adsorbed on the ore surface were without flagella, a finding, which was noted by Le Roux,et al. (14) during the investigations on the bacterial oxidation of pyrite.

The following finding will give an idea about the total quantity of the ferrobacilli, adsorbed on the ore surface and swimming in the solutions. The leaching solution passes through the dump in about eleven hours and during that time all the ferrous iron, combined in it (3-4 g/ℓ) turns into Fe^{3+}. This is a rate which cannot be reached even under laboratory conditions and that by using specially selected strains. Regardless of the ferrobacilli count, this high rate is possible only when the oxdiation process is carried out in a thin film. In fact, the difference between the oxidation rates under laboratory conditions and in the dump is much more marked, because the reduction of Fe^{3+} takes place during the oxidation of suphides in the dump. On the other hand, the catalytic effect of cupric ions during the oxidation of ferrous iron, based on the reaction (1)

$$Fe^{2+} + Cu^{2+} \longrightarrow Fe^{3+} + Cu^{+} \qquad [1]$$

is not significant.

T. thiooxidans was isolated much more rarely and always in a smaller number than the ferrobacilli. Its maximal quantity in the leaching solution was 2.10 cells/mℓ; more frequently its number varied between 5.10^1 - 4.10^2 cells/mℓ. These bacteria were found mainly in the dump effluents. It is possible that their growth depends on the elemental sulphur, produced by the chemical and electrochemical processes. The thiobacilli did not oxidize the sulphide minerals.

Both *T. thioparus* and *T. denitrificans* have been isolated sporadically from different points in the dump (mainly from sections, where covellite and chalcocite are the dominating copper minerals). The synergistic action of the different thiobacilli during the leaching of sulphide minerals was successfully demonstrated under laboratory conditions. That was the classical case of such kind of action. *T. thioparus* (or *T. denitrificans*) oxidizes the elemental sulphur (but not the sulphides) in an alkaline medium of a pH range of 6 to 9. As the oxidation proceeds and water-soluble metal sulphates are formed, both the hydrolysis and the formation of sulphuric acid lower the pH. As the pH drops to 4.0 *T. ferrooxidans* continues the oxidation of minerals. The inability of our strains of *T. thioparus* and *T. denitrificans* to oxidize the copper sulphide minerals as well as their small cell number suggest that they do not play an important role in the leaching of copper in Vlaikov vrah.

The presence of the heterotrophs and even of *T. thiooxidans* itself had an unfavorable effect on the physiological activity of the ferrobacilli. This is shown in Table III, which presents the results from the leaching of sulphide minerals by means of a pure culture of autochthonic ferrobacilli as well as by means of a mixed microbial culture. In both cases the shake-flask technique was used.

The harmful action of the other microbial species was due to their competitive participation in the consumption of oxygen and some useful ions, and to their precipitation on the mineral surface. In some cases the influence was specific - thus some protozoa (*Amoeba*) used the ferrobacilli as food and *Caulobacter* inhibited them by means of organic compounds secreted in the medium.

TABLE III   *Leaching of Sulphide Minerals by means of a*
Pure Culture of Ferrobacilli and a Mixed Microbial Culture

Mineral	Pure Culture	Mixed culture
	copper, leached in 30 days, %	
Chalcopyrite	5.9	3.5
Covellite	67.1	52.1
Chalcocite	93.6	77.0

*Both cultures contained 10^6 ferrobacilli per ml each in the
beginning of the experiment; the mixed culture contained also six
other species of microorganisms in quantities and ratios, characteristic of the leaching solution.*

In some sections of the dump the number of *Amoebae* was
particularly high (up to 500 per ml). Like Ehrlich (15) we have
not been able to subculture the *Amoebae* in 9K medium or in other
artificial media for ferrobacilli, but these protozoa grew very
well in acid drainage waters from the dump. In their presence
the rate as well as the final degree of leaching decreased. After
90 days of leaching, the number of viable ferrobacilli in the
experiment in the presence of *Amoebae* was only 5.10^3 cells/ml in
comparison with 10^8 cells/ml in the absence of *Amoebae*.

The sulphate - reducing bacteria were most numerous in the
deeper levels of the dump, where the rH_2 was low. It is possible
that the H_2S, formed by these bacteria, is used by *T. thioparus*
as a source of energy. It is not known with certainty whether
the sulphate-reducing bacteria are capable of reducing sulphates
at a low pH (16).

The quantity of the microbial species, concomitant to *T.
ferrooxidans* (expressed as dry weight of their biomass), depends
on the ecological parameters - mainly on the pH and on the concentration of copper ions in the solution. Under certain special

conditions (pH over 3.5-4.0 and copper concentration below 1 g/ℓ) the percentage of these microorganisms was close to 30% of the total amount of the biomass (Table IV).

The experiments to inhibit the harmful microflora by lowering the pH were conducted under laboratory conditions. It was found that the change in the pH from a previous 3.2 to 2.4 resulted in increasing the percentage of the ferrobacilli from 76% to 93%, reaching 97% at pH 2.1.

Similar experiments have been carried out and in the industrial plant itself. The lowering of pH in an experimental section of the dump by adding sulphuric acid in the leaching solution resulted in increasing the ferrobacilli number as well as the rate of copper leaching (Table V). On the other hand, the addition of ammonium and phosphate ions as well as of surface active agents, including Tween 20, had no positive effect. The greater part of these ions were adsorbed firmly on the ore surface and their concentration in the circulating solution was much smaller than expected. When the leaching was carried out mainly by strains, possessing strong direct mechanism, i.e. adsorbing

TABLE I V   *Percentage of the Ferrobacilli in the Total Dry Weight of Biomass*

Analyzed point	ferrobacilli, % of the total biomass
Pregnant solution, pH 2.4 and 1.4 g/ℓ Cu	94
Pregnant solution, pH 2.4 and 1.2 g/ℓ Cu	91
Pregnant solution, pH 2.8 and 1.4 g/ℓ Cu	83
Pregnant solution, pH 3.1 and 1.4 g/ℓ Cu	76
Solution taken from the cementators, pH 3.6	80
Solution taken from the cementators, pH 3.9	74
Solution taken from the cementators, pH 4.3	71
Collection pond II effluent, pH 4.0	74
Regeneration pond effluent, pH 2.5	85
Feed to dump leaching solution, pH 3.6	87

TABLE V   *The Effect of pH of Leaching Solution on the Copper and Ferrobacilli Content in the Pregnant Solution*

Period and Sampling No.		Leaching Solution		Pregnant Solution		
		pH	T.f.	pH	Cu	T.f.
I	1	4.45	2.10^5	2.68	1.05	4.10^9
	2	4.37	5.10^5	2.77	1.12	6.10^5
	3	4.09	8.10^5	2.59	1.14	7.10^5
II	1	3.27	2.10^6	2.44	1.21	4.10^6
	2	3.09	4.10^6	2.37	1.39	7.10^6
	3	3.22	4.10^6	2.33	1.32	6.10^6
III	1	2.51	5.10^6	2.30	1.49	9.10^6
	2	2.40	3.10^6	2.26	1.40	1.10^7
	3	2.49	6.10^6	2.31	1.52	2.10^7
	1	4.55	6.10^6	2.49	1.35	8.10^6
	2	4.40	1.10^6	2.57	1.18	2.10^6
	3	4.43	8.10^5	2.60	1.01	9.10^5

Duration of each period was 30 days.   The samplings were carried out at 10, 20 and 30th day, respectively.   The copper concentration is shown in g/ℓ, and T. ferrooxidans *count – in cells/mℓ.*

themselves in great number on the mineral surface, the addition of phosphate (over 10 mg/ℓ PO_4^{3-} in the leaching solution) was found to be particularly harmful to the speed of the process. These data are valid only for the copper sulphide ore from Vlaikov vrah. The phosphate requirement for maximum rate and extraction during the bacterial leaching of other substrates is quite different (17).

The form (Fe^{2+} or Fe^{3+}) under which the iron was present in the leaching solution caused   little effect on the ferrobacilli number and rate of leaching. It is known that the ratio of these two ions determines the Eh of solution [but see also Bhappu, et al. (18)] and, in principle, has a significant influence on the leaching. It seems, however, that the activity of microflora is a more important factor in a leaching process, which takes place in a big dump, containing mainly $CuFeS_2$.

The characteristic pecularities of the microflora of each separate stage of the complex production process must be noted. The dump effluents were characterized by relatively uniform composition of their microflora. It was due to their low pH (usually below 2.5) and a high content of heavy metal ions. The heterotropic microorganisms were represented only in small numbers. On the other hand, the ferrobacilli count is not high in comparison with that observed in other smaller dumps in Bulgaria and abroad. There are various reasons for that,but mainly it may be the dump size itself. The size determines to a great degree, the availability of oxygen and carbon dioxide in the different parts of the dump.

After many years of effort, it seems doubtful that the ferrobacilli number in all dump effluents can be increased. The addition of sulphuric acid was efficacious, but only in small dump regions, in which the secondary sulphides are well represented and where the acid-consuming processes have been accomplished. It is also doubtful that the increasing of the swimming ferrobacilli will indicate an increasing of total ferrobacilli number and the rate of leaching. There are indications that an equilibrium exists between the adsorbed and swimming bacteria in the leached systems. In various systems that equilibrium is achieved at different ratios of these two groups. The increasing of swimming bacteria (which are ferrous iron oxidizers) may cause a decreasing of the activity of adsorbed bacteria, many of which directly oxidize the sulphides. In such cases the Fe^{3+} (for instance $CuFeS_2$) would be diminished. On the other hand, it is not certain that an increase of the number of adsorbed bacteria would result in an increase in the number of swimming ferrobacilli. Unfortunately, only the numbers of swimming bacteria are easily controlled.

In the precipitation plant the ferrobacilli count suffered interesting changes. In the front cementation chambers it decreased (by up to twenty fold in separate cases), which was due

to the increasing of pH and formation of hydrogen in the
system:

$$H_2SO_4 + Fe \longrightarrow FeSO_4 + H_2 \qquad\qquad [2]$$

In the back chambers, however, the ferrobacilli count increased (up to 9.10^4 cells/mℓ), which was connected with the relatively long holding time of the sheet iron in the cementation
chambers and with smaller formation of H_2. The precipitation
plant effluents were rich in heterotrophic microorganisms.
Among the latter, the bacteria and moulds dominated in numbers,
and yeasts and protozoa were found only sporadically. Sometimes
many microscopic algae were isolated. The heterotrophic microflora, concomitant to *T. ferrooxidans*, exhibited a certain
stability in the time.

The microflora of the collection pond II (the latter collects
the precipitation plant effluents; on the other hand, the collection pond I collects the dump effluents) was very varied, up to
14 different species of microorganisms were isolated from some
samples. The main reason for that was the relatively high pH
(about 3.5-4.0) of the solutions.

The composition of microflora became again much more uniform
and the ferrobacilli count-higher in the regeneration pond, where
favorable conditions for ferrous iron oxidation existed. It is
necessary to note that only a part of the precipitation plant
effluents pass through the regeneration pond. In the Vlaikov
vrah, the significance of the regeneration stage consists not in
the production of Fe^{3+}, but in the precipitation of iron salts out
of dumps. In part, such role is played also by the collection
ponds because of relatively long holding time of the solutions
here. That holding time is much shorter during the winter months
to avoid the cooling of solutions.

It is obvious that the problem of maintenance of a suitable microflora in the leached material becomes of increasing importance for industrial leaching operations. For that purpose, a laboratory selected strain, possessing a high leaching activity versus Vlaikov vrah ore has been introduced in a small part of the dump. That part contained about 100,000 tons of ore and was partially separated from the main ore mass by elevations of lay. The strain (marked here as Pmv) has been obtained by UV-irradiation of another selected strain-Pm. The latter has been obtained as a result of gradual selection of strain P, isolated in 1968 from the same place, where the introduction of strain Pmv has been made subsequently. The gradual selection was carried out by consecutive platings of ferrobacilli on the 9K silica gel medium or on the Manning's solid medium (19) (each of the media has been prepared with 3 g/ℓ Fe^{2+}) and by testing of the activity of single colonies, bred on the solid surface. Everytime, the most active colonies were subcultured.

Both wild strain P and strain Pm oxidized tetrathionate by the metabolic pathway, suggested by Tuovinen and Kelly (20). The oxidation via that pathway is described by the equation:

$$2K_2S_4O_6 + 7O_2 + 6H_2O = 2K_2SO_4 + 6H_2SO_4 \qquad [3]$$

i.e. each of the four sulphur atoms of tetrathionate is oxidized to sulphate.

The strain Pmv oxidized tetrathionate with formation of trithionate [like the strain of Sinha and Walden (21)], but had a unique ability to oxidize the internal sulphur atom of trithionate to sulphate. Differences between the strains have been demonstrated clearly by using of sulphur compounds, containing labelled sulphur atoms (^{35}S).

$$^2{}^{35}S - SO_3{}^{2-} \longrightarrow O_3S - S^X - S^X - SO_3{}^{2-}$$

with branches to $S^X O_4{}^{2-}$ (upper left and upper right) and $SO_4{}^{2-}$ (lower branches)

pathway, suggested by Tuovinen and Kelly (20) (trithionate has not been formed)

$$^2{}^{35}S - SO_3{}^{2-} \longrightarrow O_3S - S^X - S^X - S^X - SO_3{}^2$$

$$\longrightarrow S^X O_4{}^{2-}$$

(possible mechanism: +2e, as it is proposed by Imai, et al. (22) for *T. thiooxidans*)

$$O_3S - S^X - SO_3{}^{2-} \longrightarrow SO_3{}^{2-} + S^X - SO_3{}^{2-}$$

$$\downarrow$$

$$SO_4{}^{2-}$$

$$S^X O_4{}^{2-}$$

pathway, used by strain Pmv   pathway, suggested by Sinha and Walden (21).

The specific pathway, possessed by strain Pmv, is a "big" mutation, caused by a post-replication repair mechanism following UV-irradiation. That metabolic change has been a stable characterisitic of the strain and the probability for back mutation has been insignificant.

This particular feature of the Pmv strain was used as a marker for its distinguishing from the wild strains of the Vlaikov vrah. These wild strains oxidize tetrathionate via the pathway, suggested by Tuovinen and Kelly. A simple test for estimation of the cell number of both selected and autochthonic wild strains in mixed population was worked out on the basis of the above-mentioned

marker. The test was the following. Aliquote samples from a
mixed population (i.e. from the leaching or pregnant or barren
solutions) were inoculated in Erlenmeyer flasks, containing iron-
free 9K medium and a labelled trithionate ($O_3S - S^X - SO_3^{2-}$) as a
sole source of energy. After a short cultivation the culture was
analyzed by ion-exchange chromatography method (23-26) for detect-
ing various sulphur compounds. The immediate advent of labelled
sulphate showed the presence of strain Pmv. The cell number of
the latter was calculated from the quantity of sulphate by
means of standard curve.

For introduction of the strain six concrete fermentors were
constructed near to the dump. Every one of them had a system for
aeration. The volume of a single fermentor was 1.5 m^3. Barren
solution has been used as a nutrient medium. Ammonium and phosphate ions have been added to the solution. Sometimes, ferrous
sulphate was added to increase the substrate concentration. Pure
culture of the selected strain Pmv, containing about 4.10^8 cells/
mℓ has been used as an inoculum. After the ferrous iron exhaustion, the latter was pumped to a definite part of the experimental dump region. Another part of the dump (possessing the same
volume and size) served as a control.

In 1974 and in 1975 the introduction was repeated 18 times.
Content of copper and iron ions as well as the ferrobacilli number
in the experimental and control dump effluents were checked regu-
larly. Mineralogical characterization of leaching material was
also carried out periodically.

We are able to present here only the most principal results
and conclusions of our experiments.

The introduction itself of the selected strain Pmv was successful. In March 1976 about 55% of the ferrobacilli, isolated
from the experimental dump, had the specific pecularity of strain
Pmv. It does not mean that all of them were decended from that

selected strain. Different ways of transfer of genetic material between the selected and wild strains are possible.

Pmv-like ferrobacilli were only about 4-5% of the total amount of these bacteria in the control dump (there was some connection between the experimental and control dumps by soaking up of solutions). In the main dump the selected ferrobacilli were found sporadically and only in regions near to the experimental dump. It is important to note that only a part of the Pmv-like isolates had been retained the striking high activity of their laboratory-bred relatives. That finding is an indication that the high oxidative activity of strain Pmv is not connected with its specific pathway for oxidation of sulphur compounds.

Leaching kinetics in the experimental dump did not increase in comparison with those of the control dump. Total ferrobacilli count (selected and wild) also did not increase. The last finding was valid for cells swimming in circulating solutions as well as for cells adsorbed on the ore (rock) surface.

III. CONCLUSIONS

The above-mentioned experiment was preceded by analogous experiments, conducted under laboratory and semi-industrial conditions. In some experiments other laboratory selected strains were used besides the strain Pmv. These strains had specific markers, which were not very stable (for example, different tolerance to heavy metal ions, organic compounds, antibiotics, temperature, etc.). Introductions were carried out into ores, leached by different methods: by shake-flask technique, in reactors with mechanical stirring, and in percolation columns. The more important conclusions from these experiments especially for leaching of copper sulphide and mixed ores, are the following:

1. The results from introduction are impossible to be predicted in advance.

2. Introduction is useful only for leaching of some ores, containing mainly chalcopyrite (i.e. sulfide, which is very stable versus oxidative agents). No positive results were obtained by strain introduction in ores, whose copper was represented mostly as secondary sulphides (Cu_2S, CuS).

3. Introduction is successful only when selected strains, possessing a pronounced direct oxidative mechanism for sulphide minerals, are used. These strains are capable of selective adsorption on the mineral surface and possess a high sulphur-oxidizing activity. On the other hand, the selected strains must oxidize Fe^{2+} faster than the autochthonic wild strains. That ability presents the possibility for barren solutions to be used as a nutrient medium during the cultivation in fermentors. When the introduction is successfully advanced in some dump region, the latter can be used effectively as a fermentor.

It has been found that the use of strains is very efficacious and they act probably by the mechanism suggested by Dugan and Randles (27). That mechanism is possible for sulphide minerals where a crystal lattice having a specific state exists. This state can be associated with a "reverse exciton" · It is known that the exciton is an excited state of a system of electrons, and that the exciton is not connected with the transfer of electric charge. A transfer of an electron from the cation to the next anion is carried out in the crystal lattice during the process of formation of an exciton. At the "reverse exciton" the transfer of an electron is carried out from an anion to the next cation. That phenomena gives the possibility for the ferrobacilli, adsorbed on the mineral surface, to oxidize mainly the ferrous atoms of the mineral. In the structure of chalcopyrite, these atoms are more available. The resulting ferric atoms accept electrons from the next cations (S^{2-}) and serve as intermediates in the electronic transport between the mineral and the bacterial cell. As the ferrous iron-oxidizing enzyme of *T. ferrooxidans* acts by the so-called Ping Pong Bi Bi mechanism (28), the oxidation will be most rapid when the transfers of electrons from S^{2-} to Fe^{3+} in the crystal lattice, on the one hand, and from the bound iron in protein moiety of Fe^{2+}-cytochrome c reductase to cytochrome c in the bacterial cell, on the other hand, are in resonance. It is very difficult to establish by experimental means that such a state exists. In practice, only the actual (observed) rate of leaching of a sulphide mineral can be accepted as a criterion for efficacy of a certain strain. The high ferrous iron oxidation rate itself does not indicate that the strain will actively oxidize the chalcopyrite.

Estimation of rhodanese activity can be used as a simple, but quite sure test if strains must be picked out to serve as an initial material of gradual selection, whose aim is the obtaining of selected strains, possessing a high oxidative activity versus $CuFeS_2$. A high rhodanese activity is a very important factor, especially in dump leaching, where the lower sulphur and sulphite-oxidizing activities of ferrobacilli permit a non-enzymatic formation of thiosulphate in the cell envelope. It is known, on the one hand, that thiosulphate inhibits the biological oxidation of ferrous iron and sulphide minerals (29), and, on the other hand, that the sensisitivy of ferrobacilli versus heavy metal ions increases during the oxidation of thiosulphate (30).

4. The most successful introduction of selected biomass is carried out by its preliminary dilution in the leaching solution to about 10^5 - 10^6 cells/ml (the most suitable leaching solution for is a leaching solution with low iron content - below 1 g/l). Introduction of that solution into the ore mass must be conducted between two long pauses in dump irrigation.

5. Use of laboratory selected strains is very efficacious for leaching of flotation concentrations, which do not contain an autochthonic ferrobacilli population. Leaching of such concentrates is well investigated (31-37) and the use of selected strains

here is only a question of time. The introduction can be rather perspective and in some ore sections, intended for *in-situ* leaching. These are sections, in which the autochthonic ferrobacilli population is not numerous or is even absent, because of unfavorable natural conditions (low rH_2, absence of free oxygen etc).

The use of laboratory selected strains of ferrobacilli for industrial leaching operations poses some new problems, already existing in the other microbiological operations (for instance, production of antibiotics). Some of those problems are: 1. Developing of methods of experimental selection of ferrobacilli as well as methods of express microbiological control in the industrial leaching operations; 2. Maintenance of the selected strains under laboratory and industrial conditions; 3. Finding the optimal ways and conditions for introduction itself to be carried out; 4. Technological realization of introduction process as well as its inserting in the complex production scheme. These problems can be successfully solved only by close interaction between the fundamentalists and the practitioners in the area of bacterial leaching.

IV. REFERENCES

1. Buchanan, R., and Gibbons, N., (eds.), "Bergey' Manual of Determinative Bacteriology", 8th Edition, The Williams and Wilkins Co., Baltimore, 1974.

2. Gilman, G., "A Manual of Soil Fungi", Iowa, 1957.

3. Lodder, J. and Kreger, N., - Van Rij, "The Yeasts - a Taxonomic Study", Interscience Publishers, New York, 1952.

4. Madgwick, J., and Ralph, B., "Round Table Conference of Leaching", Braunschweig, Germany, 1977.

5. Beck, J., *Biotechnol. Bioeng.* 9, 487 (1967).

6. Brock, T., Brock, K., Belly R., and Weiss, R., *Arch. Microbiol.*, 84, 54 (1972).

7. Brierley,C.L., and Brierley, J., *Can. J. Microbiol.* 19. 183 (1973).

8. Brierley,C.L., and Murr, L.E. *Science,* 179, 488 (1973).

9. Zavarzin, G., *Microbiology* (Moscow), 41, 369 (1972).

10. Tabita, R., and Lundgren, D., *J. Bacteriol.* 108, 334 (1971).

11. Leathen, W., Kinsel, N., and Braley, S., *J. Bacteriol.* 72, 700 (1956).

12. Guay, R., Silver, M., and Torma, A., *Rev. Can. Biol.*, 35 61 (1976).

13. Brock, T., and Gustafson, J., *Appl. Environ. Microbiol.* 32, 567 (1976).

14. Le Roux, N., North, A., and Wilson, J., "Tenth International Mineral Processing Congress", London 1973.

15. Ehrlich, H., *Bacteriol. J.*, 86, 350 (1963).

16. Tuttle, I., Dugan P., Macmillan, C., and Randles, C., *J. Bacteriol.* 97, 594 (1969).

17. Bruynesteyn, A., Annual Meeting of the AIME, Denver, Colorado 1970.

18. Bhappu R., Johnson, P., Brierley, J. and Reynolds, D., *Trans. AIME*, 244, 307 (1969).

19. Manning, H., *Appl. Microbiol.*, 30, 1010 (1975).

20. Tuovinen, O., and Kelly, D., *Arch. Microbiol.*, 98, 351 (1974).

21. Sinha, D., and Walden, C., *Can. J. Microbiol.* 12, 1041 (1966).

22. Imai, K., Okuxumi, M., and Katagiri, H., *Koso Kagaku Shimposiumu*, 17, 132 (1962).

23. Iguchi, A., *Bull. Chem. Soc.*, Japan, 31, 597 (1958).

24. Iguchi, A., *Bull. Chem. Soc.*, Japan, 31, 600 (1958).

25. Trudinger, P., *Biochem. J.*, 78 680 (1961).

26. Trudinger, P., *Austral. J. Biol. Sciences*, 17 446 (1964).

27. Dugan, P., and Randles, C., in: "Acid Mine Drainage Formation and Abatement", The Ohio State University Research Foundation 1971.

28. Din, G. and Suzuki, I., *Can. J. Biochem.*, 45, 1547 (1967).

29. Razzell, W., and Trussell, P., *J. Bacteriol.*, 85, 595 (1963).

30. Tuovinen, O., Niemelä, S., and Gyllenberg, H., *Antonie van Leeuwenhoek*, 37, 489 (1971).

31. Torma, A., Walden, C., and Branion, R., *Biotechnol. Bioeng.*, 12, 501 (1970).

32. Bruynesteyn, A., and Duncan, D., *Can. Metallurg. Quart.*, 10, 57 (1970).

33. Duncan, D., Walden, C., Trussell, P., and Lowe, E., *Trans. SME*, 238, 122 (1967)

34. Torma, A.E., Walden, C., Duncan, D., and Branion, R., *Biotechnol. Bioeng.* 14, 777 (1972).

35. Torma A.E., and Subramanian, K., *Intr. J. Miner. Proc.* 1, 125 (1974).

36. Sakaguchi, H., Silver, M., and Torma, A.E., *Biotechnol. Bioeng.*, 18, 1091 (1976).

37. Rossi, G., Morra, A., and Savoini, G., Personal communication (1977).

ON THE MECHANISM OF BACTERIAL LEACHING

Kazutami Imai

Faculty of Agriculture
Okayama University
Tsuchima, Okayama-shi, Japan

Manganese dioxide is not soluble in sulfuric acid, but it is solubilized in the growing culture of Thiobacillus thiooxidans. *The leaching mechanism is postulated that manganese dioxide is firstly reduced with hydrogen sulfide or sulfite which is formed as an intermediate during the oxidation of sulfur to sulfuric acid by the bacterium, and then solubilized with sulfuric acid. Intact cells or cell-free extracts of* T. ferrooxidans *are able to oxidize chalcocite in the absence of iron. It suggests the existence of direct oxidizing mechanism, and the mechanism has been investigated. The rate of covellite-leaching in the growing culture of* T. ferrooxidans *is enhanced by a pre-treatment (shaking in water for 10 days at 30°C) of the covellite. The phenomenon is also observed in the leaching experiments using cells of the bacterium suspended in iron-free 9K medium.*

I. INTRODUCTION

Interest in the action of autotrophic bacteria on inorganic metal compounds has been increased since the early 1950's. However, biological reactions on inorganic metal compounds have not been studied intensely in the field of biochemistry, and the mechanisms of the reactions are generally complicated with incidental chemical reactions.

This paper describes the mechanism of biological contribution of *Thiobacillus thiooxidans* and *T. ferrooxidans* in the leaching of manganese dioxide and copper sulfides, respectively.

II.   MATERIALS AND METHODS

A.   Microorganisms and Growth Conditions

For the leaching of manganese dioxide, a sulfur-oxidizing chemoautotrophic bacterium was used. The strain was isolated from a sewage and identified as *Thiobacillus thiooxidans* (1). The bacterium was cultured at 30°C under shaking in a shaking flask containing 100 ml of a basal medium which had the following composition: KH_2PO_4, 0.4 g; $MgSO_4$, 0.03 g; $CaCl_2$, 0.025 g; $FeSO_4 \cdot 7H_2O$, 0.001 g; $(NH_4)_2SO_4$, 0.2 g; Sulfur, 1 g, Water, 100 ml; pH was adjusted to 2.0.

For the leaching of chalcocite (Cu_2S) and covellite (CuS), an iron-oxidizing chemoautotrophic bacterium, *Thiobacillus ferrooxidans* was used. The strain (2,3) was isolated from acid water of a mine at Yanahara (Japan) and cultured as in the case of *T. thiooxidans,* except the culture medium. The bacterium was cultured in 9K medium reported by Silverman and Lundgren (4). The composition of 9K medium is as follows: $(NH_4)_2SO_4$, 3.0 g; KCl, 0.1 g; K_2HPO_4, 0.5 g; $MgSO_4 \cdot 7H_2O$, 0.5 g; $Ca(NO_3)_2$, 0.01 g; distilled water, to 700 ml; 10 N H_2SO_4, 1.0 ml; $FeSO_4 \cdot 7H_2O$, 300 ml of a 14.74% (w/v) solution.

B.   Materials

In the experiments of manganese-leaching, a manganese dioxide ore (ground to pass through a 100 mesh sieve) or pure manganese dioxide powder was used. The chemical composition of the ore was as follows: (%) MnO_2, 10.6; SiO_2, 55.0; Fe_2O_3, 25.0; MgO, 5.32; and trace of Ca, Al, and S.

In the leaching experiments on chalcocite, a chemical reagent of Cu_2S was used. The sample was ground to pass through a 200 mesh sieve, treated with 1 N HCl for 30 minutes at room temperature, washed throughly with water, dried, and treated with acetone.

In the leaching of covellite, fine powder of analytical grade of CuS (99.99% pure, obtained from Mitsuwa Chemicals Co., Ltd.) was used.

C. Assay Methods

Sulfate formed in culture media was determined by the method of Fritz and Fruland (5). Ferrous iron was determined colorimetrically by a modification of the o-phenanthroline method. The iron- or chalcocite-oxidizing activity of cells or cell-free extracts of *T. ferrooxidans* was determined by oxygen-uptake in a Warburg manometer.

The solubilized manganese and copper were determined by chelate titration and Atomic Absorption Spectrometry (Model AA-1 of Jarrel Ash), respectively. The growth of both bacteria in culture media was determined counting the number of cells with Toma's Haematometer.

III. RESULTS AND DISCUSSIONS

A. Leaching of Manganese Dioxide with *T. thiooxidans*

It is well known that four valent manganese, for example manganese dioxide, is not soluble in sulfuric acid. However, it was found in our laboratory (6,7) that manganese is solubilized as manganese sulfate from a manganese dioxide ore (see MATERIALS AND METHODS) by the action of *T. thiooxidans*.

As shown in Fig. 1, when the culture medium was added with 3% of the ground ore, inoculated with the bacterium, and cultured under shaking, about 1,300 ppm of manganese was solubilized as manganese sulfate in nine days. In this case, a manganese-acclimatized bacterium was used as inoculum. The acclimatization was proceeded as follows: the original strain was subcultured three to four times in 1% ore containing medium, and then successively transferred to 1.5 and 3% ore containing media. On the other hand, manganese was not solubilized in the medium not inoculated (Fig. 1).

Then, in order to explore the leaching mechanism of manganese dioxide, the effect of several additions on the extraction rate of manganese was examined. In this case, pure MnO_2 powder (0.5 g/ 100 ml medium) was used as the substrate. As shown in Table I, the addition of metal sulfides enhanced the leaching rate of

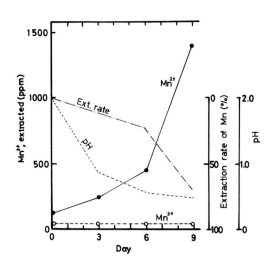

Fig. 1. *Extraction of manganese from manganese dioxide ore in the growing culture of* Thiobacillus thiooxidans.

MnO_2 content of the ore, 10.6%; Pulp density of the ore, 3.0%.

●———● *inoculated;*   ○– – – –○ *uninoculated.*

TABLE I   Effect of Metal Sulfides on the Extraction
of Manganese from MnO_2 in the Growing Culture of T. thiooxidans

Addition (1.5 mmoles/100 ml)	Sulfuric acid, formed (g/100 ml)	Mn^{2+}, extracted (ppm)
FeS	0.7	1110
CuS	0.3	214
ZnS	0.2	1850
–	0.3	170

MnO_2 (0.5 g/100 ml) and metal sulfides were added on the second day after inoculation, and sulfuric acid and Mn^{2+} were determined in 72 hours after the addition.

manganese dioxide, and FeS gave good effect on both of the leaching of manganese and bacterial growth. ZnS was effective for the former, but ineffective for the latter. On the other hand, CuS was not so effective for both of them.

The leaching process of manganese in the culture with added FeS is shown in Fig. 2. Addition of ferrous sulfate also gave good effect on the leaching of manganese, but the effect of ferric sulfate was little (Fig. 3.). From such observations, it is suggested that the leaching mechanism of manganese dioxide with T. thiooxidans is as follows: At the outset, four valent manganese is reduced to two valency with some reducing substance (or substances) which is formed from sulfur as an intermediate during the oxidation of sulfur to sulfuric acid. Then, the two valent manganese is extracted with sulfuric acid formed from sulfur by the action of the bacterium. The reducing intermediate may be thought to be hydrogen sulfide or sulfite, because the formation of hydrogen sulfide in the cultures of T. thiooxidans has been observed (8,9) and the addition of FeS enhances the leaching rate of manganese (Table I and Fig. 2). And it has also been established by many investigators that sulfite is the important intermediate in the oxidation of sulfur by T. thiooxidans (10).

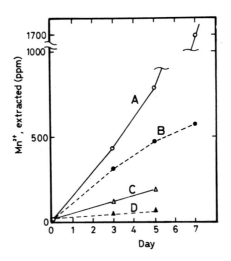

Fig. 2.   Effect of FeS on the leaching of manganese dioxide in the growing culture of T. thiooxidans.

		Additions (%)	
	MnO_2	*FeS*	H_2SO_4
A *(inoculated)*	0.5	0.1	-
B *(uninoculated)*	0.5	0.1	3.0
C *(inoculated)*	0.5	-	-
D *(uninoculated)*	0.5	-	3.0

Such postulation on the leaching mechanism of manganese dioxide will be supported by the following chemical reactions:
(1) On introducing hydrogen sulfide gas, manganese dioxide suspended in dilute sulfuric acid is solubilized immediately.
(2) Manganese dioxide suspended in water is also easily solubilized upon the introduction of sulfur dioxide gas.

Although the chemical equations to reactions (1) and (2) have not been established yet, the following equations may be inferred.

$$MnO_2 + H_2S + H_2SO_4 \longrightarrow MnSO_4 + S + 2H_2O \qquad [1\text{-}a]$$

$$4MnO_2 + H_2S + 3H_2SO_4 \longrightarrow 4MnSO_4 + 4H_2O \qquad [1\text{-}b]$$

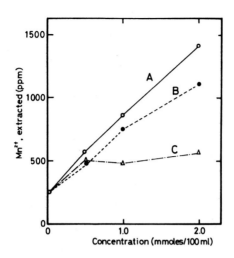

Fig. 3. Effect of the concentration of additions on the extraction of manganese in the growing culture of T. thiooxidans. *A) FeS; B) FeSO4; C) Fe2(SO4)3. Extracted manganese was determined on the seventh day after inoculation. Pulp density of* MnO₂ *was 0.5%.*

$$2MnO_2 + 4SO_2 + H_2O \longrightarrow Mn_2(SO_3)_3 + H_2SO_4$$

$$Mn_2(SO_3)_3 \longrightarrow MnS_2O_6 + MnSO_3 \qquad [2\text{-}a]$$

$$MnO_2 + H_2SO_3 \longrightarrow MnSO_4 + H_2O \qquad [2\text{-}b]$$

Among these equations, [1-b] and [2-b] will be preferable to reactions (1) and (2), respectively.   The formation of sulfur is not detected during reaction (1), and the product of reaction (2) is mainly manganese sulfate.

The results indicated in Fig. 3 suggest that ferrous iron is also effective in reducing four valent manganese.

B.   Oxidation of Chalcocite (Cu₂S) by *Thiobacillus ferrooxidans.*

It has been reported (11) that there are two mechanisms in bacterial leaching of metal sulfides.   One is indirect, that is,

metal sulfides are oxidized and dissolved chemically with ferric
iron and sulfuric acid, and ferrous iron formed in this reaction
is re-oxidized by the action of bacteria. The other is direct,
that is, metal sulfides are oxidized directly by cells of the
bacteria in the absence of acid-soluble iron. However, details
of the latter mechanism have not been clarified yet. Landesman,
et al. (12) as well as Beck and Brown (13) reported that when
pyrite or chalcopyrite is used as a substract, both iron and
sulfur moieties are oxidized simultaneously by $T.$ $ferrooxidans.$

Torma (14) reported that the bacterium is able to oxidize
iron-free analytically pure cobalt, nickel, and zinc sulfides in
the absence of acid-soluble iron. Nielsen and Beck (15) reported
that when chalcocite is used as a substrate, the bacterium
oxidizes the Cu^+ moiety, but not the sulfur moiety. In order to
detect the direct mechanism and to explore the details of the
mechanism, we carried out the following experiments:

1. *Oxidation of Chalcocite with Intact Cells in the Absence
    of Acid-Soluble Iron*

As shown in Fig. 4, chalcocite was oxidized and the copper
was extracted by the action of cells of $T.$ $ferrooxidans$ under the
condition without iron. From every value of oxygen-uptake and
copper-extraction indicated in the figure, the value of respective
control using boiled cells has been subtracted. The optimum pH
for the oxidation of chalcocite is about 2.3 and it coincides
with that for the extraction of copper. The molar ratio of
extracted copper to consumed oxygen is about 2.0 at every pH
(Fig. 4), and sulfur is not detected during the oxidation. These
facts suggest that the reaction proceeds according to the following equation:

$$2Cu_2S + 2H_2SO_4 + O_2 \longrightarrow 2CuS + 2CuSO_4 + 2H_2O$$

When ferric iron is added to the reaction mixture, the rates
of oxygen uptake and copper extraction were both accelerated as

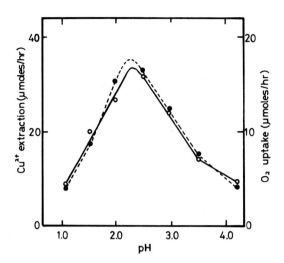

Fig. 4.   *Extraction of copper from chalcocite by intact cells of* Thiobacillus ferrooxidans. *Reaction conditions: Cu_2S, 100 μmoles; glycine-H_2SO_4 buffer, 500 μmoles; cells, 2 mg protein; total volume, 3 ml; gas phase, air; temp., 30°C.*

o————o Cu^{2+} *extraction;*   ●─ ─ ─ ─● O_2 *uptake*

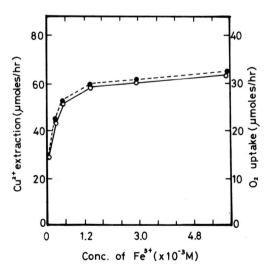

Fig. 5.   *Effect of Fe^{3+} on the leaching of chalcocite with cells of* T. ferrooxidans. *Reactions were carried out at pH 2.5.*

o————o Cu^{2+} *extraction;*   ●─ ─ ─ ─● O_2 *uptake*

indicated in Fig. 5. Also in this case, the molar ratio of
extracted copper to consumed oxygen was 2.0 at every concentration of ferric iron.

From the results shown in Figs. 4 and 5, it may be thought
that *T. ferrooxidans* has a direct leaching mechanism to chalcocite, because resting cells of the bacterium are able to oxidize
chalcocite in the reaction mixture without iron.   However,
if the cells were contaminated with a trace of insoluble iron
coming from the culture medium, it leaves room for the operation
of indirect mechanism. Therefore, we attempted to demonstrate
the existence of a direct mechanism using cell-free extracts of the
bacterium. As the cells are disrupted by sonic oscillation in
Tris-H_2SO_4 buffer (0.1 M) of pH 7.0 and then centrifuged, the
cell-free extracts cannot contain any free iron.

2.   Oxidation of Chalcocite with Cell-Free Extracts

As indicated in Table II, cell-free extracts of *T. ferrooxidans* possessed high activity of iron oxidase and also showed
chalcocite-oxidizing activity without the addition of iron, though
it was weak.   When ferric iron was added to chalcocite, the
oxidizing activity of the extracts was enhanced remarkably. On
the other hand, addition of quinacrine inhibited the activity of
iron oxidase strongly, but it did not affect the oxidation of
chalcocite (Table II).

These results suggest that oxidation of chalcocite by the
extracts without the addition of iron proceeds through direct
mechanism, since the rate of chalcocite oxidation must depend on
the rate of ferrous iron oxidation in indirect mechanism.

In fact, the rate of chalcocite oxidation of the crude
extracts increased with the addition of ferric iron, but decreased
to the level of chalcocite alone with further addition of
quinacrine (Table II). On the other hand, when the crude cell-
free extracts were dialyzed against 0.01 M Tris-H_2SO_4 buffer

TABLE II   *Oxidation of Chalcocite with Cell-Free Extract of* T. ferrooxidans

| | | Activity (μl O_2/60 min/ 40.7 mg protein) on | |
Condition	Cu_2S	Cu_2S + $Fe_2(SO_4)_3$	$FeSO_4$
1. Crude cell-free extract	26	115	337
2. Crude cell-free extract + quinacrine (5 x 10^{-3}M)	28	35	50

Composition of the reaction mixture: each substrate, 100 μmoles; β-alanine-H_2SO_4 buffer (pH 2.5), 150 μmoles; cell-free extract, 40.7 mg protein; total volume, 3.0 ml; 30°C.   From every value indicated in the Table, the value of respective control using boiled cell-free extracts has been subtracted.

TABLE III   *Effect of Dialysis on Chalcocite-Oxidizing Activity of Crude Cell-Free Extracts of* T. ferrooxidans

| | Activity (μl O_2/60 min/ 35.5 mg protein) | |
Addition	Before dialysis	After dialysis
None	25	4
$Fe_2(SO_4)_3$ (6 x 10^{-3}M) + quinacrine (5 x 10^{-3}M)	32	32

Reactive conditions were the same as given in Table II.

(pH 7.0) for 41 hours in the cold, the activity of chalcocite oxidation decreased markedly.

However, it was recovered by the addition of ferric iron even in the presence of quinacrine (Table III). These results suggest that a direct mechanism on chalcocite oxidation exists in cell-free extracts of *T. ferrooxidans*, and the enzyme system needs trace amounts of iron as a co-factor (16).

Corrans, et al. (17) suggested that both of the indirect and direct mechanisms function in the oxidation of chalcocite by *T. ferrooxidans*, but indirect mechanism is prominent under usual conditions, that is, in the presence of ferrous or ferric iron. From the results indicated in Fig. 5 and Table II, we also approve of this suggestion.

C. Leaching of Covellite (CuS) with *T. ferrooxidans*

Regarding to bacterial leaching of covellite many papers have been published (17-24). However, in order to clarify the detailed aspect of bacterial contribution to the leaching of covellite, we carried out some basical experiments. In this study, chemical reactions which are thought to participate in bacterial leaching of covellite were considered.

3. *Auto-Oxidation of Ferrous Iron in 9K Medium*

100 ml of sterilized 9K medium in cotton pluged shaking flask was shaken aseptically for 12 days at 30°C, and the rate of auto-oxidation of ferrous iron was determined. In this experiment the concentration of ferrous sulfate in 9K medium was varied as 1, 2, 3, and 4% (as $FeSO_4 \cdot 7H_2O$). At the same time, four other media containing cupric sulfate in addition to ferrous sulfate were prepared and shaken under the same condition. The content of Cu^{2+} in each of the media was the same (1000 ppm), but the concentration of ferrous sulfate was varied as indicated above.

Auto-oxidation of ferrous iron in all of the media was small even after 12 days, and the effect of cupric sulfate on the auto-oxidation of ferrous iron was not observed (Fig. 6).

4. *Chemical Leaching of Covellite with Free Sulfuric Acid*

Iron-free 9K medium (excluded merely ferrous sulfate from 9K medium) was prepared and its pH was respectively adjusted to

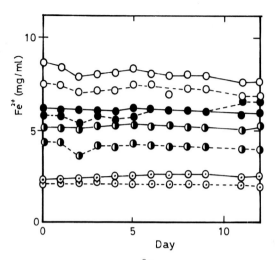

Fig. 6. Auto-oxidation of Fe^{2+} in 9K medium.   Concentration of FeSO$_4$·7H$_2$O in the medium:

○ 4%,   ◐ 3%,   ◑ 2%,   ⊙ 1%

-----, Cu^{2+} (1000 ppm) was added further.

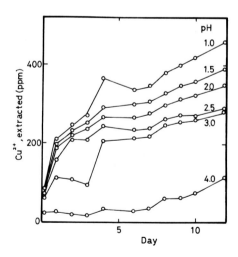

Fig. 7. Effect of initial pH of iron-free 9K medium on the chemical leaching of covellite.

4.0 - 1.0 with sulfuric acid. Each medium was sterilized, added
with covellite (1%), and shaked aseptically for 12 days at 30°C.

As shown in Fig. 7, the amount of copper extracted was
smaller t pH 4.0 compared with that at other lower pH values. In
the range of pH 1.0 to 3.0, 300 to 500 ppm of copper was extracted,
and a correlation was observed between the rate of copper extraction and the pH of the medium.

5. *Effect of Ferric Concentration on the Chemical Leaching of
    Covellite in 9K Medium.*

Ferrous sulfate and ferric sulfate were added to iron-free
9K medium in the following ratio (1% of $FeSO_4 \cdot 7H_2O$: % of
$Fe_2(SO_4)_3$ in the medium): A. 5:0; B. 4:1; C. 3:2; D. 2:3; E.1:4;
and F. 0:5.

After sterilization, 1% of covellite was added to each of
them, shaken aseptically for 12 days at 30°C, and the amount of
extracted copper was determined. As shown in Fig. 8, copper
extraction occurred rapidly in the initial stage of the reaction
(within one hour), and the rate of copper extraction was nearly
proportional to the initial concentration of ferric sulfate in
the medium. However, in the final stage of the reaction, the
amount of extracted copper became larger in the media with the
mixture of ferrous and ferric sulfate (B and C. in Fig. 8).

6. *Leaching of Covellite in Growing Culture of* T. ferrooxidans

9K medium with 1% of covellite added was inoculated with
T. ferrooxidans and cultured for 8 days at 30°C under shaking.
During the culture, cell-growth, iron oxidation, and copper
extraction were determined. As shown in Fig. 9, the growth of
the bacterium and the rate of iron oxidation in the medium were
almost similar to those in normal 9K medium (without covellite).
However, the rate of copper extraction in inoculated
medium was higher than in uninoculated medium.   It implies

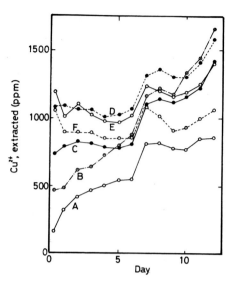

Fig. 8. Effect of the initial concentration of Fe^{2+} and Fe^{3+} in 9K medium on the chemical leaching of covellite. % $FeSO_4 \cdot 7H_2O$: % $Fe_2(SO_4)_3$ in 9K medium: A. 5:0; B. 4:1; C. 3:2; D. 2:3; E. 1:4; F. 0:5.

that the leaching of covellite is accelerated by the action of the bacterium, especially in the early phase of its growth.

7. Effect of Ferrous Iron Concentration on the Leaching of Covellite with the Bacterium

The concentration of ferrous sulfate in 9K medium was varied respectively as 2, 3, and 4% (as $FeSO_4 \cdot 7H_2O$), and to each medium 1% of covellite was used. These media were inoculated with *T. ferrooxidans* and cultured at 30°C under shaking. At the same time, control media which had the compositions corresponding to the respective test media, but not inoculated were shaken under the same conditions.

As shown in Fig. 10, the rate of copper extraction in inoculated media was not affected by a change in the concentration of ferrous sulfate in the range of 2 to 4%, and the amount of

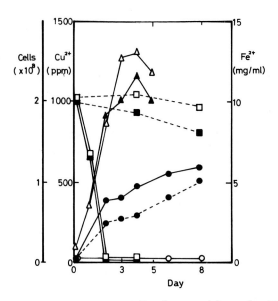

Fig. 9.   Culture process of T. ferrooxidans in 9K medium.

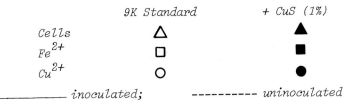

	9K Standard	+ CuS (1%)
Cells	△	▲
Fe^{2+}	□	■
Cu^{2+}	○	●

_____ inoculated;   ---------- uninoculated

extracted copper was much larger in inoculated media than in uninoculated, except one case. The exceptional medium (A in Fig.10)
was one of the uninoculated controls, and the copper extraction was
comparable to that in inoculated media. Exclusively in this
medium, a sample of covellite powder which had been exposed to
the air for a long time was used accidentally. The color of
fresh powder of pure covellite is usually deep indigo blue, but
that of the sample was rather black.

Thereupon, fresh powder of pure covellite was suspended in
water (7 g/100 ml) and shaked aseptically for 10 days at 30°C
using a cotton pluged shaking flask. After the treatment, the

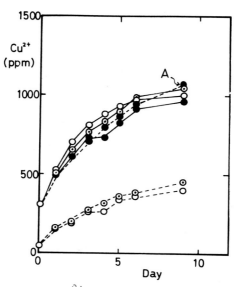

Fig. 10. Effect of Fe^{2+} concentration on the leaching of covellite in the growing culture of T. ferrooxidans. *Concentration of FeSO$_4$·7H$_2$0:*

○ *2%,*   ⊙ *3%,*   ● *4%*

——————— *inoculated,*   - - - - - - - - - *uninoculated*

covellite was filtered, washed with water, and dried. The treated
covellite powder was used as a substrate for leaching experiments,
and the rate of copper extraction was compared with that of untreated fresh covellite.

8. *Comparison of Copper Extraction from Treated - and Untreated - Covellite*

9K medium added with 1% of treated- or untreated-covellite
was inoculated with *T. ferrooxidans*, and shaken for 12 days at
30°C.

As shown in Fig. 11, the rate of copper extraction was higher
in the culture using treated-covellite than in that using untreated covellite, and the phenomenon was also observed in uninoculated

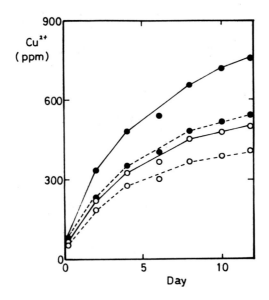

Fig. 11. Extraction of copper from treated- and untreated-CuS in the growing culture of T. ferrooxidans.

● *Treated CuS,*   ○ *Untreated CuS*

———— *inoculated,*   --------- *uninoculated*

control media. In the next place, the leaching of both kinds of the covellite with resting cell-suspension of *T. ferrooxidans* was then examined.

Cells of the bacterium which were obtained from the culture in 9K medium and washed thoroughly were suspended (2×10^8 cells/ ml) in iron-free 9K medium containing 1% of the treated- or untreated-covellite. The suspensions were shaken for 15 days at 28°C. As the results, copper was extracted from both kinds of the covellite, but the rate of the extraction was also higher in the treated-covellite (Fig. 12).

The fundamental change which occurs in covellite during the pretreatment has not been clarified yet, but it is indicated that the treatment gives good effect on the leaching of covellite.

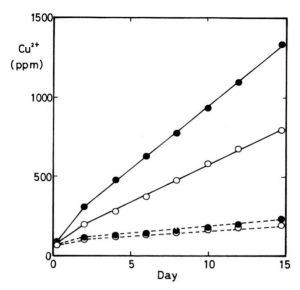

Fig. 12.   Extraction of copper from treated- and untreated-
CuS with cells of T. ferrooxidans.

● Treated CuS,   ○ Untreated CuS

──────── inoculated   ---------- uninoculated

As widely known, the mechanicsm of bacterial leaching is
complicated with many reasons. However, I think that the most
important reason is that many series of chemical and biological
reactions occur and they affect one another in complex pattern
during the leaching process.

Furthermore, as I described above, the effective reactions
are oxidative in most of the leaching processes, especially in the
leaching of metal sulfides, but in certain cases, for example the leaching of manganese dioxide, some reducing reactions play an important
role.

From such observations and others, it is evident that the
mechanism of bacterial leaching is not similar, but diverse depending
upon the kind of metal compounds, leaching condition, sort of
bacteria, and so on.

IV. REFERENCES

1. Imai, K., Okuzumi, M., and Katagiri, H., *J. Ferm. Technol.*, 42, 755 (1964).

2. Imai, K., Sugio, T., and Tano, T., in "Abstracts of Papers, Fifth International Fermentation Symposium, Berlin, 1976", p. 453.

3. Imai, K., in "World Mining and Metals Technology", (A. Weiss, Ed.), Vol. 1, p. 321. American Institute of Mining, Metallurgical, and Petroleum Engineers, Inc., 1976.

4. Silverman, M.P., and Lundgren, D.G., *J. Bacteriol.*, 78, 326 (1959).

5. Fritz, I., and Fruland, M., *Anal. Chem.*, 26, 1593 (1954).

6. Tano, T., and Imai, K., *J. Agr. Chem. Soc. Japan*, 37, 576 (1963).

7. Imai, K., Tano, T., and Noro, H., U.S. Patent, 3433629, March 1969.

8. Starkey, R., *J. Bacteriol.*, 33, 545 (1937).

9. Imai, K., Okuzumi, M., and Katagiri, H., in "Symposium on Enzyme Chemistry, Japan", Vol. 17, p. 132 (1962).

10. Peck, Jr. H.D., *Ann Rev. Microbiol.*, 22, 489 (1968).

11. Silverman, M.P., *J. Bacteriol.*, 94, 1046 (1967).

12. Landesman, J., Duncan, D.W., and Walden, C.C., *Can. J. Microbiol.*, 12, 959 (1966).

13. Beck, J.V., and Brown, D.G., *J. Bacteriol.*, 96, 1433 (1968).

14. Torma, A.E., *Rev. Can. Biol.*, 30, 209 (1971).

15. Nielsen, M.M., and Beck, J.V., *Science*, 175, 1124 (1972).

16. Imai, K., Sakuguchi, H., Sugio, T., and Tano, T., *J. Ferm. Technol.*, 51, 865 (1973).

17. Corrans, I.J., Harris, B., and Ralph, B.J., *J. South African Inst. of Mining and Metallurgy*, 72, 221 (1972).

18. Bryner, L.C., Beck, J.V., Davis, D.B., and Wilson, D.C., *Ind. Eng. Chem.*, 46, 2587 (1954).

19. Ito, I., *J. Ferm. Assoc.*, Japan, 25, 1 (1967).

20. Watanabe, A., Uchida, T., and Furutani, S., *J. Ferm. Assoc.*, Japan, 25, 21 (1967).

21. Sakaguchi, H., Torma, A.E., and Silver, M., *Appl. Environ. Microbiol.*, 31, 7 (1976).

22. Silver, M., and Torma, A.E., *Can. J. Microbiol.*, 20, 141 (1974).

23. Torma, A.E., Gabra, G.G., Guay, R., and Silver, M., *Hydrometallurgy*, 1, 301 (1976).

24. Torma, A.E., in "Abstracts of Papers, Fifth International Fermentation Symposium, Berlin, 1976", p. 455.

POTASSIUM RECOVERY THROUGH LEUCITE BIOLEACHING:

POSSIBILITIES AND LIMITATIONS

Giovanni Rossi

University of Cagliari
Cagliari, Sardinia, Italy

Leucite is a framework silicate $(2SiO_2 \cdot AlO_2)^{-1} \cdot K^{+1}$ with a specific gravity of 2.5, Mohs hardness about 6, and vitreous luster. It is tetragonal (pseudocubic) and can be differentiated from orthoclase because of its lower silica content (four molecules in leucite vs. six in orthoclase). It is more basic and more susceptible to acid attack and weathering than orthoclase. In the volcanic belt of Central Italy, comprised between Upper Latium and the Alban Hills and in nearby areas, there exist large occurrences of leucitic rocks containing estimated reserves of nine billion tons of potassium oxide (K_2O) equivalent and the same amount of aluminum oxide (Al_2O_3). Numerous attempts aimed at developing industrial processes of potassium extraction from leucite ores date as far back as 1856, but so far they have not been successful. Microbial solubilization of potassium and aluminum from leucite using heterotrophic microorganisms from collections or isolated from soils of leucitic areas and thiobacilli was studied in static conditions (batch cultures). Leaching with thiobacilli, even in the presence of pyrite, and with some strains of soil microorganisms did not give good results. In 150 days, strains of Aspergillus niger *(van Tieghen),* Scopulariopsis brevicaule *(Sacc. Bain) and* Penicillium expansum *(Link) leached between 21 and 27% of the potassium.*

I. INTRODUCTION

Leucite is a potassium aluminum silicate (a framework sili-
cate) with the formula $(2SiO_2 \cdot AlO_2)^{-1} \cdot K^{+1}$, or $KAl(SiO_3)_2$, specific
gravity 2.5, Mohs hardness 5.5 to 6.6, which generally occurs as
white or gray trapezohedral crystals (icositetrahedrons or "leu-
citohedrons") or rounded grains disseminated in eruptive igneous
rock, with sizes comprised between a few millimeters and a few
centimeters (1). It is tetragonal (pseudocubic) and can be
differentiated from orthoclase because of its lower silica content
(four molecules in leucite vs. six in orthoclase) (2) (Fig. 1).

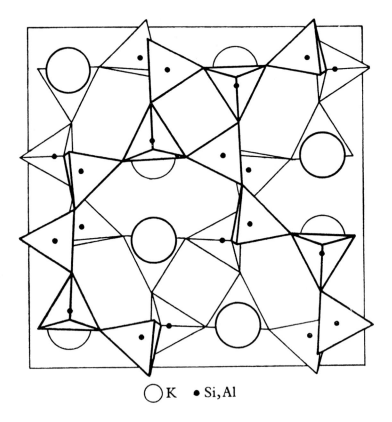

$\bigcirc$ K $\bullet$ Si,Al

Fig. 1. Lower half of the unit cell of leucite (2).

Leucite has vitreous luster, is more basic and more susceptible to acid attack and weathering than orthoclase. It contains
21.58% K_2O and 23.4% Al_2O_3.

Leucite occurs abundantly in the rocks of the volcanic belt
of Central Italy, between Acquapendente in Upper Latium and the
Alban Hills and, further South, in the province of Caserta as well
as in the Sessa Aurunca area and in the vicinity of the Vesuvius
volcano (Fig. 2).

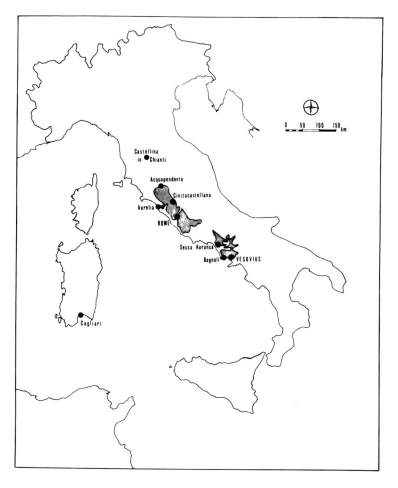

Fig. 2. Map of Italian leucite deposits.

On the basis of estimates made by Washington, who extensively investigated the Italian leucite occurrences (3), potash equivalent (K_2O) and alumina (Al_2O_3) reserves contained in the above mentioned deposits can be calculated in more than nine billion tons each (4).

The potassium mineral salt deposits discovered in the fifties in Sicily and exploited since then are insufficient to satisfy the potash demands of the Italian economy, and the situation is even worse as far as aluminum metal is concerned, since the major Italian bauxite deposits are depleted, and in the O.E.C.D. area the only bauxite producer is France.

The development of an industrial process of potash and aluminum extraction from leucite rocks could therefore have a significant effect on the Italian economy, and this is the reason why several investigators and specialists in this field have tackled the problem.

Furthermore, one cannot ignore the presence of considerable leucite deposits in other areas of the world, for instance in the United States.

The first attempts [made by Bickell (5)] to recover potash from leucite date as far back as 1856. Since then, several processes have been developed and tested all of them based on attack by a strong inorganic acid or on alkaline treatment. In every case the major difficulty consisted in the elimination of the silica and the iron.

Among the processes which have been developed up to the pilot plant scale the following should be mentioned (6,7):
 - those based on hydrochloric acid attack, with formation of potassium chloride and aluminum chloride, subsequent separate precipitation of the latter and final decomposition of the hydrated aluminum chloride by heating at about 350°C; among these, the most important was that developed by Prof. Blanc, which was

realized in a plant located at Aurelia, 70 km North of Rome, in
the thirties (8,9);

    - those based on nitric acid attack, similar, in principle,
to the previous process;

    - those based on sulfuric acid attack; most noteworthy is
that patented by Prof. Gallo, of Pisa University, according to
which leucite is attacked with sulfuric acid at 52° C, the silica
is filtered and the pure alum thus formed is purified, crystalliz-
ed, dehydrated and calcined at about 800°C with recovery of sulfur
trioxide, which is recycled; the calcined residue is treated with
water: the soluble alumina is then separated and the potassium
sulfate solution is finally evaporated. This process was tested
in a pilot plant built at Castellina in Chianti, in Tuscany, by
the Italian Company S.A.L.P.A. in the forties;

    - the processes based on alkaline attack, they consisted in
principle in the reaction of leucite with limestone with formation
of calcium silicate which was then separated, and of the water-
soluble potassium aluminate. Among these processes should be
mentioned the Messer-Schmidt process, tested by S.A. Vulcania, the
Jourdan process, tested by S.I.P. (Società Italiana Potassa), and
the process developed and tested on a pilot plant scale by S.I.P.
in the late forties in a plant located at Bagnoli (Naples).

All of the above mentioned processes, though technologically
sound, were too expensive due to the fact that the formation on
the leucite particles of an insoluble inhibiting layer required,
for the process to be rendered commercially practicable, expensive
auxiliary operations, and the products, though excellent as far as
commercial requirements are concerned, were not competitive.

For these reasons, all of the industrial attempts to exploit
leucite for aluminum and potash extraction have been abandoned up
to now.

II. THE BIODEGRADATION OF ALUMINOSILICATES

The success of the bioleaching techniques developed in the
past decade in the field of sulfide minerals has prompted the investigation of the possibilities of biological solubilization of
aluminum and potassium from leucite.

The biodegradation of aluminosilicates and carbonates has up
to now constituted a field of research mainly for agricultural
microbiologists. From literature, it would seem that the investigations carried out up to the present time have followed two lines
of approach: the first utilized mainly in the Western hemisphere
and in Europe, adopted the conventional procedures of agricultural
microbiology, and aimed at ascertaining the action of the known
soil microorganisms on aluminosilicates and carbonates; the second,
mainly followed by the Slav-speaking countries, under the leader-
ship of the Russian School, aimed at the research of microbial
strains specifically suitable for aluminosilicate biodegradation.

Among the first - one may very likely say "the first" -
investigators who tackled the problem can be considered the
Italians De Grazia and Camiola (10). The object of their investigation was to understand the process by which plants obtain the
soluble potassium supply needed for their nutrition from the insoluble volcanic rock. They worked on batches of very finely
ground museum-grade leucite crystals mixed with 1,000 ml of culture
medium each in 2,000 ml static flasks which were inoculated with
various strains of moulds. A culture yielded, in 70 days, more
than 70% dissolution of K_2O equivalent, though the sterile control
yielded results quite high too (30%).

Shortly after, Bassalik (11) reported on an investigation of
orthoclase biodegradation by means of several strains; the best
results were obtained with strains of *Bacillus extorquens*, n.sp.,
with yields, expressed in terms of percent weight of the initial
feldspar sample dissolved, lower than 4% in 68 days in the most

favourable case, but more than 40 times higher than in the sterile controls.

Investigations aimed at throwing light on the problem have recently been undertaken by the research workers of the "Bureau de Recherches Géologiques et Minières" (B.R.G.M.) in France, in the framework of a wide program concerning the understanding of the processes of geologic rock alteration (12,13). They experimented with several rocks (olivine from basalt nodules of the Srmci region in Northern Bohemia, dunite from New Caledonia, "green rocks" (serpentine) from the Alps and mica muscovite) with various strains of moulds and bacteria, both of collection and isolated from soils. The best results were obtained with *Aspergillus niger*, which, in 18 days, produced a 26.8% solubilization of the potash contained in the muscovite samples. Evidence of transformations induced in biotite crystals by the same strain was also produced by Boyle, Voigt and Sawhney (14). As has already been pointed out in the introduction, the philosophy underlying the research conducted in this field by the Russian School has been somewhat different from the one outlined so far, its objective being the isolation of microorganisms specifically capable of aluminosilicate destruction. The Russian Academician V.I. Vernadskii had already discussed, as far back as 1934, the possibility of silicate destruction by the soil microorganisms which liberate potassium together with other silicate-forming elements. In his study on alumina and silica he expressed the view that the decomposition of aluminum silicate ($Al_2Si_2O_7$) could only take place through biochemical action (15).

These investigations were taken over, in the forties, by Alecsandrof, whose objective was to find and isolate soil bacteria capable of decomposing silicates and their utilization for improving the agricultural output as a substitute for potash fertilizers. According to the results reported by Alecsandrof in three papers published in 1947 and in 1949, the bacteria isolated from

Russian black soil ("cernoziom") were able to transform the potassium contained in the latter into a form not susceptible to assimilation by upper plants into an assimilable form (16,17,18).

These findings were confirmed by a subsequent work published in 1950 Alecsandrof and Zac, who, culturing the silicobacteria on "silicated agar", obtained the solubilization of 75.9% and 67.0% of the aluminosilicate present in the culture medium (19). Unfortunately, they did not indicate the chemical composition nor the crystalline structure of the aluminosilicate subjected to biodegradation in their tests.

Even though the existence of silicobacteria was taken for granted by some researchers, who devoted a remarkable amount of work to their characterization (20-26), Tesic and Todorovic (27-29) have come to the conclusion that they really belong to a genus known for some time under the name of *Bacillus circulans* Jordan (1890), and therefore, the bacteria discovered by Russian scientists can probably be considered as ecotypes of an already known species of silicobacteria; furthermore according to them, it does not seem to have been entirely proved that these are bacteria specific for silica. Alecsandrof himself, in later papers, (30-35) seems to share their opinion.

Bacteria having characteristics similar to those described by Alecsandrof have been reported also, in the paper already mentioned, by Goni, Gugalski and Sima (13), whereas Heinen assessed that this type of microorganism assimilates silica in the place of phosphorus (36,37).

III.   MATERIALS AND METHODS

A.   The Mineral Substrates

The leucite sample used in the experiments is a concentrate obtained by magnetic separation from a volcanic rock ("leucito-firo") of the Civitacastellana area, described in detail by

Ventriglia (38), with procedures, described elsewhere (39,40). It was dry-ground in a "Vibratom" vibrating mill (Siebtechnik, G.m.b.H., Mühlheim, German Federal Republic) with a grinding charge of ceramic pebbles. Its composition is reported in Table I and its size distribution analysis in Table II.

The pyrite was the same as that described in a previous work (41).

TABLE I   Chemical Data on Leucite Concentrate
Utilized as Solid Substrate

Component	Weight %	Component	Weight %
SiO_2	55.44	MgO	0.14
Al_2O_3	22.58	SO_3	0.08
Fe_2O_3	0.50	Na_2O	0.82
CaO	0.40	K_2O	19.00

TABLE II   Size Distribution Analysis
of Leucite Concentrate

Size, μm

Screen fractions		Weight %
	+208	11.83
-208	+147	8.58
-147	+104	10.53
-104	+ 74	9.75
- 74	+ 61	21.96
- 61	+ 53	26.90
- 53	+ 43	4.03
- 43	+ 38	0.19
- 38		6.24

B. The Microorganisms

The *Thiobacillus ferrooxidans* microorganisms used for the
bioleaching experiments of mixtures of leucite and pyrite, were
originally isolated from acid mine waters in Southern Sardinia
(42) and maintained by weekly transfers into Silverman and
Lundgren 9K medium (43).

For leaching tests on leucite alone, microorganisms both from
collections and isolated from Italian soils were used. The
former were purchased from Centralbureau voor Schimmelcultures
(CBS), Barn, The Netherlands, and are:

- *Scopulariopsis brevicaule* (Sacc.), CBS 467.48;
- *Penicillium expansum* Link, CBS 146.45;
- *Aspergillus niger* van Tieghen, CBS 131.52

The bacteria were isolated from a sample of soil collected in
the vicinity of a volcanic rock quarry close to Civitacastellana.
The samples were kept sealed in polythene bags for 24 months; at
the end of this time interval, the bacteria were "revived" by
humidifying the soil samples and keeping them for three days in a
thermostat at 30°C.

0,1 gram batches of the above soil samples were then suspended in 1,000 ml of water previously deionized and sterilized, and
the suspensions thus obtained were used as inocula of petri dishes
containing "silicate agar" according to Alecsandrof (Table III).

TABLE III Silicate Agar According to Alecsandrof (17)

Agar-agar (purified of potassium)	2.00%
Saccharose	0.50%
$(NH_4)_2SO_4$	0.10%
Na_2HPO_4	0.20%
$MgSO_4$	0.05%
$FeCl_3$ *(1 drop of 1% solution per 100 ml of aluminum and potassium silicate)*	

From the colonies thus developed, two were selected which had a
liquid gelatinous appearance with a prominent cupola and were
labeled Al and A2. The strain Al organisms are squat rods, about
1 μm long, characterized by a subterminal clostridium-shaped
deformation and provided with a capsula about 6 μm in diameter.
The strain A2 organisms are similar in size to those of strain Al,
but have no capsula (aerobic *Bacillus* sp.).

C. The Cultures

For the experiments with mixtures of leucite and pyrite the
9K Silverman and Lundgren medium was used (43).

For the experiments with *Scopulariopsis brevicaule*, *Penicillium expansum* and *Aspergillus niger* a culture medium proposed for
bioleaching tests on aluminosilicates by Goni, Gugalski and Sima
(13) was used, which is described in Table IV.

Finally, for the leucite bioleaching tests with bacteria
isolated from Italian leucitic soils, the Alecsandrof medium was
used (17) and its composition is reported in Table V.

TABLE IV   Culture medium of Goni, Gugalski and Sima (13)

Deionized Water	*1,000 g*
Maltose	*10 g*
Peptone	*1 g*

TABLE V   Culture medium of Alecsandrof (17)

Distilled water	*1,000 g*
Saccharose	*0.75 g*
Ammonium sulphate	*0.15 g*
Disodium phosphate	*0.30 g*
Magnesium sulphate	*0.75 g*
Ferric chloride	*(1 drop of 1% soln.)*

In order to evaluate the extent of the leucite degradation in the presence of organic acids produced by certain microorganisms (like *Aspergillus niger*) a 50-50 sterile mixture of M/10 oxalic acid and M/10 citric acid solutions were prepared.

All the tests were carried out in 2,000 ml Florence flasks, each containing 1,000 ml of culture medium (or acidic solution) and 3 grams of ground leucite. The bottles were kept in a thermostatic chamber at 32°C and samples of about 5 ml were taken aseptically at regular time intervals.

D.   The Analytical Methods

The potassium content of the samples was determined using a Gio De Vita "Flammelectron II" flame photometer; analyses for aluminum were effected by means of a Perkin-Elmer 306 atomic absorption spectrophotometer. The potassium, aluminum and silica content of the dried biomasses were determined by means of a Philips PW1212 automatic X-ray fluorescence spectrometer; some dried biomasses were also investigated using a General Electric XRD5 X-ray unit.

The pH of the media were measured with a Beckman   Century S-S pH-meter.

IV.   RESULTS

A.   The Experiments with Pyrite and Leucite Mixtures

The microorganism of the genus Thiobacillus was not capable of leaching pyrite in presence of leucite and, one month after the inoculum, it appeared inhibited; in the meantime the culture medium pH had slightly increased from the initial value of 2.35 to 2.68.

B.   The Experiments with Moulds

The moulds grew in the respective flasks in different ways:
the *Aspergillus niger* and *Penicillium expansum* strains formed
aggregates, up to some centimeters in size, which, already at the
end of the second month, had covered the whole free surface of
the culture medium without spreading downwards and coming
into direct contact with the mineral powder which had settled on
the bottom; on the contrary the *Scopulariopsis brevicaule* formed a
bulky biomass, constituted of several superposed layers (at least
seven), the surface of each layer being continuous and practically
covering the whole cross-section of the flask perpendicular to its
vertical axis at the level at which it was formed.

Direct contact of the biomass with the mineral powder thus
occurred. No deposit on the inner wall of the flask was noticed.

The initial and final pH of the culture media are summarized
in Table VI. The increase in pH is, thus, a feature common to all
of the culture media: it seems, however, remarkably more pronounced in the inoculated media and in the sterile mixture of
organic acids, whereas the opposite appears to take place in the
sterile control. It was preferable, in order to evaluate the
evolution with respect to time of the leucite bioleaching process,
to assay the culture media for potassium rather than for aluminum,
since the latter, being amphoteric, forms hydroxides the solubility of which is pH-dependent, as shown by the diagram in Fig. 3,
taken from a paper by Correns (44).

The results of the assays are illustrated by the curves of
Figs. 4, 5, and 6 in which the curves of the sterile control as
well as those of the organic acids mixture were also plotted.

During the first month, the leucite solubilization proceeded
in a similar way both in the acids mixture and in the cultures,
since the rate of increase of potassium ion in solution was
practically 2.65 mg per liter per day for all of them.

TABLE VI   *Variation of the pH during the
time of the experiments*

Flask contents	Initial pH	Final pH
Aspergillus niger	5.65	6.85
Scopulariopsis brevicaule	5.75	8.10
Penicillium expansum	6.00	7.98
Sterile control for the *cultures of moulds*	7.20	5.81
Sterile organic acids mixture	1.62	1.65
Culture A1	3.98	5.91
Culture A2	5.42	4.62
Sterile control for cultures *A1 and A2*	4.35	6.72

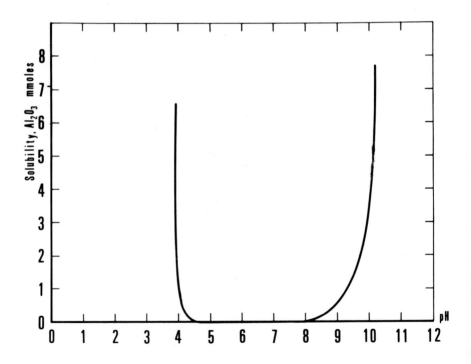

*Fig. 3.   Solubility of aluminum hydroxide gel as a function
of pH (after Correns).*

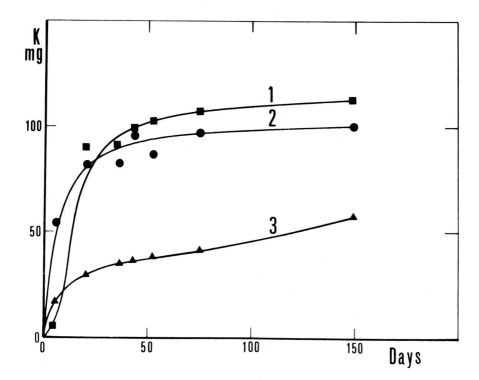

Fig. 4. Potassium solubilization from leucite vs. time –
Curve 1: sterile acids mixture. Curve 2: Penicillium expansum
culture. Curve 3: sterile control.

The solubilization rate underwent an abrupt slowing down at
the end of this initial period, and the curves then followed
different trends: those of the acids mixture and of the flask
inoculated with Penicillium expansum (Fig. 4) tended to become
parallel to the axis of the abscissae; the potassium contents of
the media thus tend asymptotically to values constant versus time.

The curves of the potassium contents in the media inoculated
with Aspergillus niger (Fig. 5) and Scopulariopsis brevicaule
(Fig. 6), after having reached a maximum at the 45th and 55th day
respectively, started decreasing. This decrease was more marked
and permanent in the case of the second strain. The sterile

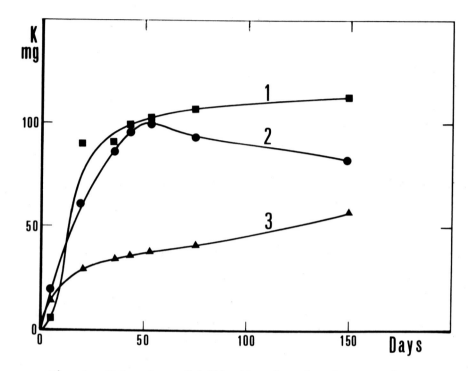

Fig. 5. Potassium solubilization from leucite vs. time -
Curve 1: sterile acids mixture. Curve 2: Aspergillus niger.
Curve 3: sterile control.

control curve runs constantly lower than those of the cultures,
though showing a slightly increasing trend with time.

At the end of the experiment the flasks were emptied, care
being taken to separate the biomasses from the residual mineral
powder; this operation was quite easy for the *Aspergillus niger*
and *Penicillium expansum* cultures, whereas it was practically
impossible for the *Scopulariopsis brevicaule*, which, as pointed
out earlier had come into close contact with the solids which had
settled on the bottom of the flask.

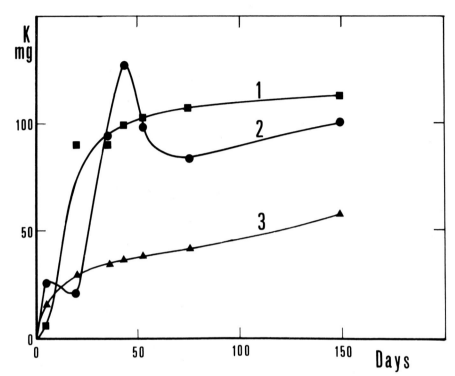

*Fig. 6.   Potassium solubilization from leucite vs. time -
Curve 1:   sterile acids mixture.   Curve 2:* Scopulariopsis
brevicaule *culture.   Curve 3:   sterile control.*

The three biomasses were dried and assayed for potassium
contents. The amount of solid residue of *Scopulariopsis brevicaule* was sufficiently large to be analyzed by X-ray diffraction,
which showed the presence of crystalline leucite.

Reliable X-ray diffraction patterns could not be obtained for
the solid residues of the other two biomasses as they were too
small; they were therefore analyzed by means of the fluorescence
spectrometer. They revealed the presence of potassium but not of
aluminum or silica. This result confirmed the observation made
during the experiment, i.e., that these biomasses never came into
contact with the leucite.

From the diagrams it appears that the potassium solubilized is comprised between at least 27% in the case of the *Scopulariopsis brevicaule* culture and 22% in the case of the *Penicillium expansum* culture. These figures are conservative, since the potassium assimilated by the organisms was not taken into account, due to the uncertainty relating to *Scopulariopsis brevicaule*.

C. Experiments with Microorganisms Isolated from Leucite Soils

In the flasks inoculated with the strains labeled A1 and A2, isolated from leucitic soils according to the procedure suggested by Alecsandrof, the leucite biodegradation occurred at a remarkably slower rate than in the flasks inoculated with moulds; after 150 days, the potassium solubilized was of the same order of magnitude as that of the sterile controls of the latter, even though the turbidity of the culture media, which was much more pronounced than that of the sterile controls, was indicative of a flourishing microbial flora.

The experiment was stopped after 24 months and during the final 19 months no samples were taken in order to not excessively affect the environmental conditions. In both flasks, but particularly in the flask containing the strain A1, the settled slime had formed clots of various sizes, which were not present in the sterile controls.

Furthermore, on the inner wall of the flask the formation of a white-gray deposit was noticed, which extended from the meniscus down to the bottom of the Florence flask. This phenomenon was not observed on the walls of the sterile control flasks.

The filtrates of both the inoculated flasks and the sterile controls assayed 41 p.p.m. of potassium. The amount of deposit formed on the inner wall was very small (about 200 mg) and it was not possible to prevent its mixture with the very fine leucite fractions. Analysis of this deposit was only possible by

means of the X-ray fluorscence spectrometer, but the results were
not conclusive, as the presence of Al, Si, and K was observed.
The initial and final pH are shown in Table VI.

V. COMMENTS AND CONCLUSIONS

The experiments indicated that the utilization of the genus
Thiobacillus and of pyrite for leucite is impracticable since
the pH of 9K medium did not drop below levels incompatible with
the viability of the organism. It must be concluded that some
unknown soluble component of the leucite mineral tested is toxic,
even in very slight amounts, for the organism.

The results of the bioattack experiments with microorganisms
isolated from leucitic soils are likewise negative: leucite is,
nevertheless, not toxic for these organisms and it is quite
probable that other strains yield a better leucite dissolution.

The results of the experiments carried out with moulds
were very encouraging; among the latter, *Scopulariopsis brevicaule* appeared to be the most efficient, having produced a
notable potassium solubilization. The other two moulds, even
though they produced a less pronounced solubilization, have
contributed to throwing light to a certain extent on the possible
mechanism of (bio)solubilization and to determine its present
limitations. A first significant indication is constituted
by the fact that in the initial phase the dissolution rates were
the same for both the moulds and the organic acids mixture
whereas, after one month, they were practically reduced to zero
or, as shown by Figs. 5 and 6, the potassium contents of the
solutions showed a decrease.

The continuous slowing down of the potassium solubilization
rate observed in both the cultures and the acid mixture might
denote that (a) leucite dissolution takes place by indirect

chemical attack, as already assumed by Goni, et al. (17) and by Keil[1] in the case of biodegradation of other minerals in the presence of heterotrophs and (b) the solubilization rate is progressively slowed down by the formation of a protective (inhibiting) coating on the surfaces of the mineral particles. The latter explanation is consistent with the results of a research by Kruger (45), who carried out decomposition experiments with leucite in an electrodialysis apparatus, demonstrating that the potassium ions at first go rapidly into solution and, after a protective residue-layer has been formed, migrate only very slowly. According to Correns (46), this thin decomposition layer of SiO_2 and Al_2O_3 measures, at pH 3 with particles of radius 1 μm, some 0.03 μm and does not increase as the experimental time is prolonged. Furthermore, since relatively little aluminum, as compared to potassium and silicon, is dissolved out, Correns assumes that the residue layer should be rich in aluminum.

The biodegradation process would thus suffer, in this case, from the same drawbacks which have so far rendered useless the efforts of the investigators who endeavoured to develop a chemical process of leucite exploitation.

It cannot be excluded that *Scopulariopsis brevicaule* gives rise to direct attack. However, attempts made so far to ascertain that this kind of attack exists have up to now been unsuccessful, because of the intimate mixture of the biomass with the mineral particles.

The maximum shown by the curves of Figs. 5 and 6 and the subsequent decreasing trends might be explained by the subtraction of potassium from the solutions by the organisms. Experiments are being carried out presently to illuminate this subject, even though it is reasonable to believe that, from the viewpoint

---

[1] *Paper to be published in the Proceedings of the "Round Table on Leaching", Braunschweig, 1977.*

of the application of the process on an industrial scale, large volumes of biomass might create practical problems.

As far as the drawback represented by the slowing down in the solubilization rate is concerned, a remedy can be found by re-grinding the ore in order to disencrust the surfaces of the parti-cles and to create new fresh surfaces. It must not be forgotten, however, that every comminution step increases substantially the overall processing cost.

Investigations are presently under way aimed at throwing light on these problems. Research aimed at the isolation of sili-cobacteria capable of leucite and, more generally, feldspar bio-degradation, is also being continued.

VI. ACKNOWLEDGMENTS

The encouragement of Prof. Mario Carta, Director of the Min-ing and Mineral Dressing Institute of the University of Cagliari, is gratefully acknowledged. The author would also like to thank Prof. Paolo Piga, Director of the Mining Institute of Rome Univer-sity for the discussions which led to the initiation of the inves-tigation and for supplying the leucite concentrate samples. It is furthermore a great pleasure to thank Prof. Antonio Spanedda, of the Microbiology Institute of the University of Cagliari, who, throughout the years of investigation, gave the author the support of his knowledge and encouragement. Thanks are also due to Prof. Giovanni Alfano, Mining and Mineral Dressing Institute of Cagliari University, for performing the X-ray fluorescence spectrometer analyses, to Prof. Antonio Massidda, Industrial and Applied Chem-istry Institute, University of Cagliari, for performing the X-ray diffraction determinations, and to Antonella Rossi for the coop-eration in performing analytical work. The investigation was financed by the Consiglio Nazionale delle Ricerche, Centro Studi Geominerari e Mineralurgici, Engineering Faculty, University of Cagliari.

VII. REFERENCES

1. Deer, W., Howie, R.A., and Zussman, M.A., "Rock forming minerals", Vol. 4, p. 276. Longmans, Green & Co., Ltd., London, 1963.

2. Andreatta, C., "Mineralogia e cristallografia", p. 853. Libreria Universitaria L. Tinarelli, Bologna, 1957.

3. Washington, H.S., "The Roman comagmatic region". The Carnegie Institution, Washington, 1906.

4. Washington, H.S., *Met. Chem. Eng.*, 18, 65 (1918).

5. Strappa, O., *Ind. Mineraria (Rome)*, 25, 258 (1974).

6. Spada, A., *Ind. Mineraria (Rome)*, 10, 11 (1969).

7. Spada, A., *Ind. Mineraria (Rome)*, 11, 711 (1960).

8. Blanc, J.A., in "Atti Congr. Naz. Chim. Ind.," Milan, 1924.

9. Blanc, J.A., in "Atti Soc. Ital. Progr. Sci., XVI Riunione,". Perugia, 1927.

10. De Grazia, S., and Camiola, G., *Le stazioni sperimentali agrarie italiane*, 39, 829 (1906).

11. Bassalik, K., *Zeitschrift für Gärungsphysiologie*, 2, 1 (1913).

12. Department Laboratories du B.R.G.M., *Bull. Bur. Rech. Geol. Minières (Fr).*, Sect. 2, 5, 35 (1971).

13. Goni, J., Gugalski, T., and Sima, M., *Bull. Bur. Rech. Geol. Minières (Fr).*, Sect. 4, 1, 31 (1973).

14. Boyle, J.R., Voigt, G.K., and Sawhney, B.L., *Science*, 155, 193 (1967).

15. Vernadskii, V.I., "Ocierchi geochimii" ("A treatise on Geochemistry") in Russian, Moscow, 1934.

16. Alecsandrof, V.G., Tesisi dokladov na 5-ii naucnoy conferenzii Kuybicescova s.-x. in-ta. 27-30 Ianvariia 1937 g. Kuybiscev, 1947.

17. Alecsandrof, V.G., *Dok. Vses. Akad. Sel'skokhoz. Nauk.*, 3, 34 (1949).

18. Alecsandrof, V.G., *Dok, Vses. Akad. Sel'skokhoz. Nauk.*, 12, 12 (1949).

19. Alecsandrof, V.G., and Zac, G.A., *Mikrobiologiya*, 19, 97 (1950).

20. Norchina, S.P., and Pumpianscaia, L.V., *Dok. Vses. Akad. Sel'skokhoz. Nauk.*, 27 (1956).

21. Surman, K.J., *Byul. Nauchn.-Tekhn. Inform. po s-ch Mikrobiol.*, 2 (1956).

22. Surman, K.J., *Docl. Vaschnil.*, 4 (1958).

23. Surman, K.J., *Byul. Nauchn.-Tekhn. Inform. po s-ch Mikrobial.*, 5 (1958).

24. Surman, K.J., *Sb. Bacterialni udobriva*, Kiev, 1959.

25. Surman, K.J., *Sb. Bacterialni udobriva*, Kiev, 1961.

26. Surman, K.J., Avtoreferat Dissert., Moscow, 1962.

27. Tesic, Z., and Tudorovic, M., *Zemljiste i biljka*, I.1, 3 (1952).

28. Tesic, Z., and Todorovic, M., in "Atti Congr. Intern. Microbiol. 2°-Rome, 1953", 6, 356 (1955).

29. Tesic, Z., and Todorovic, M., *Zemljiste i biljka*, VIII, 233 (1958).

30. Alecsandrof, V.G., *Tr. Konf. po Vopr. Poch, Microbiol. Kompleksa Dokuchaeva-Kostycheva-Vil'yamsa, Inst. Mikrobiol., Akad. Nauk SSSR*, Moscow, 1951, 271 (1953).

31. Alecsandrof, V.G., *Tr. Sovesh. po Vopr. Bacter. Udobr.*, 1956, Kiev, Izv. Akad. Nauk SSSR, 62 (1958).

32. Alecsandrof, V.G., Burangulova, MN., and Solov'eva, E.P., *Tr. Odessk. Sel'skokhoz. Inst.*, 13, 91 (1958).

33. Alecsandrof, V.G., and Ternovskaia, M.J., "A methodic manual for liquid preparation of silicate bacteria", *Tr. Odessk. Sel'skokhoz. Inst.*, 14 (1961).

34. Alecsandrof, V.G., Ternovskaia, M.J., and Blagodir, R.N., *Vest. Sel'skokhoz. Nauki, Vses. Akad. Sel'skokhoz. Nauk SSSR*, 11, 95 (1963).

35. Alecsandrof, V.G., and Kodra, V.A., *Dokl. Akad. Sel'skokhoz Nauk SSSR*, 1, 10 (1965).

36. Heinen, W., *Arch. Mikrobiol.*, 37, 199 (1960).

37. Heinen, W., *Arch. Mikrobiol.*, 45, 172 (1967).

38. Ventriglia, U., *Ind. Mineraria (Rome)*, 1, 371 (1950).

39. Piga, P., *Boll. Tecn. Ing. e Arch. Sardi*, 3-4 (1967).

40. Fontanive, F., and Massacci, P., *Ind. Mineraria (Rome)*, 9, 279 (1968).

41. Rossi, G., *Resoconti Ass. Miner. Sarda*, (Iglesias), 76, 5 (1971).

42. Rossi, G., *Resoconti Ass. Miner. Sarda*, (Iglesias), 74, 5 (1969).

43. Silverman P., and Lundgren, D.G., *J. Bacteriol.*, 77, 642 (1959).

44. Correns, C.W., *Kolloid-Z*, 34, 341 (1924).

45. Krüger, G., *Chem. Erde*, 12, 236 (1939).

46. Correns, C.W., *Clay Minerals Bull.*, 4, 249 (1961).

OPTIMUM CONDITIONS FOR LEACHING OF URANIUM AND OXIDATION OF LEAD SULFIDE WITH *Thiobacillus ferrooxidans* AND RECOVERY OF METALS FROM BACTERIAL LEACHING SOLUTION WITH SULFATE-REDUCING BACTERIA

Noboru Tomizuka
Mitsuo Yagisawa

Fermentation Research Institute

Chiba, Japan

The optimum conditions for leaching of uranium and oxidation of lead sulfide in the presence of T. ferrooxidans *and the recovery of metals from bacterial leaching solution with sulfate-reducing bacteria were investigated on a laboratory scale.*

The optimum leaching of uranium occurred when the oxidation-reduction potential increased to above 0.75V and the pH decreased to below 2.0. The continuous culture and the continuous leaching of uranium were carried out in an apparatus for single stage continuous culture. 'Wash out' and the highest output of cells and ferric irons were observed at the dilution rate 0.10/hr and 0.062/hr respectively and over 80% of the uranium could be leached at a dilution rate below 0.011/hr.

In the virtual absence of iron, the microbiological oxidation of galena was limited. With addition of iron to the medium, the rate of oxidation and the percentage oxidation could be increased until other factors became limiting. Iron in amounts corresponding to 10% of the galena on a molar basis was found to be most effective.

A mixed culture of sulfate-reducing bacteria exhibited superior ability to recover metals from bacterial leaching solution. A recovery treatment employing continuous culture was effective. Batch culture was impossible. The optimum pH was 6.0, which agreed with that for growth. The maximum rate of metal removal for copper, zinc and iron was 47.5mg/ℓ/hr, 18.1mg/ℓ/hr and 25.7mg/ℓ/hr, respectively. Content of copper, zinc and iron in the black precipitate was about 6.9, 38.0 and 1.0 times that of the original sulfide ore, respectively.

I. INTRODUCTION

Heap or dump leaching and in-place leaching have become today standard methods in extractive metallurgy. Hydro-metallurgical processes are playing an increasingly important role in the extraction of both the common and the rare metals. On the other hand it is now widely recognized that certain kinds of microorganisms play an important role in dissolution of metals from ore and in precipitation of metals from metal solution.

The microorganisms involved in these phenomena are currently used or have been suggested for use in two hydro-metallurgical processes, namely, the dissolution of the desired metal from ore into solution and the recovery of the desired metal from the metal solution or mine drainage. The first of those processes called bacterial leaching is used in the following extractive processes of metallurgy: 1) in-place leaching, 2) dump or heap leaching, 3) leaching in stirred or agitated reactors.

The present paper is a summary of a number of published articles and largely concerned with the establishment of the optimum conditions for bacterial uranium leaching and bacterial lead oxidation on a laboratory scale. Brief mention of recovery of metals from a bacterial leaching solution with sulfate-reducing bacteria is also made.

II. METHODS, RESULTS, DISCUSSION, AND CONCLUSION

A. Bacterial Leaching of Uranium (1,2)

The investigation was initiated in order to find those microorganisms which are able to leach significant quantities of

Table I   Isolation of Iron and/or Sulfur
Oxidizing Microorganisms

Strain	No. 4Fe	No. 11Fe	No. 19Fe	No. 2S	No. 11S
Size (micron)	0.5 X 1.5	0.5 X 2.0	0.5 X 1.5	0.5 X 1.0	0.5 X 2.0
Opt. temp. (°C)	27	27 - 30	30	27	27 - 30
Opt. initial pH		2.5 - 3.0	2.5	2.0 - 3.0	3.0
Gram reaction		−	−	−	−
Flagellum		−	−	+	+
NaNO$_3$	+	+	+	−	++
(NH$_4$)$_2$SO$_4$	+++	+++	+++	+++	+++
Liquid medium					
FeSO$_4$ 7H$_2$O	+	++	++	−	−
S	−	++	+	++++	++++
Solid medium					
FeSO$_4$ 7H$_2$O	+	+	+	−	−
Na$_2$S$_2$O$_3$ 5H$_2$O	+	+	+	+	+

uranium from its ores and to clarify which of them are applicable
to in-place leaching of uranium.

The Ningyo-toge uranium deposit used for this study is of the
sedimentary type and is composed of conglomerate and arkose sandstone just above the unconformity plane. The ore contains a tetravalent uranium mineral called ningyoite (CaU(PO$_4$)$_2$ nH$_2$O) which
occurs coating the surface of pebbles and cementing the matrices
of the conglomerates. Pyrite is intimately associated in concentration with ningyoite and is usually contained at less than 1.0%.
The grade of the uranium ore is generally 0.03 to 0.15% U$_3$O$_8$.

1. Isolation and Selection of a Microorganism which Has Efficient
    and Stable Uranium Leaching Ability

Isolation of sulfur and/or ferrous iron oxidizing microorganisms from ores and mine waters of the Ningyo-toge mine was mainly
carried out in a modified 9K medium (composition of medium
(g/L): FeSO$_4$·7H$_2$O, 50.0; KH$_2$PO$_4$, 0.5; MgSO$_4$ 7H$_2$O, 0.5; KCl, 0.1;
(NH$_4$)$_2$SO$_4$, 3.0; Ca(NO$_3$)$_2$, 0.01; pH 3.0). Some of the properties
of the five strains which were isolated are shown in Table I. No.
11Fe had the most efficient and stable uranium leaching ability
among them and exhibited many characteristics which were similar

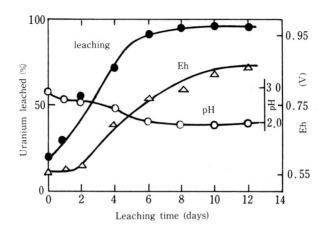

Fig. 1. Uranium leaching by T. ferrooxidans *at 25°C. The incubation flask contained 10g ore and 100ml of 0.03% ammonium sulfate solution (pH 3.0) and was shaken on a rotary shaker.*

to those of *Thiobacillus ferrooxidans*.

Figure 1 shows a typical uranium leaching curve of strain No. 11Fe on a medium in which 0.03% ammonium sulfate alone was dissolved.

2. *Role of the Microorganism on Uranium Leaching*

As shown in Fig. 1, over 90% of the uranium was leached when the pH decreased to below 2.0 and the oxidation-reduction potential increased to above 0.75V. According to the Eh - pH diagram of $U-O_2-H_2O-(S-Na-CO_2)$, the area of the pH and the Eh is that of uranyl sulfate.

Dialysis cultivation was carried out to test whether direct microbiological action on the ores was necessary for uranium leaching or not. In this cultivation, no microorganisms but only the metabolic products were allowed to diffuse through a cellulose membrane from the culture chamber to the ore containing reservoir. As shown in Fig. 2, the pH was slightly decreased by the addition of sulfuric acid to a range suitable for bacterial growth. The decrease in the pH caused by the metabolic product was accompanied

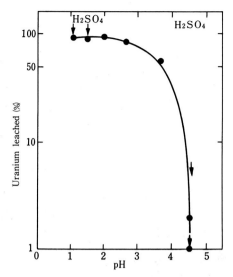

Fig. 2. *Uranium leaching on the modified 9K medium by* T. ferrooxidans *in the dialysis flask. The culture chamber contained 100ml medium (pH 4.5) and was inoculated with 2.0ml of* T. ferrooxidans *maximum culture. The ore containing reservoir contained 80g ore and 700ml of energy source-free medium (pH 4.5). By the addition of sulfuric acid the pH was slightly decreased to a suitable pH range (below pH 4.0) for the growth. Arrows show the points at which sulfuric acid was added. Cellulose membrane was used as the semipermeable membrane.*

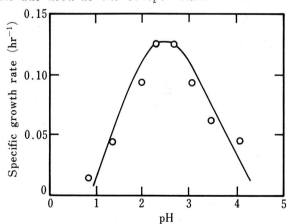

Fig. 3. *Effect of pH on specific growth rate. Cultivation was conducted on the modified 9K medium at 30°C and at 8.5 X 10^{-7} g mole O_2/ml min atm.*

by an increase in the leached uranium concentration of the solution. This result shows that the microorganism does not directly attack uranium mineral, but rather that it sets up chemical conditions suitable for the dissolution of uranium.

From these results it is suggested that the microorganism plays an indirect role in the supply of ferric iron and sulfuric acid and these oxidizing agents oxidize tetravalent uranium to the acid-soluble hexavalent state. Thus, the leaching processes of uranium mineral from the Ningyo-toge ore is thought to be as follows:

$$CaU(PO_4)_2 + Fe_2(SO_4)_3 + H_2SO_4 + 2H_2O$$
$$\longrightarrow UO_2SO_4 + 2FeSO_4 + 2H_3PO_4 + CaSO_4$$

$$4FeS_2 + 15O_2 + 2H_2O \xrightarrow{Bacteria} 2Fe_2(SO_4)_3 + 2H_2SO_4$$

$$4FeSO_4 + 2H_2SO_4 + O_2 \xrightarrow{Bacteria} 2Fe_2(SO_4)_3 + 2H_2O$$

3. *Optimum Culture Conditions and Continuous Culture of* T. ferrooxidans

All experiments were carried out in a 2-liter fermentor with 1-liter working volume on the modified 9K medium.

In batch culture it was observed that there was a parallel relationship between the increase of ferric iron concentration and of cell numbers. As shown in Figs. 3, 4 and 5, for the specific growth rate, the optimum pH, temperature and aeration were between 2.3 and 2.7, between 29 and 34 °C and above 5.0×10^{-7} g mole O_2/ml min atm, respectively. Ammonium salts, urea and certain kinds of amino acids were used as nitrogen sources. Of these, ammonium sulfate was the most effective for the growth and its optimum concentration was 0.3%. The optimum concentration of ferrous sulfate ($FeSO_4$ $7H_2O$) as the energy source was 5%. The organism was capable of growing on glucose in the presence of an inorganic energy source. The growth was inhibited by externally added keto acids having relatively small pKa value such as pyruvic acid, α-

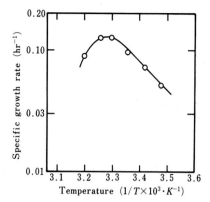

Fig. 4.   Effect of temperature on specific growth rate.
Cultivation was conducted on the modified 9K medium at pH 2.7 and
at 8.5 X 10⁻⁷ g mole O_2/ml min atm.

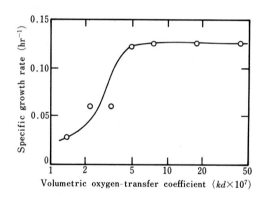

Fig.5.   Effect of aeration on specific growth rate.   Cultiva-
tion was conducted on the modified 9K medium at pH 2.7 and at 30°C.

keto-glutaric acid and maleic acid.

Continuous culture was conducted in a single stage system.
The fermentor was first run batchwise for a few days and then was
shifted to continuous operation.   Fig. 6 presents the effect of
the dilution rate on cell growth and the oxidation of ferrous iron.
'Wash out' and the highest output of cells and ferric iron were
observed at the dilution rate of 0.10/hr and 0.062/hr, respectively.   The curves illustrated in Fig. 6 show that it is not possible
to predict continuous culture behavior from batch culture data.

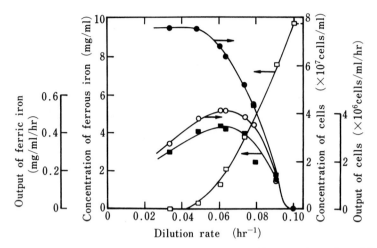

Fig. 6. Effect of dilution rate on cell growth and oxidation of ferrous sulfate. Cultivation was conducted on the modified 9K medium. Temperature, pH and aeration were 30°C, 2.7 and 8.5 X 10⁻⁷ g mole O₂/ml min atm, respectively. ● concentration of cells; ○ output of cells; □ concentration of ferrous iron; ■ output of ferric iron.

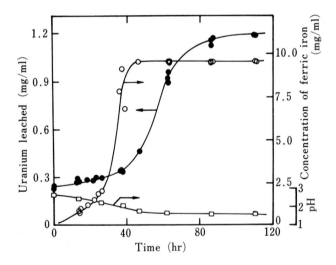

Fig. 7. Relationship between ferrous oxidation and uranium leaching. Leaching of uranium was conducted on the modified 9K medium with 10% ore at 30°C. ○ oxidation of ferrous iron; □ pH; ● concentration of uranium leached.

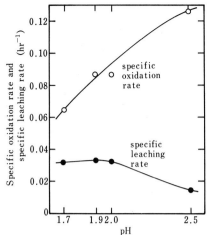

*Fig. 8. Effect of pH on specific oxidation rate and specific
leaching rate. Leaching of uranium was conducted on the modified
9K medium with 10% ore at 30°C. ● specific leaching rate of
uranium; ○ specific oxidation rate of ferrous iron.*

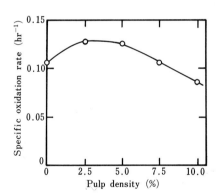

*Fig. 9. Effect of pulp density on specific oxidation rate
of ferrous iron. Leaching of uranium was conducted on the
modified 9K medium at 30°C and at pH 2.0.*

The reason has not yet been determined.

4. *Optimum Conditions for Bacterial Leaching of Uranium and
Continuous Leaching of Uranium*

The relationship between the characteristics of the uranium
ore and the necessary amount of pyrite or ferric iron for uranium
leaching could not be determined. The ore used in this study
contained less pyrite than necessary in order to carry out effi-

cient uranium leaching in a short incubation time. In general the addition of 5%(W/V) ferrous sulfate ($FeSO_4 \cdot 7H_2O$) was sufficient to induce bacterial uranium leaching from the Ningyo-toge uranium ore. Therefore, the modified 9K medium was used throughout the tests. Fig. 7 shows the time course curves of uranium leaching and ferrous oxidation. The increase of leaching ratio was not directly related to the ratio of ferrous oxidation, but rather, as already demonstrated in Fig. 1, it was found that low pH and high oxidation-reduction potential are essential to the uranium leaching.

Effect of pH on specific oxidation rate of ferrous iron and specific leaching rate of uranium is illustrated in Fig. 8. The optimum pH for uranium leaching was about 2.0 and there was absence of a parallel relation between uranium leaching and ferrous oxidation in the terms of specific rate.

Effect of pulp density on specific oxidation rate of ferrous iron is presented in Fig. 9. At low pulp densities, uranium ore contributed in a positive manner to the oxidation of ferrous iron added. The reason why the addition of ore has this promoting effect is not clear.

Continuous leaching of uranium was carried out with 10% ore at pH 2.0 and at 30°C. Data on the continuous leaching of uranium at different dilution rates is shown in Fig. 10. The highest output of uranium leached was observed at the dilution rate of about 0.060/hr, which was numerically, nearly equal to the dilution rate of the highest output of cells and ferric irons (Fig. 6). But, to leach over 80% of the uranium contained in ore employed, it was necessary to operate the continuous leaching at a dilution rate of under 0.011/hr. However, from Figs. 6 and 8, it is seen that continuous leaching with high recovery ratio of uranium at higher dilution rate may be feasible by conducting the continuous leaching in two-vessels with the continuous culture carried out in a first vessel and the continuous leaching in a second vessel.

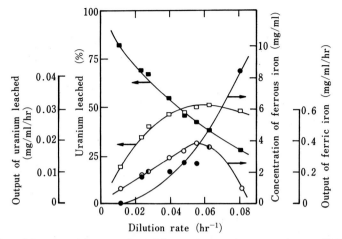

Fig. 10. Continuous leaching of uranium on the modified 9K medium. Leaching of uranium was conducted with 10% ore at pH 2.0 and at 30° C. ■ *ratio of uranium leached;* □ *output of uranium leached;* ● *concentration of ferrous iron;* ○ *output of ferric iron.*

5. Conclusion

It has been demonstrated that *Thiobacillus ferrooxidans*, which has efficient and stable uranium leaching ability and is naturally present in the Ningyo-toge uranium deposit, is capable of continuously leaching uranium in stirred and/or agitated reactors. However, no consideration has yet been given to the problem of large-scale operation. At any rate, it is not believed that in-place leaching and dump (heap) leaching are practically applicable to Ningyo-toge uranium ore. The reasons are as follows;

(1) The Ningyo-toge uranium deposit is of small scale.

(2) The ore contains less pyrite than necessary in order to carry out efficient uranium leaching.

(3) There are many sandstone layers in the uranium deposit which are impervious to the leaching solution.

(4) The ore is so rich in clay that it is difficult to supply air into the deposit or an artificial ore heap.

(5) At low pH there was observed decomposition of the clay

cementing cracks of the unconformity plane. This means the leaching solution would escape through the cracks.

B. Microbiological Oxidation of a Lead Sulfide (3)

The oxidation of lead sulfide (galena) in the presence of *T. ferrooxidans* has been investigated. Information regarding the biological oxidation of lead sulfide is scarce and conflicting conclusions have been reported.

In this study, lead sulfide (galena) was obtained from the carboniferrous limestone deposits of South Wales (U. K.) and was crushed to pass a 300-mesh B. S. screen. The sample was found to contain 84.5 − 86.2% lead and less than 0.1% iron. Oxidation tests were carried out in Erlenmeyer flasks on 100ml of the modified 9K medium with lead sulfide replacing ferrous sulfate as an energy source. The temperature was maintained at 35 C and the pH was manually controlled by daily adjustment back to an initial pH value with sulfuric acid or sodium hydroxide during the oxidation process. Oxidized lead was measured by atomic absorption spectrophotometry after it had been dissolved into the aqueous phase by treating samples with 10% diethylenetriamine (DETA) solution at room temperature.

1. *Oxidation of Galena in an Iron-free Medium*

The ability of *T. ferrooxidans* to use galena as an energy source in a medium in the virtual absence of iron was measured by monitoring the oxidation of galena and the bacterial growth. Fig. 11 shows typical results with different initial concentrations of bacteria. The results illustrate the ability of *T. ferrooxidans* to participate in the oxidation of galena and use it as an energy source for growth. It can be seen that the rate of oxidation is growth associated and the oxidation curve shows a logarithmic increase with culture time.

Optimum conditions for galena oxidation were as follows:

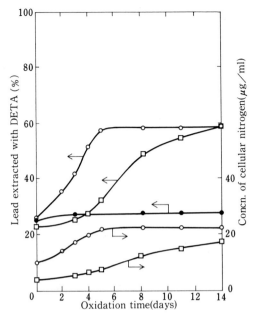

Fig. 11. *Oxidation of galena by* T. ferrooxidans *and its growth in an iron-free medium. Reaction conditions were as follows: volume of medium, 100ml; galena, 2.5g; pH, 2.5; inoculum as cellular nitrogen,* □ *0.49mg,* ○ *0.98mg,* ● *0mg.*

(1) A particle size for the galena of less than 50 microns, i. e. -300 mesh B. S. screen.

(2) A solids concentration lower than 4 grams/100ml of medium.

(3) A pH value for the tests of 2.0.

As shown in Fig. 12, the pH of the culture was found to have a pronounced effect. At pH values lower than 1.3 the percentage oxidation of galena was found to be the same in the inoculated flask as in the control.Although the inoculum size could also be seen to have a definite effect on the oxidation process, the final percentage oxidation was the same for all inoculum sizes.

Under these conditions when an inoculum of 0.0068mg/ml of medium as cellular nitrogen was used, the maximum percentage oxidation obtained was 60.0% and the cell yield per gram of galena oxidized was 1.6mg as cellular nitrogen. The doubling time of the microbe was found to be 4.2 days and consequently the specific

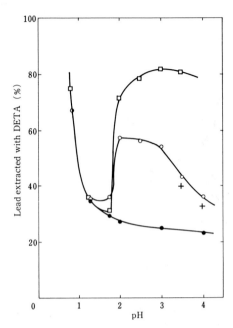

Fig. 12. Effect of pH on final percentage oxidation in an iron-free medium and in an iron added medium. Reaction conditions were as follows: volume of medium, 100ml; galena, 2.5g; □ *in a 0.001 moles iron added medium with microbes (as cellular nitrogen, 0.48mg);* ○ *in an iron-free medium with microbes (as cellular nitrogen, 0.79mg);* ● *in an iron-free medium without microbes. Crosses show the results are those obtained after the incubation for 8 days.*

oxidation rate and the specific growth rate were 6.9 X 10^{-3}/hr.

2. Oxidation of Galena in an Iron Added Medium

The marked influence of iron on the oxidation response of galena is illustrated by the oxidation curves in Fig. 13. They show that, once, the maximum percentage oxidation of galena was obtained in the absence of iron, further oxidation of the galena was produced when iron was added. Addition of ferrous sulfate to the control produced no increase in the amount of galena oxidized.

The optimum concentration of iron was found to be 0.001 moles of iron per 100ml of medium. Iron additions in excess of the optimum concentration were inhibitory. The reason may be that, as the

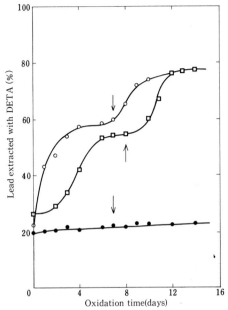

Fig. 13. *Effect of iron on oxidation of galena. Reaction conditions were as follows: volume of medium, 100ml, ○ pH 2.0, 2.3g galena and 2.64mg microbes as cellular nitrogen; □ pH 3.0, 2.5g galena and 0.66mg microbes as cellular nitrogen; ● pH 2.0, and 2.3g galena without microbes. Arrows show the points at which 0.00l moles ferrous sulfate was added.*

iron is oxidized to the ferric form, a large amount of it precipitates as ammonium ferric sulfate which contributes to the oxide coating of the galena surface limiting the oxidation process.

Figure 14 illustrates the relationship observed between the galena oxidation and the oxidation of iron from the ferrous to the ferric form. The rate of oxidation of galena was enhanced and the final percentage oxidation was much higher than in corresponding tests without iron additions. The amount of extracted lead was found to increase logarithmically while ferrous iron was oxidized with the increasing oxidation rate but was not completely oxidized until after the galena had ceased to oxidize.

Typical galena oxidation curves and the growth curves on the medium with iron additions are shown in Fig. 15. The oxidation of

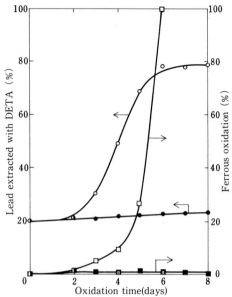

Fig. 14.   Relationship between galena oxidation and ferrous oxidation.   Reaction conditions were as follows; volume of medium, 100ml; galena, 2.3g; pH, 3.0; iron, 0.001 moles ferrous sulfate; inoculum as cellular nitrogen, 0.44mg, ○ , □ with microbes, ● , ■ without microbes.

galena was also found to be associated with bacterial growth in a similar manner to oxidation tests performed with iron-free medium. The final percentage oxidation was approximately 82.0% for most of the tests performed.

The optimum conditions in the iron added medium at 35°C were nearly the same as observed in the iron-free medium, except that optimum pH was about 3.0 as shown in Fig. 12. For a typical test performed at pH 3.0 with 2.5g of galena and an inoculum size of 0.0088mg of cellular nitrogen per ml of solution in the presence of 0.001 moles of ferrous sulfate, the maximum percentage oxidation obtained was 83.5%, the doubling time of the microbe was 1.6 days (specific growth and oxidation rate were 1.8×10^{-2}/hr) and the cell yield per gram of galena oxidized was 1.8 mg as cellular nitrogen.

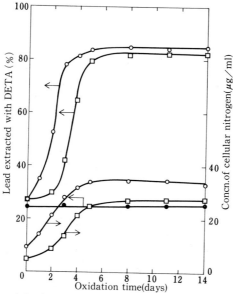

Fig. 15.   Oxidation of galena by T. ferrooxidans *and its growth on iron added medium.   Reaction conditions were as follows: volume of medium, 100ml; galena, 2.5g; pH, 3.0; iron, 0.001 moles ferrous sulfate; inoculum as cellular nitrogen,* □ *0.43mg,* ○ *0.86mg,* ● *0mg.*

3.   Mechanism of Oxidation of Galena

As shown in Table II, the formation of sulfur as predicted by equation (1) was noted by an extraction (with carbon disulfide) of residue obtained from control tests with galena in the presence

$$PbS + H_2SO_4 + 1/2O_2 \longrightarrow PbSO_4 + H_2O + S \qquad (1)$$

and absence of iron. The incomplete chemical oxidation of galena is probably due to the formation of lead sulfate and sulfur reaction layers masking the reactive sulfide surface. In oxidation tests performed upon galena in the presence of bacteria under experimental conditions identical to those of the control tests, it was found that no sulfur was detected in the tests without iron additions and only very small quantities in the tests with iron additions.

The role of the bacteria in the oxidation of galena in the

*Table II Formation of Sulfur During Oxidation of Galena
in the Presence and Absence of* T. ferrooxidans

Reaction conditions				Oxidation ratio[4] (%)	Sulfur (mg)	
Galena[1]	Microbe[2]	Iron[3]	pH		Expected[5]	Found
2.5g	−	−	2.0		23.6	19.8
2.5g	+	−	2.0	59.1	154.0	0
2.5g	−	+	3.0	25.4	42.9	32.6
2.5g	+	+	3.0	79.1	219.0	8.4

*(1) Galena sample less than 45 micron (minus 400-mesh) in
particle size of which 12.3% had been oxidized to lead
sulfate during preparation.*
*(2) Inoculum size used in tests with microbes 0.88mg as cellular
nitrogen;*
*(3) Concentration of ferrous sulfate employed 0.001 moles/100ml
medium;*
(4) Incubation period, 10 days;
(5) Sulfur as calculated from equation (1) and (2).

absence of iron would appear to lie in bacterial oxidation of any
sulfur formed during the reaction expressed by equation (1) thus
maintaining the acid level and thus removing sulfur from the sulfide surface allowing further oxidation of sulfide to proceed.
Therefore the extent of final percentage oxidation of galena
depends mainly on the acid reaction expressed by equation (1).

The precise reason why the rate of oxidation and the final
percentage oxidation of galena in the presence of iron with bacteria were much higher than those tests without iron is not immediately apparent. However, it is suggested that one of the roles of
iron in the medium may be explained as the role of enhancing chemical oxidation of the galena producing elemental sulfur, as

$$PbS + Fe_2(SO_4)_3 \longrightarrow PbSO_4 + 2FeSO_4 + S \qquad (2)$$

expressed by equation (2), and it may also be involved in increasing the activity of the bacterial oxidizing enzyme system. It is
also suggested that ferric iron produced by the oxidation of ferrous iron in the medium acts as an 'oxygen carrier' which promotes

the oxidation of the less accessible surface area of the galena particles which is unable to react with sulfuric acid and oxygen in the manner expressed by the reaction in equation (1).

4. *Conclusions*

It has been demonstrated that *T. ferrooxidans* can be actively employed in the enhancement of the oxidation of lead sulfide in relatively weak acid solutions and that the presence of iron also improves both the oxidation rate and final percentage oxidation. It is possible that such reactions may be useful as a preliminary step in the recovery of lead from sulfide concentrates via an amine extraction route.

C. Recovery of Metals from a Bacterial Leaching Solution with Sulfate-reducing Bacteria (4)

The study is related to the recovery of metals from bacterial leaching solution and acid mine drainage, which are characterized by high concentrations of sulfate, metals and hydrogen ion.

The mixed culture of sulfate-reducing bacteria used in this study was isolated from a soil sample obtained near our Institute. The basal medium was as follows (g/L): Na_2SO_4, 1.0; $CaCl_2$ $6H_2O$, 0.1; $MgSO_4$, 1.0; $(NH_4)_2SO_4$, 1.0; $FeSO_4$ $7H_2O$, 0.11; Na-glycerophosphate, 0.1; yeast extract, 0.1. Sulfate was supplied by Na_2SO_4 salt solution or the supernatant of bacterial leaching solution which was prepared from copper sulfide ore with *T. ferrooxidans* (medium (g/L): $(NH_4)_2SO_4$, 1.0; K_2HPO_4, 0.1; $MgSO_4$ $7H_2O$, 0.1; KCl, 0.05). Sodium lactate was used as an energy source and the amount thereof added to inoculated solutions was calculated from the following reaction:

$$2CH_3CHOHCOO^- + HSO_4^- \longrightarrow HS^- + 2CH_3COO^- + 2CO_2 + H_2O$$

The cultures were incubated in a 1-liter fermentor with 0.5-liter working volume at 30°C throughout the experimental work.

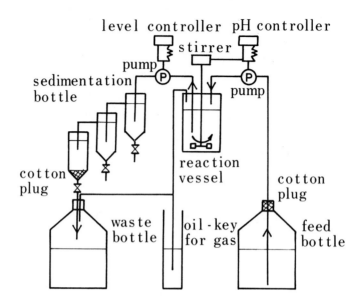

Fig. 16. Test procedure for continuous recovery of metals from bacterial leaching solution.

1. Continuous Culture of Sulfate-reducing Bacteria

Batch culture of sulfate-reducing bacteria on the supernatant of bacterial leaching solution was impossible, because the low pH and high concentration of metals of the supernatant were obstructive to growth. However, the bacteria grew in batch culture on basal medium and the pH of the incubated solution gradually increased during the growth.

Continuous culture was conducted in a single stage system as shown in Fig. 16. The fermentor was first run batchwise with the basal medium and then was shifted to continuous operation with the supernatant of bacterial leaching solution. Namely, the addition of the supernatant for the adjustment back to an initial pH value during the cultivation means the continuous feeding of fresh

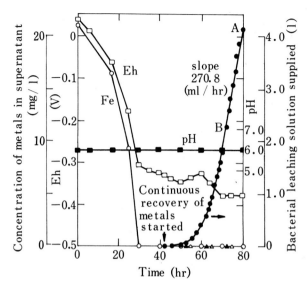

Fig. 17. Continuous recovery of metals from bacterial leaching solution with sulfate-reducing bacteria. The supernatant contained 55.0ppm Cu, 32.0ppm Fe and 20.5ppm Zn. Continuous recovery was conducted at 30°C. O Fe; △ Cu; ▲ Zn.

medium to the incubated solution. Agitation with a magnetic stirrer was carried out only during the period of pH adjustment. To maintain constant volume in the reaction vessel, the cells and products were removed at the same rate as the feed rate.

Figure 17 shows the results obtained in one example of continuous culture. The steady state was achieved within 40 hours. The maximum feed rate of bacterial leaching solution was 270.8ml/hr, which was calculated from the slope between A and B.

2. *Effect of pH and Metal Concentration on Growth Rate and Rate of Removal of Metals*

The specific growth rate and the rate of removal of metals from bacterial leaching solution were found to be strongly influenced by the pH of the culture. As shown in Fig. 18, the optimum pH for metal removal was 6.0, which agreed with the optimum pH for specific growth.

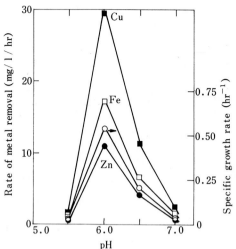

Fig. 18. Effect of pH on metal removal rate and specific growth rate. Cultivation was conducted at 30 C. The solutions with different pH values each contained Na-lactate (10.3g/l) and yeast extract (0.1g/l) and had a metal concentration of 55.0ppm Cu, 32.0ppm Fe and 20.5ppm Zn. ■,●,□ rate of metal removal; ○ specific growth rate.

Figure 19 shows the influence of metal concentration of bacterial leaching solution on the rate of metal removal and the specific growth rate. In the continuous culture at optimum pH, the maximum rate of metal removal was obtained at 40% metal concentration of the original bacterial leaching solution (Cu, 270.0mg/l; Zn, 102.5mg/l; Fe, 135.0mg/l) and for copper, zinc and iron the maximum rate of metal removal was 47.5mg/l/hr, 18.1mg/l/hr and 25.7mg/l/hr, respectively.

The maximum specific growth rate observed in the continuous culture was 0.541/hr. The value was faster than the value which was observed in batch culture. The reason for this phenomenon is not apparent. It is suggested, however, that the bacterial growth rate was accelerated as a result of the formation of harmless metal sulfide from hydrogen sulfide and metal ions which exert a deleterious influence on microorganisms and as a result of the fact that a mixed culture was used.

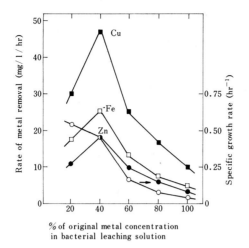

*Fig. 19. Effect of metal concentration of bacterial leaching
solution on metal removal rate and specific growth rate.
Cultivation was conducted at 30°C and each diluted solution
contained Na-lactate (10.3g/l) and yeast extract (0.1g/l).
Metal concentration of original supernatant of bacterial leaching
solution was 270.0mg/l Cu, 102.5mg/l Zn and 135.0mg/l Fe.
■,●,□ rate of metal removal; ○ specific growth rate.*

3. Characteristics of Precipitate and Conclusion

There was obtained a black precipitate containing copper,
zinc and iron at 19.96%, 6.13% and 10.95%, respectively. The
ratios among the metals in the precipitate were similar to those
among the metals in the supernatant of the bacterial leaching
solution and the concentrations of the metals in the precipitate
were about 6.9, 38.0 and 1.0 times those of the original copper
sulfide ore.

Although no data was obtained concerning larger systems, it
seems possible to conclude from these results that, if the problem
concerning energy source can be solved, it should be possible to
use sulfate-reducing bacteria for the recovery of certain kinds of
metals from bacterial leaching solution and acid mine drainage.

III. REFERENCES

1. Tomizuka, N., and Takahara, Y., in " Proc. IV. Intern. Ferment. Symp. (Kyoto), Ferment. Technol. Today " (G. Terui, Ed.), p. 513-520. Society Fermentation Technology, Osaka, Japan, 1972.

2. Tomizuka, N., Yagisawa, M., Someya, J., and Takahara, Y., *Agr. Biol. Chem.*, 40, 1019 (1976).

3. Tomizuka, N., *Report of the Fermentation Research Institute*, 48, 51 (1976).

4. Yagisawa, M., Murakami, Y., Kato, Y., Tomizuka, N., Yamaguchi, M., and Ooyama, J., *Journal of the Mining and Metallurgical Institute of Japan*, 93, 447 (1977).

BIOGENIC EXTRACTION OF URANIUM

FROM ORES OF THE GRANTS REGION

Corale L. Brierley

New Mexico Bureau of Mines and Mineral Resources
Socorro, NM 87801 U.S.A.

Uranium ores from the Anaconda Co. Jackpile and Paute orebodies were leached using Thiobacillus ferrooxidans, Thiobacillus thiooxidans, *enrichment cultures of mesophilic, iron-oxidizing bacteria,* Sulfolobus acidocaldarius, *and a* Sulfolobus-*like organism. The latter two organisms are extremely thermophilic acidophiles. The four test ores ranged between 0.13% and 0.56% U_3O_8 with uranium mineralization occurring as coffinite, oxidized uranium complexes, uraninite, and organo-uranium complexes. The ores varied in pyritic and organic matter content with some acid-consuming gangue present. Leaching tests were conducted in airlift percolator columns, batch reactor flasks, and drip-leach columns in which biologically-generated leach solutions were applied to the uranium ores. Eh measurements, iron oxidation kinetics, and uranium dissolution suggested that leaching was not enhanced by bacteria in direct contact with the ore. Supplementing inoculated ores with ferrous iron did not increase biogenic U_3O_8 dissolution. Growth and manometric studies indicated that bacterial growth and function were suppressed in the presence of some ore samples. Bacterial suppression by the ores may have resulted from the presence of toxic agents or the strong reducing environment.*

I. INTRODUCTION

Bacterial leaching of pyritic uranium ores is both feasible
and economical. The role of *Thiobacillus ferrooxidans* in the
extraction of uranium is probably indirect and confined to the
generation of the oxidizing agent, ferric sulfate, and the
solvent, sulfuric acid, accordingly:

$$2FeS_2 + 7\tfrac{1}{2}O_2 + H_2O \longrightarrow Fe_2(SO_4)_3 + H_2SO_4$$

Reduced uranium is oxidized by ferric sulfate to acid soluble
hexavalent uranium:

$$UO_2 + Fe_2(SO_4)_3 \longrightarrow UO_2SO_4 + 2FeSO_4$$

T. ferrooxidans then regenerate ferric sulfate as follows:

$$2FeSO_4 + \tfrac{1}{2}O_2 + H_2SO_4 \longrightarrow Fe_2(SO_4)_3 + H_2O$$

This bacterial process has been used as a scavenger operation
(1, 2, 3, 4) to obtain uranium values from mined-out areas.
The Agnew Lake Mines Ltd. in northern Ontario is now using
this bacterial process as the principal means for *in-situ*
recovery of uranium (5).

The uranium ore of the Grants uranium belt in New Mexico
is less amenable to bacterial leaching because of the low pyrite
content. The primary mineralization is coffinite ($USiO_4 \cdot nH_2O$),
uraninite (UO_2), organo-uranium complexes, and some oxidized
forms of uranium (6). These uranium deposits are associated
with highly insoluble organic material (7) which has an infrared
spectra resembling coal (W. L. Blankenship, unpublished data).

This paper describes manometry experiments and leach studies
on uranium ores from the Anaconda Co. Jackpile and Paute Mines
near Grants, New Mexico. *T. ferrooxidans*, *T. thiooxidans*, an
iron-enrichment culture obtained from the Jackpile-Paute Mines,
Sulfolobus acidocaldarius, and *Sulfolobus*-like organisms were

used in the studies. The latter two organisms are the extremely thermophilic, acidophilic bacteria which inhabit hot, acid springs (8, 9) and have been shown to enhance metals extraction from recalcitrant ores (10, 11).

II. MATERIALS AND METHODS

A. Ores

Five ore types -- J11, J27, P6B, P92, and SP46 -- from the Anaconda Co. Jackpile and Paute Mines, located 35 miles east of Grants, New Mexico, were sized to -1 mm + 425 μm (-16 + 35 mesh) and analyzed for U_3O_8 content. J-designated ores were collected from the Jackpile Mine; P-designated ores were from the Paute Mine, and the SP46 ore was collected from a stockpile of blended Jackpile and Paute ores.

B. Bacteria

Used in the studies were *Thiobacillus thiooxidans* (American Type Culture Collection # 8085), *Thiobacillus ferrooxidans* (ATCC # 19859), a mesophilic iron-oxidizing enrichment culture from the Jackpile-Paute mining district, *Sulfolobus acidocaldarius* (8), and a *Sulfolobus*-like bacterium (9).

C. Leach Solutions

Bryner and Anderson medium (12) and 9K medium (13), made up with distilled water rather than iron solution, were used as leach solutions. The initial pH was 2.5.

Energy sources and supplements included: 78 mM flowers of sulfur, sterilized separate from the medium by intermittent steaming for 30 min on 3 consecutive days; 36 mM acidified, $FeSO_4 \cdot 7H_2O$, sterilized separate from the medium at 18 psia, 121°C for 15 min; and 0.02% and 0.2% yeast extract (Difco) for *Sulfolobus*-like bacteria and *Sulfolobus acidocaldarius*, respectively.

Chemical control experiments were sterilized with 0.2 mM
panacide (2,2-methylene bis (4-chlorophenol))(ICN).

D. Analyses

U_3O_8 analysis of the leached ore (tails) was completed by
tributyl phosphate extraction (TBP) (14). Spectrophotometric
interferences were encountered when leach solutions were analyzed
by the TBP method.

Ferrous iron was determined by dichromate titration (15),
and total iron was assayed by a Perkin Elmer 303 atomic
absorption spectrophotometer.

pH and Eh were measured with a Corning Model 7 pH meter.

The Folin-phenol method (16) was used for protein determination, and bacterial numbers were estimated with the 3-tube
most probable number (MPN) technique (17). Oxygen uptake was
measured with a Gilson Differential Respirometer according to
standard methods (18). One-g ore samples were reacted with 5 ml
leach solution for 8-12 hr before inoculation and oxygen uptake
monitoring to allow for carbonate neutralization and CO_2
dissipation. Ores and ores supplemented with 36 mM Fe (II) were
tested.

E. Leaching Experiments

The leachability of the ores was tested in 250-ml
Erlenmeyer flasks with 100 ml leach solution and a 1% to 5% pulp
density of ore. Batch reactors were incubated at 25°C for the
mesophilic bacteria and 60°C for thermophilic bacteria up to
32 days. After the incubation period, the reactor contents were
examined microscopically, and the pH and Eh values of the solutions were measured. In some cases Fe (II) concentrations were
determined, and U_3O_8 analysis was completed on the tails.

Percolator columns, 3.8 cm by 25 cm with 100 g of unsterilized ore, were operated at room temperature for experimental
studies. Recycled over the ore was 250 ml Bryner and Anderson
medium (12) with 36 mM Fe(II). Test columns were inoculated with
the iron-enrichment culture, and control columns were sterilized
with Panacide. Samples were collected weekly for Eh and pH
measurements and Fe(II) analysis. After completion of the
experiment, the tails were analyzed for residual U_3O_8.

Spent media from the growth of *T. thiooxidans* and *T.
ferrooxidans* on Bryner and Anderson medium (12) were applied in
500-ml aliquots at an approximate rate of 1.0 ml per min to
100-g samples of uranium ore in 3.8 cm by 25 cm columns.
Influent solutions (spent media) and effluent solutions from the
columns were analyzed for Eh, pH, number of bacteria, and Fe(II)
when iron was initially used in the medium. The tails from the
"drip-leach" columns were assayed for U_3O_8 content.

III. RESULTS

A. Ores

The -1 mm + 425 μm (-16 + 35 mesh) ores had the following
U_3O_8 content: J11, 0.28%; J27, 0.56%; P6B, 0.13%; P92, 0.20%;
and SP46, 0.41%. Non-sulfate sulfur content ranged from 0.22%
for high carbon ores to 0.1% for refractory ores (J. Grunig,
personal communication). Total iron content in the ores was as
follows: J11, 0.7%, J27, 1.1%; P6B, 0.4%; P92, 0.8%; and
SP46, 0.8%.

B. Manometry Experiments

Figure 1 shows Q_{O_2} (μl oxygen uptake/mg protein/hr) by
T. ferrooxidans in the presence of J11 and P6B ores. Oxygen
uptake was greatly reduced when these ores served as sole energy

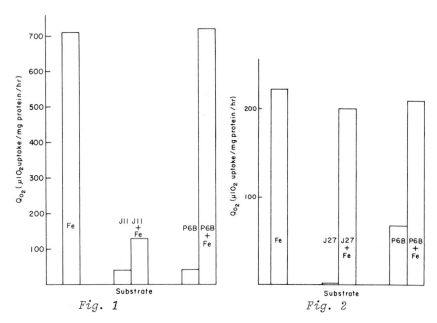

Fig. 1. The effect of uranium ores, J11 and P6B, and iron-supplemented ores on oxygen uptake by T. ferroxidans.

Fig. 2. The effect of uranium ores, J27 and P6B, and iron-supplemented ores on oxygen uptake by Sulfolobus.

sources. If J11 and P6B ores were supplemented with Fe(II), oxygen uptake by *T. ferrooxidans* in the presence of P6B was equivalent to oxygen uptake observed when Fe(II) was the energy source. Oxygen uptake by *T. ferrooxidans* remained depressed when J11 ore was supplemented with iron.

Figure 2 illustrates results obtained on the oxygen uptake of a mixed population of *Sulfolobus acidocaldarius* and *Sulfolobus*-like organisms in the presence of J11 and P6B ores. When J27 served as the sole energy source, no oxygen uptake was noted; however, when P6B ore was the sole energy source, some oxygen uptake was observed. If J27 and P6B ores were supplemented with Fe(II), oxygen uptake was nearly equivalent to that

observed with iron as the only energy source; the presence of
the ores did not suppress iron oxidation.

TABLE I   pH, Eh, and Microscopic Data
from Control and T. ferrooxidans-Inoculated Uranium
Ores after 10-Day, Shake-Flask Test

		Control			T. ferrooxidans	
Ore	pH	Eh(+mv)	Micro^a	pH	Eh(+mv)	Micro^a
J11	3.8	634	0	3.7	736	0
J27	3.4	656	0	3.4	744	0
P6B	3.5	574	0	3.5	736	0
P92	3.7	613	0	3.7	741	0

^{a}Subjective microscopic evaluation -- 0 to +4

C. Batch Reactor Studies

The following results summarize data collected from
stationary and shake-flask leach studies.

1. *T. ferrooxidans*

Table I displays Eh, pH, and microscopic data collected
after 10 days of *T. ferrooxidans* growth on uranium ores in shake
flasks with Bryner and Anderson medium. The pH increased in
both control and inoculated flasks, and an increase in Eh was
observed for inoculated ores. After 10 days of incubation, no
T. ferrooxidans could be observed microscopically in inoculated
shake flasks.

If the uranium ores and Bryner and Anderson medium were
supplemented with Fe(II) (Table II), a pH increase was again
observed, and higher Eh increases were noted in *T. ferrooxidans*-
inoculated shake flasks than in comparable iron-poor ores. An
exception was the P92 ore.

*TABLE II   pH, Eh, and Microscopic Data
from Control and* T. ferrooxidans-*Inoculated, Iron-
Supplemented Uranium Ores after 10-Day, Shake-Flask Test*

	Control			T. ferrooxidans		
Ore	*pH*	*Eh (+mv)*	*Micro*[a]	*pH*	*Eh (+mv)*	*Micro*[a]
J11	*3.4*	*563*	*0*	*3.6*	*817*	*+3*
J27	*3.2*	*564*	*0*	*3.4*	*819*	*+3*
P6B	*3.2*	*565*	*0*	*3.5*	*814*	*+3*
P92	*3.3*	*563*	*0*	*3.3*	*516*	*+3*

[a]*Subjective microscopic evaluation -- 0 to +4*

A microscopic examination of inoculated leach solutions after 10 days of incubation revealed large numbers of organisms.

When pyrite was added as an energy supplement, *T. ferrooxidans* growth was noted, and Eh increased over control reactors. The pH values of both control and inoculated reactors were between 3.5 and 4.0.

*TABLE III   pH, Eh, and Microscopic Data
from Control and* T. ferrooxidans-*Inoculated, Pyrite-
Supplemented Uranium Ores after 10-Day, Shake-Flask Test*

	Control			T. ferrooxidans		
Ore	*pH*	*Eh (+mv)*	*Micro*[a]	*pH*	*Eh (+mv)*	*Micro*[a]
J11	*3.8*	*575*	*0*	*3.9*	*724*	*+2*
J27	*3.5*	*607*	*0*	*3.7*	*756*	*+2*
P6B	*3.7*	*586*	*0*	*3.9*	*749*	*+2*
P92	*3.8*	*574*	*0*	*3.6*	*753*	*+3*

[a]*Subjective microscopic evaluation -- 0 to +4*

*TABLE IV   pH, Eh, Fe(II) and Microscopic Data
from Control and Enrichment-Inoculated Uranium
Ores after 32-Day, Shake-Flask Test*

		Control				Iron-enrichment		
Ore	pH	Eh(+mv)	Micro[a]	Fe(II) (mM)	pH	Eh(+mv)	Micro[a]	Fe(II) (mM)
J11	2.6	360	0	0.36	2.6	535	+2	0.36
J27	2.6	410	0	0.18	2.6	510	+1	0.36
P6B	2.6	370	0	0.36	2.6	535	+2	0.18
P92	2.8	345	0	0.54	2.7	425	+2	0.90

[a]*Subjective microscopic evaluation -- 0 to +4*

2. Iron-enrichment Culture

An iron-oxidizing enrichment culture, obtained from the Jackpile-Paute uranium district, grew on uranium ores with no supplemental iron added (Table IV). Eh values were only somewhat higher than those observed for uninoculated controls, and soluble Fe(II) concentrations were similar for inoculated and uninoculated reactors.

If Fe(II) was added to the leach solutions and uranium ores, good growth of the iron-enrichment culture occurred; the Eh was greatly increased over control reactors, and added iron was oxidized (Table V).

3. *Sulfolobus* Organisms

Mixed cultures of *Sulfolobus acidocaldarius* and *Sulfolobus*-like organisms were grown on uranium ores not amended with additional energy sources (Table VI). The pH values were variable for the ores with J11 exhibiting the greatest pH increase. Eh values for inoculated reactors were not substantially higher than Eh values for control reactors. Uranium was not extracted from J11 and P6B ores, and uranium extraction values

TABLE V   pH, Eh, Fe(II), and Microscopic Data
from Control and Enrichment-Inoculated, Iron-Supplemented
Uranium Ores after 32-Day, Shake-Flask Test

	Control				Iron-enrichment			
Ore	pH	Eh(+mv)	Micro[a]	Fe(II) (mM)	pH	Eh(+mv)	Micro[a]	Fe(II) (mM)
J11	2.3	350	0	40.32	2.3	630	+3	1.08
J27	2.3	355	0	35.84	2.3	645	+3	0.18
P6B	2.3	355	0	39.07	2.3	640	+3	1.08
P92	2.3	350	0	40.50	2.3	655	+3	1.43

[a]Subjective microscopic evaluation -- 0 to +4

from J27 and P92 ores were nearly the same for inoculated and control reactors.

When the ores and 9K medium were supplemented with Fe(II), (Table VII), the pH values of control and inoculated reactors were maintained between pH 2.3 and 3.1. Eh values did not increase for inoculated reactors, and Fe(II) concentrations were similar for both control and inoculated flasks. The percents U_3O_8 extracted were nearly the same for both control and Sulfolobus-inoculated reactors.

Additional leaching experiments with Sulfolobus organisms included supplementing the uranium ores and solutions with elemental sulfur, yeast extract, yeast extract and iron, pyrite, and yeast extract, and pyrite. None of these supplements or combinations of supplements measurably enhanced biological U_3O_8 extraction.

TABLE VI   pH, Eh, Microscopic and U_3O_8
Extraction Data from Control and Sulfolobus-Inoculated
Uranium Ores after 10-Day, Stationary, Batch Reactor Test

Ore	Control				Sulfolobus			
	pH	Eh(+mv)	Micro[a]	%U_3O_8[b]	pH	Eh(+mv)	Micro[a]	%U_3O_8[b]
J11	6.5	564	0	0	6.6	514	+2	0
J27	2.6	604	0	33	2.6	614	+3	33
P6B	2.9	574	0	0	2.7	624	+2	0
P92	3.0	544	0	10	3.5	544	+3	25

[a] Subjective microscopic evaluation -- 0 to +4
[b] Percent U_3O_8 extracted from ore

D. Percolator Leach Column Studies

Uranium ores were leached in percolator columns with unsterilized ore inoculated with the iron-enrichment culture; control columns were sterilized with Panacide. When P6B ore was

TABLE VII   pH, Eh, Fe(II), and U_3O_8 Extraction Data
from Control and Sulfolobus-Inoculated, Iron-Supplemented
Uranium Ores after 10-Day, Stationary, Batch Reactor Test

Ore	Control				Sulfolobus			
	pH	Eh(+mv)	Fe(II) (mM)	%U_3O_8[a]	pH	Eh(+mv)	Fe(II) (mM)	%U_3O_8[a]
J11	3.1	529	ND[b]	11	3.0	549	ND[b]	18
J27	2.5	574	1.21	55	2.5	574	1.16	56
P6B	2.4	564	1.48	39	2.3	564	1.31	23
P92	2.8	ND[b]	1.48	45	2.6	ND[b]	0.90	45

[a] Percent U_3O_8 extracted from ore
[b] Not determined

leached, it was noted that iron was not oxidized by the enrichment culture until approximately 32 days after the leaching
experiment was initiated (Fig. 3). Bacterial counts at day 21
indicated about 10^3 iron-oxidizing microbes per ml of slurry.
At this time Eh increased in the inoculated column. Iron was
not oxidized in the control column, and the Eh remained at about
+600 mv for the entire experiment. The pH values for both columns were about 3.0. After 54 days of leaching the percents
U_3O_8 extracted were 92% and 95% for control and inoculated
columns, respectively. Similar results were noted for pH, Eh,
Fe(II) concentrations, and U_3O_8 extraction for control and
enrichment-inoculated P92 uranium ore.

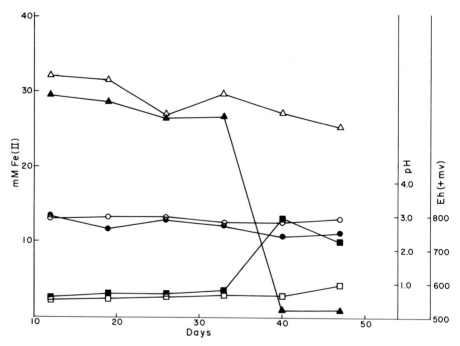

*Fig. 3. Changes in pH, Eh, and ferrous iron concentrations
during column leaching of iron-supplemented P6B ore. Iron-
enrichment culture:* ▲-Fe, ●-pH, ■-Eh; *Control:* △-Fe, ○-pH,
□-Eh.

Figure 4 illustrates the pattern for pH, Eh, and Fe(II)
concentrations during control and enrichment- culture leaching
of SP46 uranium ore. The Eh values in the control and inoculated
columns remained between +500 mv and +550 mv, and the pH
stabilized at about 3.0 for both columns. At day 9,counts
indicated that about 10 iron-oxidizing bacteria per ml were
present. The Fe(II) concentration decreased in both columns at
day 14 with the decrease occurring at a slightly lower rate in
the control column. The final Fe(II) concentration in both
columns was about 18 mM. The U_3O_8 extraction after 42 days of
leaching was the same for both columns -- 96%.

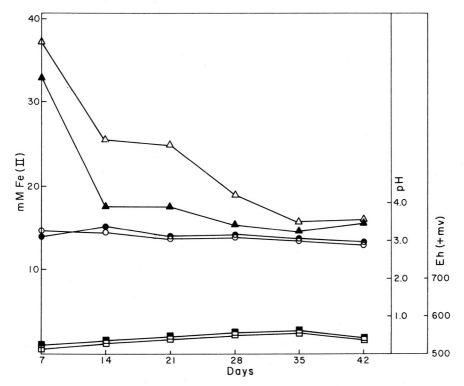

*Fig. 4. Changes in pH, Eh, and ferrous iron concentrations
during column leaching of iron-supplemented SP46 ore. Iron-
enrichment culture: ▲-Fe, ●-pH, ■-Eh   control: △-Fe, o-pH,□-Eh.*

In leach column tests of 2 weeks duration similar pH, Eh, and Fe(II) oxidation patterns were observed. The U_3O_8 dissolution ranged from 65% to 95% with little variation between inoculated and control columns.

E. "Drip Leach" Column Tests

Ferric iron and sulfuric acid leach solutions, generated by *T. ferrooxidans* and *T. thiooxidans*, were applied to columns containing 100 g of uranium ores. The results from these leach experiments were compared with results from experiments in which the same volumes of uninoculated leach solutions were applied to the ores.

Table VIII, part A summarizes data collected on the generated lixiviants which were applied to two columns of J11 ore. *T. ferrooxidans* generated a Fe(II) concentration of 18 mM, and 1.6×10^3 organisms per 100 ml (MPN) of solution were present. The Eh in the inoculated solution had not increased substantially. No Fe(II) was present in the control leach solution. Table VIII, part B summarizes data from the column effluent after reaction of the leach solutions with J11 ore and the percent U_3O_8 extracted from the ore. Most notable were the increases in pH and the 100% extraction of U_3O_8 from the ore leached with the biologically-generated lixiviant.

Data from similar experimentation with *T. thiooxidans*-generated leach solution are summarized in Table IX. The inoculated lixiviant had a somewhat lower pH than the control solution (Table IX, part A). Using the inoculated leach solution 41% of the U_3O_8 was extracted from J11 ore as compared with 14% when control medium was applied to the ore (Table IX, part B).

The application of *T. ferrooxidans*-generated leach solution with 34 mM Fe(II) (Table X, part A) to SP46 ore yielded 90% U_3O_8. The release of U_3O_8 from SP46 ore leached with control medium

TABLE VIII   Drip Leach Results (A) Chemical and Biological Data of T. ferrooxidans-*Generated and Control Influent Leach Solutions, and (B) Chemical and Biological Data of Column Effluent Solutions and Percent U_3O_8 Extraction from J11 Ore*

(A)	pH	Eh(+mv)	Fe(II) (mM)	Fe(TOT) (mM)	MPN/100 ml
Control	2.3	610	36	36	0
T. ferrooxidans	2.5	644	18	36	1600

(B)	pH	Eh(+mv)	Fe(II) (mM)	Fe(TOT) (mM)	MPN/100 ml	%U_3O_8[a]
Control	3.6	555	31	52	0	39
T. ferrooxidans	2.8	660	8	31	15	100

[a]*Percent U_3O_8 extracted from ore*

containing 2 mM Fe(II) was 63%.

IV.   DISCUSSION AND CONCLUSIONS

Manometric studies with *Sulfolobus* on uranium ores indicated limited or no oxygen uptake (Fig. 2), however, if the ores were amended with Fe(II), oxygen uptake by *Sulfolobus* was equivalent to that observed with Fe(II) only. Although the ores appeared to be deficient in oxidizable energy sources, the ores were apparently not toxic to these organisms in short term experiments. *Sulfolobus* did grow on unamended uranium ores, but uranium extraction was not enhanced (Table VI). This suggests that the increased temperature and acid medium are sufficient to leach some of the uranium. If the ores were amended with Fe(II), it was found that *Sulfolobus* cultures neither actively oxidized the iron nor was uranium extraction enhanced in inoculated batch

TABLE IX   *Drip Leach Results (A) Chemical and Biological Data of* T. thiooxidans-*Generated and Control Influent Leach Solutions, and (B) Chemical and Biological Data of Column Effluent Solutions and Percent* U_3O_8 *Extracted from J11 Ore*

(A)	pH	Eh(+mv)	MPN/100 ml	
Control	2.6	636	0	
T. thiooxidans	2.3	629	250	
(B)	pH	Eh(+mv)	MPB/100 ml	$\%U_3O_8{}^a$
Control	3.9	605	0	14
T. thiooxidans	3.8	539	10	41

^{a}*Percent* U_3O_8 *extracted from ore*

TABLE X   *Drip Leach Results (A) Chemical and Biological Data of* T. ferrooxidans-*Generated and Control Influent Leach Solutions, and (B) Chemical and Biological Data of Column Effluent Solutions and Percent* U_3O_8 *Extracted from SP46 Ore*

(A)	pH	Eh(+mv)	Fe(II)(mM)	Fe(TOT)(mM)	
Control	2.3	573	34	36	
T. ferrooxidans	2.7	794	4	36	
(B)	pH	Eh(+mv)	Fe(II)(mM)	Fe(TOT)(mM)	$\%U_3O_8{}^a$
Control	3.8	683	30	35	63
T. ferrooxidans	2.7	474	3	28	90

^{a}*Percent* U_3O_8 *extracted from ore*

reactors (Table VII). Since the *Sulfolobus* organisms are mixotrophic, it is possible that sufficient oxidizable organic matter was present to provide growth, and iron was not oxidized by the organisms.

Manometric studies with *T. ferrooxidans* indicated that J11 and P6B ores did not contain a sufficient source of energy for demonstrable oxygen uptake (Fig. 1). If ferrous iron was added to the ores, *T. ferrooxidans* showed oxygen uptake comparable to oxidation of iron only, but oxygen uptake with J11 ore was greatly suppressed. This indicates that J11 ore may contain substances inhibitory to iron oxidation by *T. ferrooxidans*. In a 10-day batch reactor study in which ores were amended with Fe(II), *T. ferrooxidans* growth was observed (Table II). It is possible that in long-term tests the organisms do adapt to conditions which may initially be inhibitory and would be observed in manometric studies.

Iron-enrichment cultures which had been isolated from uranium mining areas were used in leach studies. Presumably these organisms would be more adapted than the ATCC *T. ferrooxidans* to uranium ores. These organisms in contrast to *T. ferrooxidans* did grow on unamended uranium ores (Table IV). When the ores were amended with iron, good growth was observed, and iron was readily oxidized (Table V). These batch reactor studies were incubated for longer time periods than comparable tests with *T. ferrooxidans*, so direct comparison is not possible. The enrichment culture was used in percolator column tests. The Fe(II) in amended P6B ore was oxidized by the culture (Fig. 3), but it required 32 days. Uranium extractions in the control and inoculated columns were the same, and similar uranium extraction results were observed for SP46 ore. These results indicate that leaching of uranium ores with the organisms associated with the ore does not appear to enhance uranium extraction. The development of the organisms or their ability to oxidize a substrate is suppressed.

When leach solutions, which had been generated by bacterial activity, were applied to ores, uranium extraction was increased over that observed when uninoculated media were applied. The increased uranium extraction resulted from the increased Fe(II) content of *T. ferrooxidans* activity and from increased acidity of *T. thiooxidans* growth.

Acid consuming material, lack of energy source, and slow development or activity of the bacteria within the ore all indicate that bacteria cannot be used effectively for leaching of the Grants uranium ores. When biologically-generated leach solutions are applied to the Grants uranium ores, increases in U_3O_8 solubilization are noted, but these laboratory studies are preliminary and do preclude extrapolation to industrial leach systems. It may be conceivable to heap or *in-situ* leach ores with biologically-generated lixiviants. Further study is necessary to determine if recycled leach solutions would carry inhibitory factors and a cost analysis would be necessary to establish the economics of the system.

V.   ACKNOWLEDGMENTS

The author would like to express her appreciation to the Anaconda Co. for their financial support of this study and their permission to publish the results.

VI.   REFERENCES

1.   MacGregor, R.A., *Min. Eng.*, 21, 54 (1969).
2.   MacGregor, R.A., *Can. Min. Metall. Bull.*, 59, 583 (1966).
3.   MacGregor, R.A., *Nucl. Appl.*, 6, 68 (1969).
4.   Fisher, J.R., *Can. Min. Metall. Bull.*, 59, 588 (1966).
5.   McCreedy, H.H., Coordinated Research Committee Meeting of IAEA, Dec. 13-15, 1976. Coventry, Eng.

6. Granger, H.C., in "Geology and Technology of the Grants Uranium Region, Mem. 15," p. 21. New Mexico Bureau of Mines and Mineral Resources, Socorro, 1963.

7. Jacobs, M.L., Warren, C.G., and Granger, H.C., *U.S. Geol. Surv. Prof. Pap.*, 700-B, B 184 (1970).

8. Brock, T.D., Brock, K.M., Belly, R.T., and Weiss, R.L., *Arch Microbiol.*, 84, 54 (1972).

9. Brierley, C.L., and Brierley, J.A., *Can. J. Microbiol.*, 19, 183 (1973).

10. Brierley, C.L., *J. Less-Common Met.*, 36, 237 (1974).

11. Brierley, C.L., *Dev. Ind. Microbiol.*, 18, 273 (1976).

12. Bryner, L.C., and Anderson, R., *Ind. Eng. Chem.*, 49, 1721 (1957).

13. Silverman, M.P., and Lundgren, D.G., *J. Bacteriol.*, 77, 642 (1959).

14. Carl, F., *Anal. Chem.*, 30, 50 (1958).

15. Blaidel, W.J., and Meloche, V.W., "Elementary Quantitative Analysis--Theory and Practice," 2nd Ed., p. 964. Harper & Row, New York, 1963.

16. Lowry, O.H., Rosebrough, N.J., Farr, A.L., and Randall, R.J., *J. Biol. Chem.*, 193, 265 (1951).

17. Collins, C.H., "Microbiological Methods," 2nd Ed., p. 404. Plenum Press, New York, 1967.

18. Umbreit, W.W., Burris, R.H., and Stauffer, J.F., "Manometric and Biochemical Techniques," 5th Ed., p. 64. Burgess, Minneapolis, 1972.

MICROBIAL LEACHING OF CU-NI SULFIDE CONCENTRATE

H. M. Tsuchiya

University of Minnesota
Minneapolis, Minnesota, U.S.A.

Mutalism between Thiobacillus ferrooxidans *ATCC 19859 and* Beijerinckia lacticogenes *ATCC 19361 has an effect upon both the rate and extent of leaching copper and nickel from ore concentrates of the Duluth Gabbro.*

All living organisms are made up of carbon, hydrogen, oxygen, nitrogen, sulfur, phosphorus, and the other elements that comprise the protoplasm. Moreover, these occur in a definite stoichiometry; these are the minimal requirements of living systems. The stoichiometry of the environment from which the microorganisms selectively extract elements in the process of growth, differs from that of the microbial protoplasm. All organisms must therefore have an energy source to maintain the concentration gradients and the stoichiometry of the elements that make up the environment and the protoplasm.

Most of the known copper and nickel reserves in the United States are of the sulfide form. Numerous investigators have established the fact that the microorganisms play a role in the oxidation of insoluble sulfides to soluble sulfates. The thiobacilli produce the lixiviant. The more thiobacilli there are, the more lixiviant is produced and the faster the rate of and greater the extent of leaching. The addition of Beijerinckia *to the leaching liquor stimulates the fixation of nitrogen. It is proposed that the additional nitrogenous compounds stimulate the population density of thiobacilli. As the numbers of thiobacilli are increased, the amount of lixiviant is increased.*

I. INTRODUCTION AND PROCEDURES

Results of leaching tests on a mixed copper and nickel sulfide concentration from the Duluth Gabbro complex are reported here. The relative efficacies of pure cultures of *Thiobacillus ferroxxidans* and mixed cultures *T. ferrooxidans* and *Beijerinckia lacticogenes* are compared as possible leaching aids.

The sulfide concentrate studied was a combined copper and nickel rougher flotation concentrate supplied by the U.S. Bureau of Mines, Twin Cities Research Center, Twin Cities, Minnesota. The ore body is located in northeastern Minnesota and is called the Duluth Gabbro complex of low grade copper and nickel sulfides. The principal minerals are chalcopyrite, millerite, pyrrhotite, and pentlandite. The concentrate assayed 4.22% copper and 0.90% nickel and was ground to -270 mesh. Detailed chemical analysis is given in Table I.

TABLE I   *Chemical Analysis of the Copper-Nickel Concentrate*

Element	Percent	Compound	Percent
Cu	4.22	SiO_2	35.62
Ni	0.90	Al_2O_3	10.54
Fe	17.13	MgO	8.65
S	7.17	CaO	5.80
C	1.33	Na_2O	1.79
Co	0.04	TiO_2	0.99
		K_2O	0.42
		other	5.40

The part played by microorganisms in metallurgy is a well known fact. Leached copper was probably first recovered in relatively large amounts from Spanish mines. It was not until 1954, however, that the active role played by bacteria in this oxidation process was demonstrated. The bacteria involved in the oxidation of insoluble metal sulfides to soluble metal sulfates belong to the genus thiobacillus.

The bacterium most studied in context of its application in mineral processing is *T. ferrooxidans*. It is an aerobic, autotrophic, acidophilic, carbon dioxide-fixing microorganism which requires sulfur or ferrous iron as energy source for its growth and maintenance.

Vishniac and Santer (1957) (10), Kuznetsov, et al. (1963) (5), Pings (1968) (7), Beck (1969) (1), Zajic (1969) (11), Malouf (1970) (6), and Duncan (1972) (4), are among some who have given excellent accounts of properties and characteristics of thiobacilli.

From all published accounts of microbial leaching, it can be concluded that the following factors are among those important in leaching with the help of bacteria.

(a) An available source of substrate (mineral sulfate or ferrous iron or pyrite if the ore body is an oxide ore body).

(b) A supply of carbon dioxide which is the sole source of carbon for the thiobacilli.

(c) A supply of oxygen for the oxidation process.

(d) A supply of essential nutrients including nitrogen.

(e) Water as a carrier for the nutrients, source of trace elements, and as a solvent for the solvated metal.

(f) An acidic environment.

It is unnecessary to point out that in dump leaching, the organisms get almost all of the essential nutrients either from the ore or from the recirculating barren solution. In addition to the above factors, it is also recognized that temperature, particle size (surface area), and additives effect the rate of leaching.

All living organisms are made up of carbon, hydrogen, oxygen, nitrogen, sulfur, and phosphorous, and other trace elements that comprise the protoplasm; these occur in a definite stoichiometry. The stoichiometry is shown in Table II.

TABLE II *Approximate Elementary Composition of the microbial cell. S. E. Luria in the Bacteria (I.C. Gunsalus and R. Y. Stanier, eds.) Vol. 1 (New York, Academic Press, 1960)*

Element	Percent of Dry Weight
carbon	50
oxygen	20
nitrogen	14
hydrogen	8
phosphorus	3
sulfur	1

These are minimal requirements of living systems. The stoichiometry of the environment from which organisms selectively extract elements in the process of growth, differs from that of the protoplasm.

Since we are interested in the interactions of microbial forms and with their common environment, we considered the possibility of furnishing an organism which would supply nitrogen to thiobacillus.

Beijerinckia, an aerobic, acidophilic, nitrogen fixing, heterotrophic organism, characteristically non-fastidious with respect to its carbon source seemed to be the organism of choice. The acidophilic nature of *Beijerinckia* would make it suitable for the low pH conditions that exist in leaching systems. Starkey and De (1939)(9), and Becking (1961a and 1961b)(2,3), have reported on the habitat, isolation, and characteristics of these organisms. It seemed reasonable to believe that if these two organisms, a *Beijerinckia* and a *Thiobacillus*, could be grown together, the former would fix dinitrogen to satisfy the need of nitrogenous compounds for itself and thiobacillus and the latter would fix carbon dioxide for its and *Beijerinckia's* organic matter in a medium devoid of added carbon and nitrogen sources. This hypothesis is represented schematically in Fig. 1.

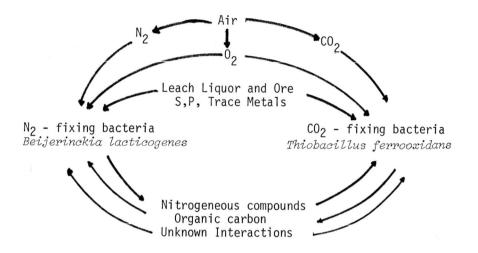

Both have total nutrient supply if
both are present.

Fig. 1.   Proposed mutualistic interaction of T. ferrooxidans
and B. lacticogenes *in a leaching environment
devoid of fixed carbon or fixed nitrogen.*

In this paper are our findings with *T. ferrooxidans,* ATCC
19859 and *B. lacticogenes* ATCC 19361. The pure cultures of
thiobacillus were maintained on the 9K medium of Silverman and
Lundgren (1959)(8). Stock cultures of *Beijerinckia* were maintained
on agar medium, the composition of which is given in Table III.

The *Beijerinckia* were systematically adapted to grow at low
pH values. This was accomplished by sequential transfers of
actively growing cells in flasks containing increasingly higher
concentrations of hydrogen ions.

Thiobacillus was also adapted to grow on the concentrate in
question in a nutrient medium consisting of the 9K basal salts of
Silverman and Lundgren (8), but not containing $(NH_4)_2SO_4$ or
$Ca(NO_3)_2$. The salts solution is shown in Table IV.

TABLE III   Composition of Agar Medium for Maintaining
Stock Cultures of the Nitrogen-Fixing Bacteria

Component	Concentration (g/l)
KH_2PO_4	0.5
$MgSO_4.7H_2O$	0.2
$CaCl_2$	0.1
$FeCl_3$	0.005
$Na_2MoO_4.2H_2O$	0.0005
Difco yeast extract	0.5
Sucrose	2.5
Agar	15.0
pH adjusted to	3.0

TABLE IV   Composition of the T Medium Used in
Leaching Tests

Component	Concentration (g/l)
K_2HPO_4	0.5
$MgSO_4.7H2O$	0.5
Na_2SO_4	0.15
KCl	0.1
$CaSO_4.2H_2O$	0.01
$Na_2MoO_4.2H_2O$	0.0005
pH adjusted to	2.5

Pure cultures of *T. ferrooxidans* and a mixed culture of *T.
ferrooxidans* and *B. lacticogenes* were thus obtained in flasks
containing no organic carbon source (only inorganic CO_2 of the
atmosphere) and no nitrogen (except N_2 of the atmosphere), NH_4^+
existing as trace impurities in chemicals, and NH_3 absorbed by
the medium at pH 2.5.

The leaching tests were carried out in Erlenmeyer flasks with
a 20% solid suspension of concentrate of 70ml of the nutrient
medium of the composition shown in Table IV.   10 ml inocula of
pure and mixed cultures were employed.   The inocula were trans-

ferred twice, prior to the inoculation, for 2 days each in the same medium as was used in the test so as to get actively growing cultures. All tests were run in duplicate and repeated to check precision. Chemical controls containing medium, but not bacteria, were run stimultaneously. The test flasks were incubated on a reciprocating shaker with a 5 inch stroke at 76 rpm at 26°C.

The experiments were monitored in leaching flasks by removing 10 ml aliquots at regular intervals. Metal analyses were carried out on a Perkin Elmer Model 303 atomic absorption spectrophotometer.

Intermittent microscopic examination of the leach flasks showed an increase of cells.

II. RESULTS AND CONCLUSIONS

Shown in Table V is the effect of % solids on leaching of copper with *T. ferrooxidans* and *T. ferrooxidans* plus *B. lacticognenes*. It can be seen that the amount of copper extracted was substantially heavier in the cultures with *Thiobacillus* and *Beijerinckia* than in the cultures with thiobacilli alone.

Shown in Table VI is the effect of solids on leaching of nickel with *T. ferrooxidans* and *T. ferrooxidans* in concert with *B. lacticogenes*. It can be seen that the amount of nickel extracted was substantially heavier in the culture with the two organisms than it was with one.

Leaching tests were carried out with unadapted pure and mixed cultures. In each case, extractions were less than achieved with adapted cultures. The tests were run in duplicate and repeated. Chemical controls were also carried out.

T. ferrooxidans oxidizes the insoluble sulfides occurring in ores to soluble sulfate. The more the number of *Thiobacillus*, owing to the nitrogen fixed by *Beijerinckia*, the greater the rate and more extensive the amount of copper and nickel leached.

TABLE V *Effect of % Solids on Leaching of Copper with*
T. ferrooxidans *and* T. ferrooxidans *plus* B. lacticogenes

% Solids	5		10		20		33	
Time (hours)	Thiob (g/l)	Thiob + Beij (g/l)	Thiob (g/l)	Thiob + Beij (g/l)	Thiob (g/l)	Thiob + Beij (g/l)	Thiob (g/l)	Thiob + Beij (g/l)
25	0.26	0.28	0.23	0.25	0.23	0.26	0.16	0.20
50	0.32	0.36	0.33	0.33	0.30	0.34	0.30	0.50
75	0.49	0.54	0.55	0.62	0.62	0.70	0.38	0.96
150	0.96	1.37	1.10	1.55	1.20	1.82	0.66	1.83
200	1.20	1.64	1.14	2.07	1.48	2.46	0.77	2.17
300	1.76	1.98	2.35	3.12	2.00	3.68	1.08	2.81

TABLE VI *Effect of % Solids on Leaching of Nickel with*
T. ferrooxidans *and* T. ferrooxidans *plus* B. lacticogenes

% Solids	5		10		20		33	
Time (hours)	Thiob (g/l)	Thiob + Beij (g/l)	Thiob (g/l)	Thiob + Beij (g/l)	Thiob (g/l)	Thiob + Beij (g/l)	Thiob (g/l)	Thiob + Beij (g/l)
25	0.20	0.22	0.27	0.03	0.30	0.34	0.31	0.50
50	0.25	0.26	0.34	0.37	0.40	0.47	0.58	0.81
75	0.33	0.35	0.50	0.54	0.63	0.62	0.82	1.17
150	0.44	0.45*	0.74	0.80	0.90	1.07	1.23	2.08
200	0.45*	0.45*	0.77	0.87	0.96	1.30	1.65	2.50
300	0.45*	0.45	0.83	0.90*	1.14	1.64	2.57	3.66

*
100% extraction

III REFERENCES

1. Beck, J.V., "*Thiobacillus ferrooxidans* and its Relation to the Solubilization of Ores of Copper and Iron", *Fermen. Adv.*, 747-771 (1969).

2. Becking, J.H., "Studies on Nitrogen-Fixing Bacteria of the Genus *Beijerinckia*. I. Geographical and Ecological Distribution in Soils", *Plant and Soil*, 14, 49-81 (1961a).

3. Becking, J.H., "Studies on Nitrogen-Fixing Bacteria of the Genus *Beijerinckia*. II. Mineral Nutrition and Resistance to High Levels of Certain Elements in Relation to Soil Type, *Plant and Soil*, 14, 297-322 (1961b).

4. Duncan, D.W. and Walden, C.C., "Microbiological Leaching in the Presence of Ferric Iron", *Develop. Ind. Microbiol.*, 13, 66-75 (1972).

5. Kuznetsov, S.I., Ivanov, M.V. and Lyalikova, N.N., "Introduction to Geological Microbiology", p. 252. McGraw-Hill New York, 1963.

6. Malouf, E.E., "Bioextractive Mining. In: SME Short Course in Bioextractive Mining". Society of Mining Engineers of AIME, New York, p. 1-45, 1970.

7. Pings, W.B., "Bacterial Leaching", *Colo. Sch. Mines Miner. Ind. Bull.*, 11(3), 1-19 (1968).

8. Silverman, M.P. and Lundgren, D.G., "Studies on the Chemoautotrophic Iron Bacterium *Ferroobacillus ferrooxidans*", *J. Bacteriol.*, 77, 642-647 (1959).

9. Starkey, R.L. and De, P.K., "A New Species of *Azotobacter*", *Soil Sci.*, 47, 329-343 (1939).

10. Vishniac, W. and Santer, M., "The Thiobacilli, *"Bacteriol. Rev.*, 12, 196-213 (1957).

11. Zajic, J.E., "Microbial Biogeochemistry", p. 345. Academic Press, New York, 1969.

COMPLEX LEAD SULFIDE CONCENTRATE

LEACHING BY MICROORGANISMS

A.E. Torma

New Mexico Institute of Mining and Technology
Socorro, New Mexico 87801 U.S.A.

Recent bacterial leaching studies indicate that zinc, copper, and cadmium values can selectively be removed from off-grade lead sulfide concentrates without producing any air pollution. The bacterial leaching technique is based on the difference in the solubility of the metal sulfates which are resulted in this process. The leach residue is an up-graded lead sulfide concentrate which is partially oxidized to sulfate and can be used directly in the classical smelting process for recovery of lead. The preliminary economic assessment presented in this study is encouraging to warrant further consideration.

I. INTRODUCTION

Lead is one of the earliest known elements by man (1). The Egyptians used it for glazing pottery as early as about 7 to 5 thousand B.C., the ancient Romans made water-pipes from it and the alchemists believed that it was associated with the planet of Saturn (2). Today, there is a variety of lead products available for use (3) including metal products (alloys, pipes, battery grids), pigments (white and colored) and chemicals (gasoline anti-knock additives).

The principal lead mineral is galena PbS (4). Its ore deposits are generally associated with varying quantities of iron, zinc, copper, silver, gold, antimony, arsenic and bismouth impurities. Other lead minerals are oxides, PbO, sulfates (anglesite, $PbSO_4$) and carbonates (cerussite, $PbCO_3$).

Metallic lead is produced from flotation sulfide bearing concentrates by the conventional smelting processes (2), which are based on the following reactions:

$$2PbS + 3O_2 \rightleftharpoons 2PbO + 2SO_2 \tag{1}$$

$$PbS + 2PbO \rightleftharpoons 3Pb^\circ + SO_2 \tag{2}$$

$$PbS + 2O_2 \rightleftharpoons PbSO_4 \tag{3}$$

$$PbS + PbSO_4 \rightleftharpoons 2Pb^\circ + 2SO_2 \tag{4}$$

The thermodynamics of these equilibria is well established (5,6). It is also possible to produce lead in one step, which can be derived from equations [1] and [2]:

$$PbS + O_2 \rightleftharpoons Pb^\circ + SO_2 \tag{5}$$

However, no industrial method exists yet working with this principle.

If lead is to be produced from off-grade lead sulfide concentrates the finely dissimilated metal values (zinc, cadmium, and copper) are transferred principally into the slag during the conventional smelting processes. The recovery of these elements from the slag is difficult and costly (7-9). Many efforts have been devoted to overcome these problems and a number of hydrometallurgical processes were suggested claiming such advantages as facility for selective recovery of the associated metal values and avoiding SO_2-air pollution. These processes include:

- Pressure leaching of lead sulfides in alkaline (10,11), sulfuric acid (12,13), and hydrochloric acid (14) media;
- Anodic dissolution of lead sulfide concentrates in hydrochloric and perchloric acid solutions (15-17); and
- Bacterial leaching of off-grade lead sulfide concentrates in sulfuric acid media (18,19).

II. MICROBIOLOGICAL LEACHING OF LEAD SULFIDES

When intergrowths of sulfides of lead, zinc, cadmium, copper, and iron occur together, the quantitative recovery of the individual minerals is rendered impossible because of fine mineralization and differences in grindability. For this type of ore the microbiological leaching appears to hold promise in recovery of the associated metal values. The microorganisms involved in metal sulfide oxidation processes are intensively studied, as it is the case, for example, for the chemolithotrophic *Thiobacillus ferrooxidans* (21-27). For other type of microorganisms, which may oxidize insoluble sulfide substrates such as thermophilic organisms (28-31), many studies are undertaken to elucidate their role in these processes. The bacterial leaching can be represented by the simplified equation as follows:

$$MS + 2O_2 \xrightarrow{\text{bacteria}} MSO_4 \qquad\qquad [6]$$

where M is a bivalent metal. Generally, the metal sulfides are insoluble in the aqueous media while the sulfates are soluble. Hence, is the metal dissolution process. Evidence has been reported that *T. ferrooxidans* is able to oxidize lead sulfides (35, 36), which oxidation was found to be accelerated in presence of iron (37). However, the oxidation of lead sulfide results in the formation of insoluble lead sulfate. The difference in the solubility of heavy metal sulfates incited investigators (32) to study the possibility of selective removal of copper from a lead blast furnace matte by microorganisms. This idea was pursued by several researchers (33,34) who reported on selective bacterial leaching

of arsenic and copper from oxidized tin concentrates. The applicability of the microbiological leaching technique for selective
extraction of cadmium, copper and zinc values from an off-grade
lead sulfide concentrate (18,19), has been demonstrated. A
schematical presentation of this method is given in Fig. 1.

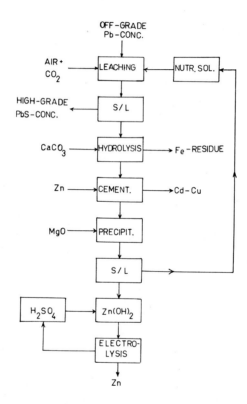

*Fig. 1. Schematical representation of a microbiological
leaching process of an off-grade lead sulfide concentrate.*

In the process outlined in Fig. 1, the finely ground lead
sulfide concentrate is contacted with a pH 2.3 iron-free aqueous
nutrient medium (38) in proportion to produce a 14% (W/V) solid
to liquid suspension, which is inoculated with an adapted active
culture of *T. ferrooxidans*. The leach suspension is agitated and
aerated with air containing 0.2% (V/V) carbon dioxide (39) and
incubated at 35°C for four days. After solid-liquid separation,

lead is recovered from the leach residue, which can be considered
to be a high-grade Pb-concentrate and the dissolved metals, copper,
cadmium and zinc, from the filtrate. A feasibility study of the
leach liquor treatment (41) indicated, that first iron has to be
precipitated by increasing the pH to 3.0, i.e. with calcite:

$$3Fe_2(SO_4)_3 + 12H_2O \rightleftharpoons 2\{H[Fe(SO_4)_2 \cdot 2Fe(OH)_3]\} + 5H_2SO_4 \qquad [7]$$
$$\text{jarosite}$$

The above partial hydrolysis reaction of ferric sulfate results
mainly hydrogen jarosite, which is easy to filter. Then, cadmium
and copper are recovered through cementation on zinc dust:

$$Cd^{++} + Zn \longrightarrow Cd + Zn^{++} \qquad [8]$$

and

$$Cu^{++} + Zn \longrightarrow Cu + Zn^{++} \qquad [9]$$

The fine Cd-Ni particles are removed as a filter-cake and can be
shipped elsewhere for further treatment. The zinc remaining in
solution can be precipitated by increasing the pH to about 7.5 by
addition of magnesium oxide:

$$Zn^{++} + 2OH^- \rightleftharpoons Zn(OH)_2 \qquad [10]$$

The zinc hydroxide can be dissolved separately by sulfuric acid
and recovered by electrolysis. The electrolyte may be used in a
cyclic process in this system.

After zinc recovery the leach solution is recycled to the
nutrient reservoir and adjusted for pH with sulfuric acid and for
the nutrient concentrations, before being reused in the leaching
step.

Semi-industrial studies (40) carried out with 30ℓ leach suspensions resulted in overall extractions higher than 94%. These
studies included releaching of the leach residue, which has been
regrounded to liberate new substrate surfaces. It has been found
that, when about 50% of extraction was achieved, the substrate

surface has been covered with basic ferric hydroxides and lead sulfate precipitates, which impeded the bacterial activity. At the end of the second leaching the leach residue has been analyzed. It contained lead in form of PbS and $PbSO_4$ in proportions of about one to one. This one to one ratio is of technological importance in the sintering and blast furnace smelting, where exothermic (according to equations [1] and [3]) and endothermic (according to equations [2] and [4]) reactions take place. If the right proportion between PbS and $PbSO_4$ existed in the feed, then the heat liberated by the exothermic reactions would provide the heat necessary for the endothermic reactions. Hence, the presence of sulfate in the feed would reduce the cyclic load used to dissipate the excess heat liberated by the exothermic reactions during conventional sintering.

III. OPTIMUM LEACH CONDITIONS

The optimum conditions for the microbiological leaching of off-grade lead sulfide concentrates are similar to those which are derived for leaching of other metal sulfides by *T. ferrooxidans*. These conditions are summarized in Table I.

TABLE I   Optimum Conditions for Lead Sulfide Concentrates Leaching

Factors	Optimum Conditions	Reference No.
1. *Temperature*	$35°C$	42
2. *pH*	2.3	42
3. *Eh*	<500 mV	27,44
4. *Substrate concentrate (-400 mesh)*	14-20% (W/V)	
5. *Bacterial inoculum*	7%	18
6. *Nutrients:* $(NH_4)_2SO_4$	3g/ℓ	38,42,43
K_2HPO_4	0.5g/ℓ	38,42,43
CO_2 *in air*	0.2% (V/V)	39
O_2	*an intensive aeration is required*	27
7. *Substrate particle size*	-325 to -400 mesh	18

IV. ECONOMICAL ASPECTS

The preliminary cost estimate described in this section deals with the microbiological leaching of an off-grade lead sulfide concentrate containing 42.9% lead, 29.6% sulfur, 16.7% iron, 7.7% zinc, 2.4% copper and 0.02% cadmium. In this process 100,000 kg of lead sulfide concentrate will be leached in suspensions containing 20% pulp density at pH 2.3 and temperature 35°C during five days. The minimum recovery for zinc, copper and cadmium is projected to be 95%.

This leaching process can be carried out in conjunction with an existing pyrometallurgical lead smelting plant. The high grade lead concentrate, containing PbS and $PbSO_4$, produced in this process can be mixed with other concentrates to provide an ideal feed for the smelter (or sinter).

The capital cost of this process is estimated to be $7,300,000 (45-47). This includes the costs of leach tanks (concrete), pumps, aeration, regrinding equipments, thickeners, filters, dryer, electrolysis equipment, their installations, and the building.

The yearly metal production is shown in Table II.

TABLE II

Metal	PbS Conc. kg		Metal Content %		Recovery %		d/Y		Yearly Metal Production kg/Y
Zn	100,000	x	7.7/100	x	95/100	x	365	≈	2,700,000
Cu	100,000	x	2.4/100	x	95/100	x	365	≈	830,000
Cd	100,000	x	0.02/100	x	95/100	x	365	≈	7,000

The yearly metal production values are rounded up to simplify the calculation. The yearly market value of metals are given in Table III.

TABLE III

Metal	Metal Production kg/y		Selling Price $/kg		Market Value $/y	Repartition %
Zn	2,700,000	x	0.76	≈	2,050,000	59.3
Cu	830,000	x	1.62	≈	1,350,000	39.0
Cd	7,000	x	8.36	≈	60,000	1.7
			Total Market Value:		3,460,000	100.0

The market values are rounded-up and the repartition percentages are based on the total market value.

The yearly expenses are given in Table IV.

TABLE IV

Direct Costs:	
Administration	
Labor	
Reactives (oxygen, limestone, hydrated lime, magnesia, H_2SO_4, K_2HPO_4, $(NH_4)_2SO_4$ and zinc powder)	
Electricity	$800,000/year
Indirect Costs:	
Maintenance	
Electricity	
Water	
Depreciation at 10%	
Interest at 10%	
Insurance	$1,600,000/year
Total:	$2,400,000/year

The repartition of the yearly expenses per metal produced is given in Table V.

TABLE V

Metal	Total Leaching Expenses $/y	Repartition (Table 3) %		Leaching Expenses $/y
Zn	2,400,000	x 59.3/100	≃	1,420,000
Cu	2,400,000	x 39.0/100	≃	940,000
Cd	2,400,000	x 1.7/100	≃	40,000
		Total:		2,400,000

The cost of metal produced with respect to the leaching expenses is given in Table VI .

TABLE VI

Metal	Leaching Expenses $/y		Metal Production (Table 2) kg/y		Cost of Metal Produced $/kg
Zn	1,420,000	/	2,700,000	=	0.53
Cu	940,000	/	830,000	=	1.13
Cd	40,000	/	7,000	=	5.71

The gain ($/kg) of metal produced is given in Table VII.

TABLE VII

	Costs of metal production $/kg		
	Zn	Cu	Cd
Costs of mining concentration (estimated)	0.11	0.14	0.39
Costs of leaching, regrinding, cementation, electrowinning (Table 6)	0.53	1.13	5.71
Total expenses (A)	0.64	1.27	6.10
Selling price (B)	0.76	1.62	8.36
Gain (B-A)	0.12	0.35	2.26

The estimated costs (48) for mining and concentration do not include any expenses for lead which make up the major part of the costs.

The realisable yearly profit of the leaching plant is given in Table VIII.

TABLE VIII

Metal	Metal produced (Table 2) kg/y		Gain realized (Table 7) $/kg		Profit $/y
Zn	2,700,000	x	0.12	$\approx$	324,000
Cu	830,000	x	0.35	$\approx$	290,000
Cd	7,000	x	2.26	$\approx$	16,000
				Total:	630,000

The yearly return of revenue of about 6.3×10^5 for an investment of 7.3×10^6 appears acceptable. The above revenue should be increased with the amount of money which would be paid for the lead concentrate produced by this leaching process.

V. CONCLUSION

The preliminary economic assessment presented in this study indicates that the microbiological leaching technique may present an economically viable alternative for treatment of off-grade lead sulfide concentrates. The leach residue of this process is an upgraded lead concentrate which can be used in the classical smelting process for recovery of lead. The dissolved metal values can be recovered by producing a reasonable profit.

Further, the microbiological leaching technique requires relatively low capital investment and low operating cost. It can be built near the mining site and used as a complementary treatment prior to the smelting process. The microbiological leaching treatment does not require high temperature nor pressure. It is easy to operate and control. This process, if properly operated does not compromise in any way the quality of the environment.

VI. REFERENCES

1.   Anonymous, in "McGraw-Hill Encyclopedia of Science and Technology", Vol. 7, p. 424. McGraw-Hill Book Co., Inc., New York, 1960.

2.   McKay, J.E., in "Kirk Ottmer Encyclopedia of Chemical Technology", Vol. 12, p. 207. John Wiley & Sons, Inc., New York, 1967.

3.   Moulds, D.E., U.S. Dept. Inter. Bur. Min., Washington, D.C., 1965.

4.   Dana, E.S., and Ford, W.E., "A Textbook of Mineralogy", p. 415. John Wiley & Sons, Inc., New York, 1932.

5.   Kellog, H.H., and Basu, S.K., *Trans. Met. Soc. AIME*, 218, 70 (1960).

6.   Kellog, H.H., *Trans. Met. Soc. AIME*, 220, 1622 (1962).

7.   Yurko, G.A., in "AIME, World Symposium on Mining and Metallurgy of Lead and Zinc", (C.H. Cotterill and J.M. Cigan, Eds.), Vol. 2, p. 330. AIME Publication, New York, 1970.

8.   Hancock, G.C., Hart, D.H., and Pelton, L.A.H., in "AIME, World Symposium on Mining and Metallurgy of Lead and Zinc", (C.H. Cotterill and J.M. Cigan, Eds.), Vol. 2, p. 790. AIME Publication, New York, 1970.

9.   Woods, S.E., and Temple, D.A., The Present Status of Imperial Smelting Process. Presented at the 8th Commonwealth Min. Met. Congr. in Melbourne, Feb. 27 - Apr. 4, 1965.

10. Vizsolyi, A., Veltman, H., and Forwards, F.A., in "Unit Process in Hydrometallurgy", (M.E. Wadsworth and F.T. Davis, Eds.), p. 336. Gordon and Breach Science Publishers, New York, 1964.

11. Gerlach, J., Pawlek, F., and Traulsen, K., *Erzbergb. Metallhüttenw.*, 18, 605 (1965).

12. Vizsolyi, A., Veltman, H., and Forwards, F.A., *Trans. Met. Soc. AIME*, 227, 215 (1963).

13. Mackiw, V.N., and Veltman, H., *Can. Min. Met. Bull.*, 60, 80 (1967).

14. Scott, P.D., and Nicol, M.J., *Inst. Min. Met. Trans. Sect. C.*, 85, C40 (1976).

15. Skewes, H.R., *Proc. Aust. Inst. Min. Met.*, 244, Dec., 35 (1972).

16. Terayama, K., Izaki, T., and Arai, K., *J. Jap. Inst. Met.*, 36, 591 (1972).

17. Kruesi, P.R., Allen, E.S., and Lake, J.L., *Can. Min. Met. Bull.*, 66, 81 (1973).

18. Torma, A.E., and Subramanian, K.N., *Internat. J. Miner. Process.*, 1, 125 (1974).

19. Torma, A.E., Can. Pat. Appl. No. 203,547 (1974).

20. Ehrlich, H.L., and Fox, S.I., *Biotechnol. Bioeng.*, 9, 471 (1967).

21. LeRoux, N.W., *New Scientist*, 43, 12 (1969).

22. Zajic, J.E., "Microbial Biogeochemistry", Academic Press, New York, 1969.

23. Trudinger, P.A., *Miner. Sci. Eng.*, 3, 13 (1971).

24. Karavaiko, G.I., Kuznesov, S.I., and Golomzik, A.E., "Bacterial Leaching of Metals from Ores", Nauka, Moscow, 1972.

25. Tuovinen, O.H., and Kelly, D.P., *Internat. Metal. Rev.*, 19, 21 (1974).

26. Schwartz, W., *Bild Wiss.*, 13, 60 (1976).

27. Torma, A.E., *Adv. Biochem. Eng.*, 6, 1 (1977).

28. Brock, T.D., Brock, K.M., Belly, R.T., and Weiss, R.L., *Arch. Mikrobiol.*, 84, 54 (1972).

29. Brierley, C.L., and Brierley, J.A., *Can. J. Microbiol.*, 19, 183 (1973).

30. deRosa, M., Gambacorta, A., and Bu'lock, J.D., *J. Gen. Microbiol.*, 86, 156 (1975).

31. Murr, L.E., and Berry, V.K., *Hydromet.*, 2, 11 (1976).

32. Corrick, J.D., and Sutton, J.A., U.S. Bur. Min., R.I. No. 7126, 1968.

33. Polkin, S.I., Karavaiko, G.I., Tauzhayanskaya, Z.A., and Panin, V.V., in "Proceedings of the 9th International Mineral Processing Congress", (N. Arbiter, Ed.), Vol. 1, p. 347. Gordon and Breach Science Publisher, New York, 1970.

34. Polkin, S.I., Panin, V.V., Adamov, E.V., Karavaiko, G.I., and Chernyak, A.S., Presented at the 11th Internat. Miner. Process. Congr. in Cagliary, Italy, 1975.

35. Lyalikova, N.N., Shlain, F.B., Unanova, O.G., and Anisimova, L.S., *Izv. Akad. Nauk SSR. Ser. Biol.*, No. 4, 564 (1972).

36. Silver, M., and Torma, A.E., *Can. J. Microbiol.*, 20, 141 (1974).

37. Tomizuka, N., *Report of the Fermentation Research Institute, Japan*, No. 48, 51 (1976).

38. Silverman, M.P., and Lundgren, D.G., *J. Bacteriol.*, 77, 642 (1959).

39. Torma, A.E., Walden, C.C., Duncan, D.W., and Branion, R.M.R., *Biotechnol. Bioeng.*, 14, 777 (1972).

40. Torma, A.E., CRM-Internal Report, Q.D.N.R., May 17, 1973.

41. Kougioumoutzakis, D., Torma, A.E., Ouellet, R., and LeHouillier, R., Presented at the Ann. Congr. Assoc. Can. Franc. Adv. Sci. in Quebec, May 8-10, 1974.

42. Torma, A.E., Walden, C.C., and Branion, R.M.R., *Biotechnol. Bioeng.*, 12, 501 (1970).

43. Tuovinen, O.H., Niemela, S.I., and Gyllenberg, H.G., *Biotechnol. Bioeng.*, 13, 517 (1971).

44. Moss, E.T., and Andersen, J.E., *Proc. Aust. Int. Min. Met.*, No. 225, 15 (1968).

45. Guthrie, K.M., *Chem. Eng.*, March 24, 114 (1969).

46. Leibson, I., and Trischman, C.A., *Chem. Eng.*, May 31, 69 (1971).

47. Parkinson, E.A., and Mular, A.L., "Mineral Processing Equipment Cost Estimation", published by Can. Inst. Min. Met., Montreal, 1972.

48. Torma, A.E., Ouellet, R., and Gabra, G.G., Presented at the Ann. Conf. Chem. Inst. Can. in Regina, Saskatchewan in June 2-5, 1974.

MICROBIOLOGICAL LEACHING OF CARBONATE-RICH

GERMAN COPPER SHALE

K. Bosecker

Bundesanstalt für Geowissenschaften und Rohstoffe
3000 Hannover 51, Federal Republic of Germany

D. Neuschütz
U. Scheffler

Friedrich Krupp GmbH., Krupp Forschungsinstitut
4300 Essen, Federal Republic of Germany

Bacterial leaching techniques have been tested to extract the copper from German copper shale. Laboratory tests were carried out in airlift-percolators and in shaking flasks. Pilot plant units were constructed using glass-fibre columns filled with 1.4 tons of copper shale. Pure cultures of T. ferroxidans *and* T. thiooxidans *were used as leaching agents. Because of the high carbonate content of the ore, bacterial leaching was feasible only after previous neutralization of the carbonate or with simultaneous addition of acid. Within 380 days 31% of the copper and 55% of the zinc were extracted. After optimization an extraction yield of 70% Cu and 80% Zn seems to be feasible. Some factors influencing the leaching process and a preliminary evaluation of the economy of bacterial leaching of the copper shale are discussed.*

I. INTRODUCTION

Within the last few decades there has been a continously
increasing world-wide demand for metallic raw materials. This
demand can be covered in the future only by the discovery of new
deposits, the improvement of conventional processing methods, and
the development of new technical operations for recovering valuable metals from low-grade ores and also from waste materials. The
German Federal Republic is the 4th largest copper consumer in the
world ranging after the USA, Japan, and the USSR. In 1973, 20.8%
of the world copper production was mined in the USA whereas the
German copper consuming industry, which depends mainly on foreign
copper resources, had to import 1.2 million tons of copper. Fig.
1 shows the imports of copper during the last few decades and
gives an impression of the increasing costs.

German copper shale - the most important copper ore in the
GFR - contains about 12 million tons of copper. Conventional
mining of copper shale was stopped in 1955 because of unprofi-
table production costs resulting from the low copper content and
from the small thickness of the copper-bearing layer.

Low-grade ores may be extracted by microbiological leaching
and these biohydrometallurgical methods are already being used
commercially for the recovery of copper and uranium (1-6). On
the whole, the biological and chemical reactions which occur
during the leaching process are known (7). But the efficiency of
such processes depends mainly on the chemical and mineralogical
characteristics of the individual ore and therefore the leaching
conditions must be established for each type of ore. In the present paper, the applicability of bacterial leaching in the
extraction of copper from the German copper shale has been investigated.

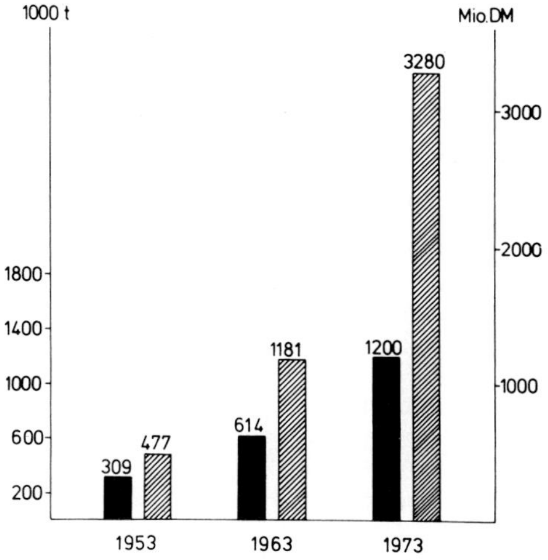

Fig. 1. German copper imports between 1953 and 1973 in tons ■ *and values* ▨

II. MATERIALS AND METHODS

Laboratory tests were carried out in air-lift-percolators and in shaking Erlenmeyer flasks. The copper shale used in these experiments was obtained from old mining dumps and contained 2.1% Cu, 3.0% Fe, 0.01% Zn, 2.5% S, and 18.1% CO_2. The main ore mineral was chalcopyrite followed by chalcocite, bornite, and pyrite.

Copper shale (200 - 500 g, crushed to a particle size of about 10 mm) was irrigated in air-lift-percolators with 150 ml of nutrient broth inoculated with pure cultures of *Thiobacillus ferrooxidans, T. thiooxidans,* or with mixed cultures of both strains.

Shaking Erlenmeyer flasks were used for submerged fermentation of fine-grained ore. Different amounts of copper shale with
a particle size of 63 - 200 µ were suspended in 90 ml of culture
medium, inoculated with 10 ml of actively growing thiobacilli,
and incubated on a rotary shaker at 200 rpm. Samples were taken
periodically from the leach suspensions to determine pH-values,
metal content, and presence of bacteria. The metal content was
determined by atomic absorption spectroscopy, surviving bacteria
were identified after inoculation into fresh culture medium.

The pure cultures of *T. ferrooxidans* and *T. thiooxidans* used
in this study were originally isolated from acid mine water in the
Harz Mountains and are cultivated in our laboratory. *T. thiooxidans* is maintained in the medium described by Starkey (8) using
elemental sulfur as an energy source. *T. ferrooxidans* is cultivated in the culture medium described by Leathen, et al. (9).

All laboratory experiments were carried out at a temperature
of 30°C. The laboratory tests were transferred to the semi-tech-
nical stage in pilot plant units, constructed of glass-fibre
columns of 2.5 m in height and 0.8 m in diameter. Each column
was filled with 1.4 tons of copper shale which had been crushed
into a grain size of 40% < 10 mm. The material was also gathered
from old mining dumps but was different in the metal content.
This ore contained 0.76% Cu, 2.3% Fe, 0.6% Zn, 1.6% S, and 19% CO_2.
At the beginning of the experiments the ore was irrigated with
diluted sulfuric acid adjusting the pH in the influent solutions
to 1.5, 1.7, and 2.0. After reaching pH-values of 2.0 to 2.5 in
the effluent solution, the columns were inoculated with mixed
cultures of *T. ferrooxidans* and *T. thiooxidans*, and the pH in the
influent suspension was maintained within a range of 1.9 to 2.1 in
one test and 2.0 to 2.2 in another with the addition of sulfuric
acid. The circulating leach suspension (about 800 to 900 l) was
collected in a big tank at the bottom of the column, pumped up

and sprayed again on top of the fill. The tank could be aerated separately.

III. RESULTS AND DISCUSSION

A. Leaching in Air-Lift Percolators

Because of the high carbonate content of the copper shale, the pH-values in the leach suspension continuously increased. The decrease in pH for different leaching experiments after inoculation with pure and mixed cultures of *T. thiooxidans* and *T. ferrooxidans* is shown in Fig. 2. Sulfuric acid was added to lower the pH, but in case of *T. ferrooxidans* (Fig. 2 b) the pH-values always rose to 7.5 followed by a complete loss of bacterial activity. After repeated addition of acid, only *T. thiooxidans* was able to maintain a pH of 4 in the leach suspension. *T. ferrooxidans* survived when mixed cultures were used (Fig. 2 c). Both strains grew rather well in the mixed culture but in spite of better growth only 0.17% of the copper was leached within 100 days. In the same period 27% of the copper was extracted by chemical leaching with 1 N H_2SO_4 (Fig. 3). In regard to these results, bacterial leaching of the copper shale seemed to be feasible only after previous neutralization of the carbonate or with simultaneous addition of acid. Strong acid conditions (6 N H_2SO_4) accelerated the process of neutralization of the ore, however the leaching process was stopped because the formation of mud in the leaching system. Using low acid concentrations (for example 1 N H_2SO_4) the neutralization was very slow and afterwards the surface of the ore was covered by a gypsum precipitate which inhibited a direct bacterial contact with the ore and diminished the leaching efficiency.

B. Leaching in Pilot Plants

Air-lift-percolators were used as a model for dump and heap leaching. However, results obtained at a laboratory scale may be

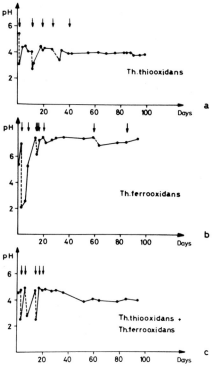

Fig. 2. Bacterial leaching of carbonate-rich German copper shale in air-lift-percolators. pH-trend after inoculation with pure cultures of T. thiooxidans *(a),* T. ferrooxidans *(b), and mixed cultures of* T. thiooxidans *and* T. ferrooxidans *(c),* ↓ *addition of 1 N* H_2SO_4.

applied on an industrial scale only with reservation. Therefore, investigations in pilot plant units seemed to be necessary. The columns we used were considered as a cut out of a mining dump in which the conditions for optimizing the leaching process were tested. Because of the high carbonate content of the copper shale, the most important facts which were taken into consideration to achieve optimum copper extraction rates are the pH-value of the leach suspension and a sufficient supply of oxygen in the fill.

The results of the percolator experiments indicated that bacterial leaching would be possible only after previous neutralization of the carbonate. Complete neutralization produces

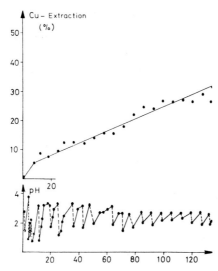

Fig. 3.   *Chemical leaching of carbonate-rich German copper shale in air-lift-percolators with 1 N H₂SO₄.   pH-trend and copper extraction.*

about 600 kg $Ca_2SO_4 \cdot 2 H_2O$ per ton of copper shale which clogs the fill and stops the leaching process. Partial neutralization is sufficient provided that the pH in the leach suspension is within a suitable range for bacterial growth. *T. ferrooxidans,* which is more sensitive to pH-changes than *T. thiooxidans,* is most active in the pH range of 1.8 to 2.5 (10-12). In order not to inhibit the bacterial activity, the pH in the influent leach suspension was maintained in the two experiments within a range of 1.9 to 2.1 and 2.0 to 2.2. At maximum flow rate, which depends on the permeability of the fill, the pH of the effluent increased by about 0.3 as a result of the partial neutralization of the carabonate during percolation. At lower flow rates, the differ - ence in the pH between the influent and effluent leaching solu- tions increased considerably. However, the oxygen supply within the fill, which is necessary for a direct bacterial oxidation of metal sulfides, is reduced at maximum irrigation. At maximum flow there is no direct oxygen supply via holes and pores in

the fill and therefore only the oxygen dissolved in the liquid
phase can be used by bacteria. An oxygen demand of 37.2 kg O_2
per ton of copper shale has been calculated for the direct bacterial oxidation of metal sulfides. Without artificial aeration
several years would be needed to transport the oxygen into the
fill. In view of these data, direct bacterial oxidation of metal
sulfides (reaction a) would seem to be negligible, and the bacterial activity is mainly restricted to an indirect oxidation of
sulfides via reoxidation of chemically reduced ferric iron (reaction b and c).

$$Me\ S + 2O_2 \xrightarrow{\textit{T. ferrooxidans}} Me\ SO_4 \tag{a}$$

$$Me\ S + Fe_2(SO_4)_3 \longrightarrow Me\ SO_4 + 2\ Fe\ SO_4 + S^0 \tag{b}$$

$$2\ Fe\ SO_4 + H_2SO_4 + 1/2\ O_2 \xrightarrow{\textit{T. ferrooxidans}} Fe_2(SO_4)_3 + H_2O \tag{c}$$

In order to accelerate the bacterial oxidation of ferrous
iron to ferric iron the effluent leaching suspension was aerated.
Precipitation of ferric iron was avoided by taking care that the
leaching suspension did not go above pH 2.3.

As shown in Table I, the pH-value in the leach suspension is
the most critical factor in the bacterial leaching of the carbonate-rich copper shale. Previous neutralization to pH 1.5
caused the most important decrease in the flow-rate and resulted
in the lowest extraction of copper and zinc. The best results
were obtained with slow partial neutralization to about pH 2.
Within 245 days, 22% of the copper and 55% of the zinc were
extracted from the copper shale. The average extraction rate
was calculated to be 280 mg Cu/h per t copper shale and 560 mg
Zn/h per t copper shale. Assuming a linear increase of the leaching intensity about 3 years are necessary to solubilize about 70%
of the copper and 1.5 to 2.5 years are calculated for the extraction of about 80% of the zinc.

TABLE I   Bacterial leaching of copper shale in pilot plants

		Column No.			
		2	3	4	5
Ore	t	1,4	1,4	1,4	1,4
particle size	mm	<10	40%<10	40%<10	40%<10
time	h	533	9231	8922	5850
leaching volume	1	700	890	820	920
flow rate, starting	1/h	120	115	125	120
stopping	1/h	1	2	13	80
pH: pre-neutralization		<0,1	1.5-2.7	1.7-2.5	ca.2.0
bacterial leaching		–	2.0-2.2	1.9-2.1	2.0-2.2
H_2SO_4 - consumption	kg	212	362	469	340
metal content: Cu	g/1	0,003	0,85	4,07	2,38
Zn	g/1	1,26	3,19	5,1	4,76
Fe	g/1	1,52	0,74	3,7	1,89
metal extraction: Cu	%	<1	7	31	22
Zn	%	10.5	34	50	55
average metal Cu	mg/h/t		60	270	280
extraction rate: Zn	mg/h/t		220	300	560
H_2SO_4-consumption per Cu	kg/kg		478	140	148
extracted metal: Zn	kg/kg		127	112	74
value of leach suspension	DM/t		6,20	15,60	13,20
H_2SO_4 - consumption (60,--DM/t H_2SO_4)	DM		15,50	20,10	14,60
deficit acid production / acid consumption	DM		-9,30	-4,50	-1,40

Data on the economics of the process are given in Tables II
and III. The most expensive component of the process is the im-
mense consumption of sulfuric acid. About 390 kg H_2SO_4/t copper
shale at a price of 60 DM/t H_2SO_4 are necessary for the neutralization of the carbonate. The total expense for the hydrometallurgical process is estimated at 41 DM/t copper shale whereas
the value of extracted metals is calculated at only 30 DM/t
copper shale. These calculations indicate that the use of per-
colating leaching is not feasible at the present time. However,
bacterial heap leaching may become of economic interest under
any of several conditions: if the content of valuable metals is
higher and the carbonate content lower than assumed, if sulfuric
acid can be obtained at lower costs, if the mining costs become
cheaper, or if metal prices increase considerably.

C.   Leaching in Shaking Cultures

Because of the low efficiency of the percolation process,
submerged fermentation of ground copper shale has been taken into
consideration. This method using fine-grained ore and shaking
cultures has been described for the extraction of concentrated
sulfide ores yielding metal extractions of about 100% (14-16).
According to our preliminary results, we feel that submerged fermentation can also be used for low-grade ores. The bacterial
attachment to the ore particles is favoured by the large surface
area and this will enhance the leaching efficiency. As shown in
Fig. 4, about 80% of the copper was extracted within 20 days.
The copper extraction depends on the pulp density in the leaching
suspension. An increase of pulp density resulted in a longer
period of neutralization and a higher consumption of sulfuric
acid. The yield of copper was about the same in all experiments.

Submerged fermentation method really appears to offer better
chances for bacterial leaching of the copper shale than heap
leaching. We are continuing work on the optimization of the

TABLE II   *Expenses of hydrometallurgical processing methods*

	costs per ton copper shale DM/t
1. Bacterial leaching	
1.1. H_2SO_4 (390 kg/t ore) (60,--DM/t H_2SO_4)	23,40
1.2. Electricity, water treatment	5,60
1.3. Labour	0,50
1.4. Capital costs	1,50
	31,--
2. Reextraction of metals from the leach suspension	10,--
Total	41,--

TABLE III   *Estimated yields of Cu and Zn at optimal leaching conditions.*

Metal	Metal content of ore %	Leaching Extraction %	Total extraction %	Total extraction kg/t ore	Metal Value DM/kg	Total yield DM/t ore
Cu	0.8	70	65	5,2	4,--	20,80
Zn	0.6	80	75	4,5	2,--	9,--
					Total	29,80

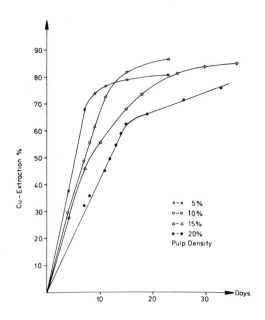

Fig. 4.   Bacterial leaching of carbonate-rich German copper shale with T. ferrooxidans *at different pulp densities.   The bacteria were added after previous neutralization, the pH was controlled manually and adjusted to 2.0.*

leaching process and hope to be able to decrease the consumption of sulfuric acid by using mixed cultures of *T. ferrooxidans* and *T. thiooxidans*.

IV.   ACKNOWLEGEMENT

These investigations are supported by the Federal Ministry of Research and Technology.

V. REFERENCES

1. Trussell, P.C., Duncan, D.W., and Walden, C.C., *Can. Min. J.*, 85, 46 (1964).

2. Beck, J.V., *Biotechnol. Bioeng.*, 9, 487 (1967).

3. Le Roux, N.W., *New Scient.*, 43, 12 (1969).

4. Malouf, E.E., *Min. Eng.*, 24, 58 (1972).

5. Guay, R., Silver, M., and Torma, A.E., *Eur. J. Appl. Microbiol.*, 3, 157 (1976).

6. Schwartz, W., *Metall*, 27, 1202 (1973).

7. Tuovinen, O.H. and Kelly, D.P., *Z. Allg. Mikrobiol.*, 12, 311 (1972).

8. Starkey, R.L., *J. Bacteriol.*, 10, 135 (1925).

9. Leathen, W.W., Mctyre, L.D., and Braley, S.A., *Science*, 114 280 (1951).

10. Razzell, W.E. and Trussell, P.C., *J. Bacteriol.*, 85, 595 (1963).

11. Golomzik, A. and Ivanov, V., *Microbiology*, 34, 396 (1965).

12. Torma, A.E., Walden, C.C., and Branion, R.M.R., *Biotechnol. Bioeng.*, 12, 501 (1970).

13. Torma, A.E., T.M.S. Paper Collection, No. A 72 - 7, 1 - 22, published by the Metallurgical Society of AIME, New York, (1972).

14. Bruynesteyn, A. and Duncan, D.W., *Can. Met. Quarterly*, 10, 57 (1971).

15. Torma, A.E., Walden, C.C., Duncan, D.W., and Branion, R.M.R., *Biotechnol. Bioeng.*, 14, 777 (1972).

16. Torma, A.E. and Subramanian, K.N., *Internat. J. Miner. Process.*, 1, 125 (1974).

STUDIES ON THE OXIDATION OF SULPHIDE

MINERALS (PYRITE) IN THE PRESENCE OF BACTERIA

A. S. Atkins

Virginia Polytechnic Institute and State University
Blacksburg, Virginia, U.S.A.

A research study on the mechanism of oxidation of pyrite has demonstrated clearly that pyrite samples from various geological origins can be leached to completion in the presence of Thiobacillus ferrooxidans. *The complete leaching of pyrite in the presence of bacteria, implicating the effect of a limiting pH value, provides a significant break-through in the mechanism of sulphide oxidation. This result demonstrates that previous speculation on the formation of a protective reaction layer on the surface of sulphide particles, which accounted for their incomplete oxidation (or the toxicity of metal in solution), is invalid for iron pyrite.*

I. INTRODUCTION

Many investigators have shown that micro-organisms are actively involved in numerous natural oxidation situations and currently 15 percent of the United States copper is produced by bacterial dump leaching operations (1, 2). Investigations into the actual role of the bacteria in these operations is comparatively recent with the isolation of a sulphur oxidizing bacterium by Colmer and Hinkle in 1947 (3).

At present bacterial leaching is widely used in several situations as a commercial scavenging process for the treatment of low grade mine wastes too low in mineral value to be treated by conventional mineral concentrating processes. Although, bacterial leaching is restricted to sulphide minerals, metals such as uranium can also be solubilized where they occur in association with sulphides.

The interest in bacterial leaching of sulphide minerals has arisen primarily from the environmental aspects in preventing air pollution from sulphur emissions associated with conventional smelting operations. The use of bacterial leaching provides an alternative extraction method to the existing pyrometallurgical and hydrometallurgical processes.

The bacteria involved in current bacterial leaching operations in acid solutions less than pH 3.0 are the species *Thiobacillus thiooxidans* and *Thiobacillus ferrooxidans* (4). These bacteria are classified as members of the genus *Thiobacillus* and are chemosynthetic autotrophic aerobic bacilli. *T. thiooxidans* having the ability to derive energy from the oxidation of inorganic sulphur while *T. ferrooxidans* can utilize both ferrous iron and/or inorganic sulphur.

Many investigators have studied the oxidation of pyrite in the presence of these bacterial types, Leathen, et al. 1953, (5) Bryner and Anderson 1957, (6) Bryner and Jameson 1958, (7) Silverman, et al. 1961, (8) Silverman, et al. 1963, (9) Silverman 1967, (10) Le Roux, et al. 1973 (11) and Silver and Torma 1974 (12). However, information concerned with the kinetics of oxidation is limited. This is probably due to an orientation of experimental study since the mid 1960's away from pyrite, (usually considered as a gangue mineral sulphide) to the treatment of commercial valuable sulphide of copper, zinc, nickel, etc.

The actual role of the bacteria in the oxidation of sulphide minerals is not yet fully resolved. However, two main mechanisms have been proposed which involve the direct bacterial oxidation of the ferrous and/or sulphide ion, and the indirect oxidation by the bacterial regeneration of ferric iron or a combination of both (13). The bacteria have also been implicated in an electro-chemical mechanism (4).

The bio-chemical conditions for the bacterial oxidation although similar with regard to kinetic factors are more critical than for a mere chemical process, because the bacteria are living organisms. Therefore mineralogical and micro-chemical environments are prerequisites to achieve optimum bacterial growth.

Bacterial leaching experiments in a batch leaching system by many investigators have shown that in most cases oxidation of sulphide mineral particles are usually incomplete (2). Parameters of tolerance to metals in solution, particle size and the formation of reaction products have been implicated as affecting both leach rate and metal extraction.

II. EXPERIMENTAL MATERIALS, BACTERIAL CULTURE LEACHING TECHNIQUE, AND ANALYSIS METHODS.

A. Materials

In this experimental study of the role of bacteria in the oxidation of pyrite, the main sample used was from Tharsis, Rio Tinto, Spain. In the previous study Atkins 1974 (14) six other pyrite concentrates reported there were also used to confirm the complete leaching of pyrite. The source, chemical analysis and brief geological description of the samples are given in Table I.

The pyrite samples used for the leaching experiments were prepared by crushing and grinding and were also concentrated when large quantities of gangue material was associated with the sample. The final concentrates were milled using a Tema Mill

TABLE I – Analysis of Pyrite Concentrates and Brief Geological Origin

Concentrate	Total iron	Pyritic iron	Non-Pyritic iron	Pyritic Sulphur Calculated	Total Sulphur	Moisture Content	Geological Origin
Avoca	42.03	40.55	1.48	46.56	47.4	1.11	S. Ireland, stratiformed syngenetic ore body. Pyrite as massive cupriferous pyrite with carbonaceous intercalations. Metamorphic.
Blaenant	40.49	36.45	4.04	41.86	43.5	0.42	Blaenant Colliery, S. Wales. Sandstone containing impregnated pyrite. Sedimentary.
Dolaucothi	44.53	43.42	1.11	49.86	50.7	0.40	W. Wales. Sheet and saddle-like bodies of gold bearing pyrite quartz. Hydrothermal.
Llanharry Surface	43.30	42.38	0.92	48.66	50.2	0.34	Llanharry Ore Mine, S. Wales. Samples from same mine but from different horizons. Hematite, geothite. Ore deposited as chemical precipitate. Sedimentary.
Llanharry Underground	39.19	32.25	6.94	37.03	43.8	0.66	
Tharsis	43.04	41.16	1.88	47.26	47.1	1.03	Rio Tinto, Spain. Massive pyrite deposits. Igneous.
Wheal Jane	42.30	40.77	1.53	46.82	49.2	0.75	Cornwall, Pyrite in close association with Arsenopyrite. Igneous.

*In a previous study, Atkins 1974 (14), these samples were shown to be iron pyrite by X-ray Diffraction Analysis.

(Tema Machinery Ltd., Banbury Oxon, manufactured in W. Germany) for a short time to produce classified concentrates of minus 350 mesh (45 µm aperature) British Standard 410 Fine Mesh Series.

B. Bacterial Culture

The culture of iron oxidizing *Thiobacillus ferrooxidans* used in the study have grown on a variety of metal sulphide concentrates and shown to be adaptive to high concentrations of 19.2 g/ℓ iron, 49.4 g/ℓ copper, 35.2 g/ℓ zinc and 4.9 g/ℓ arsenic (14, 15, 16).

C. Leaching Technique

Shake Flask - This consists of culturing the bacteria with the pyrite concentrate in a 500 mℓ Erlenmeyer flask containing 100 mℓs of suspension in an orbital incubator (Gallenkamp Model I. H. 400) maintained at 35°C and 250 revs/minute.

The mineral concentrate was dispensed into the Erlenmeyer flasks followed by 95 mℓs of substrate free Silverman and Lundgrens 1959 '9k' (17) salts medium acidified to pH 2.0 (0.01N H_2SO_4).

The pH of the medium was adjusted to pH 2.0 by the addition of 6N H_2SO_4 and the flasks stoppered with a cotton wool bung and incubated.

To achieve stable conditions prior to inoculation with bacteria, it was necessary in some cases, to make frequent adjustments of pH by acid addition. The time required for conditioning varied with the different concentrates and the pH adjustments were repeated until the desired pH remained constant for at least 24 hours. An inoculum datem which was verified experimentally (18) of 1 x 10^9 cells/mℓ suspension/1% (w/v) pulp density was added and the flask made up to 100 mls with a small volume of 0.01N H_2SO_4 (pH 2.0).

The volume of sulphuric acid required to adjust the pH was small and its effect upon the solid to liquid ratio was negligible.

Correction for evaporation during adjustment period, was made prior to inoculation. Correction for evaporation during the experiment was made half an hour prior to sampling by the addition of 0.01N H_2SO_4 and was proportionally reduced to compensate for suspension volumes lost in prior sampling. The correct volume of 0.01N H_2SO_4 was calculated using the mean of two 2000 hours evaporation tests performed in the incubator at the same standard condition. The rate of evaporation was found to be fairly constant over the duration of the test with a mean value of 0.0365 ml/hr. The emphasis placed on evaporation is important in considering the experiment usually lasts for 600 hours and at the 0.0365 ml/hr would mean an extraction error of 21.9%.

In the sterile control tests (no bacteria) a volume of DIOXIN (a brand of Dimethoxane manufactured by Givenden & Co. Ltd., Surrey) equal to the volume of inoculum was added to prevent bacterial growth.

D. Analysis

In this study known samples of growth suspension were rapidly removed while agitating the suspension to ensure that the sample was representative. The leaching of the pyrite gives rise to insoluble reaction products and in this study 1 ml of growth suspension was treated with 1 ml of 5N HCL for 30 minutes at room temperature (19). The metal concentrate in the extract will be referred to as total metal concentration. Metal concentration in untreated extracts will be referred to as soluble metal extracts. Clear extracts for analysis were prepared (soluble and total) by diluting x 10 using 0.01N H_2SO_4 (pH 2.0) and centrifuged at

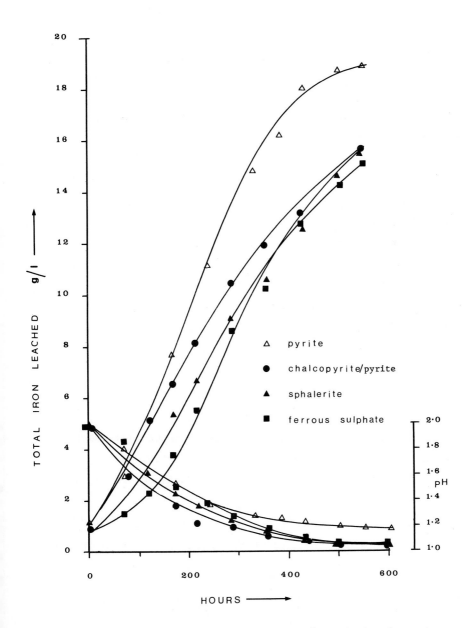

Fig. 1. Oxidation of pyrite at 6% (w/v) pulp density using cell inocula grown on a variety of substrates.   Temperature 35°C, Initial pH 2.0.

2750 x $\underline{g}$ for 30 minutes using M.S.E. super minor. The extractions were diluted as required prior to analysis using dilute 0.01N H_2SO_4 (pH 2.0).

Analysis for iron, copper, zinc and other trace elements were performed using an E.E.L. Model 240 Atomic Adsorption Spectrophotometer (A.A.S.); pH measurements were made using a Corning rugged combination electrode fitted to a Model 12 Corning E.E.L. research pH meter.

III. RESULTS

A. Oxidation of Pyrite Using Bacteria Grown on Different Substrates

To establish the effect of seed inoculum grown on a variety of substrates on the bacterial oxidation of pyrite, a series of tests using 6% (w/v) solid concentrations of Tharsis pyrite were conducted using cell inocula harvested from substrates of pyrite, chalcopyrite/pyrite, sphalerite and ferrous sulphate as indicated by Figure 1.

As expected the cell inoculum from the pyrite substrate produced the highest metal release, while chalcopyrite/pyrite, sphalerite and ferrous sulphate produced comparable metal release curves. This was indicative of the bacterial adaption, which is influenced by the amount of iron present in the respective samples together with the important end pH value of the respective systems.

In this study seed cultures grown on the chalcopyrite/pyrite concentrate were used because of the comparative nature of the experimentation, giving sufficient yield of free cells present in the supernatant for harvesting procedure, coupled with a short growth phase of 12-16 days compared to 25 days for pyrite.

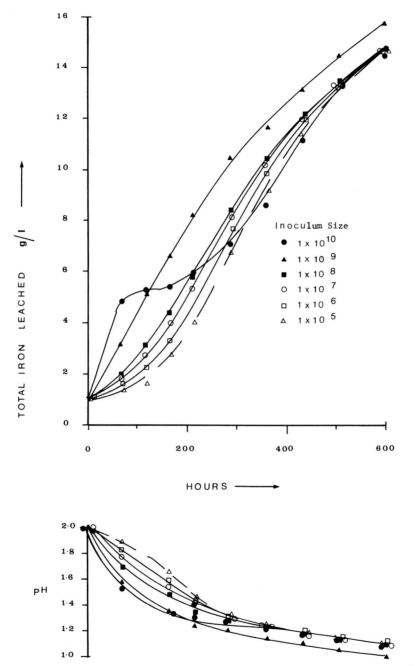

Fig. 2. The effect of varying the inoculum size from 1 x 10⁵ to 1 x 10¹⁰ cells/ml./1% pulp density on the leaching of pyrite. Pulp density 6% (w/v), temperature 35°C, Initial pH 2.0.

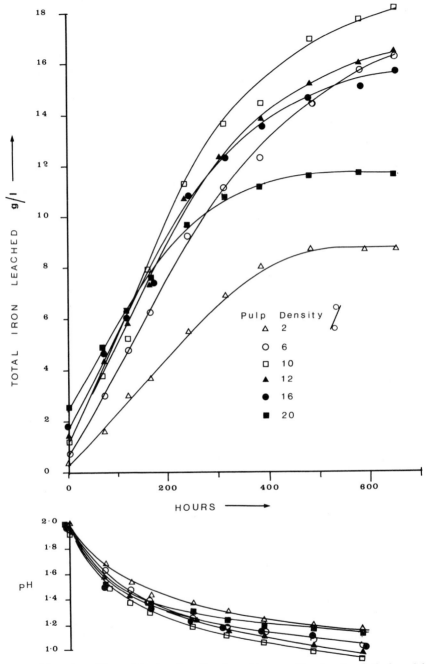

Fig. 3.   *The effect of pulp density on the bacterial leaching of pyrite.   Temperature 35°C, Initial pH 2.0.*

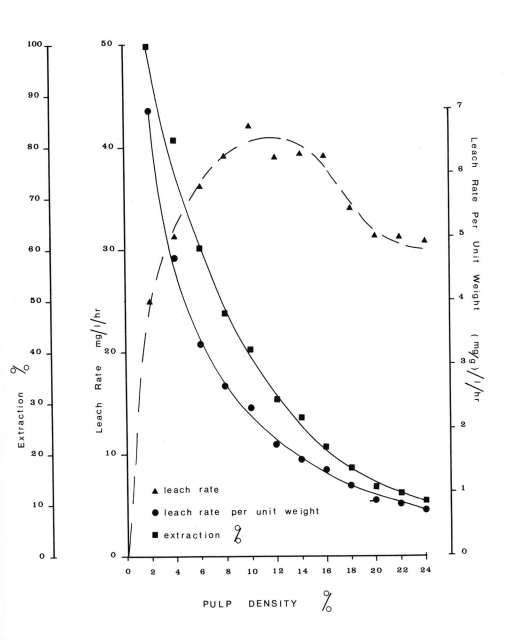

Fig. 4. The effect of pulp density on iron leach rate and extraction from pyrite. Temperature 35°C, Initial pH 2.0.

B. Effect of Inoculum Size on the Rate of Pyrite Oxidation

To investigate the importance of inoculum size on the
bacterial oxidation of pyrite a series of 6% (w/v) Tharsis
pyrite flasks were inoculated with cell concentrations varying
from 1 x 10^5 cells to 1 x 10^{10} cells per ml per 1% (w/v) pulp
density. Figure 2 represents the metal release curves of the
respective Tharsis pyrite concentrations and indicates that
varying the inoculum size produces a difference in the lag phase
of the typical logarithmic leach curves.

C. Effect of Solid Concentration on Oxidation Rates.

To demonstrate the affect of pulp density which is defined
in the study as

$$\frac{\text{Weight of Solids (g) x 100}}{\text{Volume of Liquid (ml)}}$$

on the metal leach rate and extraction during the bacterial oxidation of pyrite, shake flasks ranging from 2% to 24% (w/v)
pulp density were prepared and leached. The tests were initiated
at the standard pH of 2.0 and with a cell inoculum of 1 x 10^9
cells/ml/1% pulp density. Care was taken at higher pulp
densities to ensure that the samples for analysis were as representative as possible. Figure 3 indicates a selected representation of the metal release curves and corresponding pH
profiles produced from these tests. Table II illustrates the
leach rate and leach rate per unit weight per 24 hours of the
concentrates corresponding to the respective pulp densities.
In Figure 4 both sets of iron leach rates have been plotted
against the corresponding pulp density values together with the
respective percentage extraction. The graph indicates that the
iron leach rates were proportional to pulp density up to approximately 4% (w/v) with an optimum pulp density at about 10% (w/v).
The graph indicates a gradual decrease in the rate above 16%
(w/v) before tailing off at 24% (w/v).

TABLE II

Pulp Density %	Leach Rate mg/1/hr	Leach Rate per unit weight mg/g/1/24 hr
2	25.0	6.97
4	31.3	4.31
6	36.4	3.39
8	38.4	2.65
10	42.2	2.35
12	38.0	1.76
14	38.8	1.54
16	38.8	1.32
18	34.0	1.05
20	31.2	0.86
22	31.5	0.79
24	30.9	0.72

D.   Complete Bacterial Pyrite Oxidation

The leach data on a series of random comparative tests on samples of 6% (w/v) Tharsis pyrite as represented in Table III indicates a relationship between iron release and pH. A computer statistical analysis of the data presented in Table III indicated a linear fit with the equation:

$$y = a + bx$$

where y is the metal concentration and x is the corresponding pH of the system with a correlation coefficient of 0.92.

An examination of the proceeding pulp density experimental results showed that only the 2% (w/v) test actually leached to completion and that the remaining leach tests also stopped at approximately pH 1.0. This implied that the remaining tests at higher pulp densities would also leach to completion if the pH was not limiting the reaction.

With this factor in view, a selection of the pulp density tests, which included the 2%, 6%, 10%, 12%, 16% and 20% from the previous section, after completing the first stage of leaching (600 hours) in which the end pH values were approximately 1.0 were prepared for a repeated leach in the following

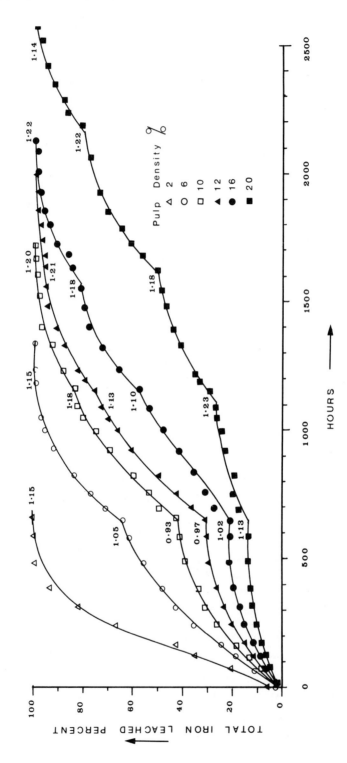

Fig. 5. The effect of re-suspending in fresh media at pH 2.0 after reaching the limiting pH value of 1.0 during the bacterial leaching of different pulp densities of pyrite.

TABLE III

*METAL RELEASE AND pH PROFILES OF A RANDOM SELECTION OF
6% (w/v) THARSIS PYRITE*

pH	Total Iron	pH	Total Iron	pH	Total Iron
2.0	1060	2.0	1080	2.0	950
1.59	3600	1.86	1800	1.57	3100
1.25	6900	1.68	3800	1.21	5100
1.29	9600	1.38	6000	1.35	6550
1.26	10800	1.28	8750	1.21	8200
1.20	12000	1.20	11500	1.21	10500
1.13	12700	1.13	12500	1.15	11600
1.09	13200	1.07	13250	1.1	13150
1.10	13500	1.10	13500	1.05	14500
				0.99	15750

pH	Total Iron	pH	Total Iron	pH	Total Iron
2.0	710	2.0	950	2.0	960
1.6	3000	1.57	3000	1.88	1450
1.43	4800	1.21	5100	1.46	2300
1.315	6300	1.35	6550	1.53	3800
1.26	9350	1.21	8200	1.33	5550
1.207	11100	1.21	10550	1.26	8600
1.156	12850	1.15	11800	1.155	10250
1.08	14500	1.10	13150	1.10	12750
1.045	15800	1.05	14350	1.05	14300
		0.99	15750	0.995	15750

manner. The leach suspension was centrifuged at 2750 x $\underline{g}$ for 45 minutes and the supernate discarded. The remaining residue was washed in dilute H_2SO_4 (pH 2.0) and re-centrifuged at 2750 x $\underline{g}$ for a further 45 minutes. The residue containing the unreacted pyrite, products of precipitation and bacterial cells were suspended in 100 ml of fresh '9k' salt solution at pH 2.0 and incubated. This procedure was repeated after each 600 hour leach phase until each respective pulp density test was leached to completion. Figure 5 illustrates the step wise leach profile, together with the final pH reading and the end of each successive leach stage of 600 hours. It must be noted that the results have been mathematically corrected to compensate for

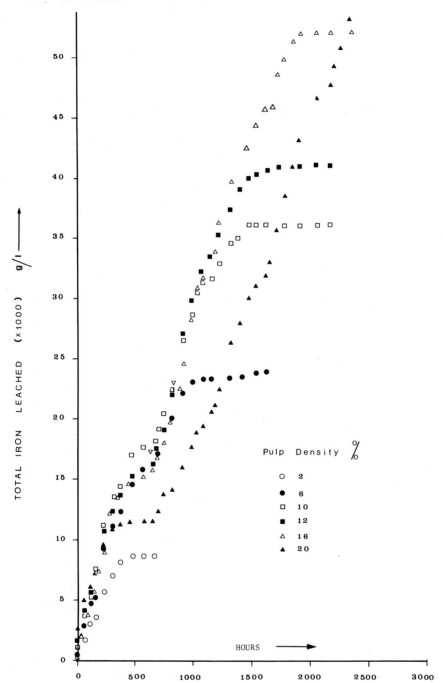

Fig. 6.   Iron release from various pulp density tests during the bacterial leaching of pyrite.

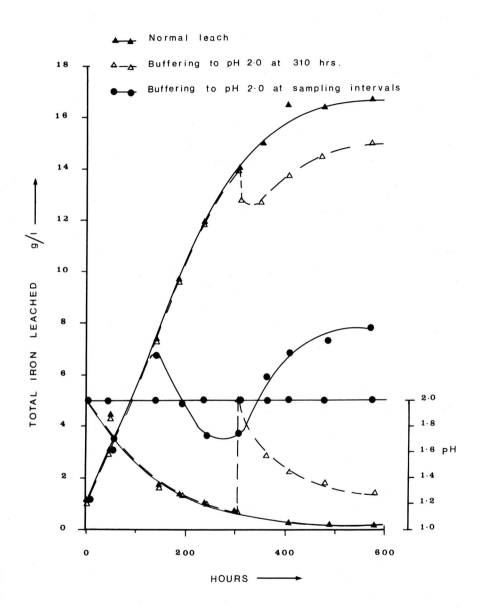

Fig. 7. The use of 5N NaOH to buffer the pH profile during the bacterial leaching of pyrite. Pulp density 6% (w/v), temperature 35°C, initial pH 2.0.

sampling loss throughout the various leach phases. The study
demonstrated that for this praticular pyrite sample, complete
dissolution could be obtained within the pulp density range of
2 to 20% (w/v) providing the pH of the suspension was maintained
by means of consecutively re-suspension of the leach at pH 2.0
after reaching the limiting pH of 1.0.

Figure 6 indicates the metal release of the 2%, 6%, 10%, 12%
16% and 20% (w/v) pulp density with respect to time. The graph
indicates that the iron release from the bacterial oxidation of
pyrite is linear with respect to time.

A series of tests were performed on the Tharsis pyrite in
which the pH of the system was regulated to overcome the need
for centrifuging and re-suspension at pH 2.0 in the following
manner:

(i) The pH of the suspension was buffered to pH 2.0
using 5N NaOH at every sampling interval

(ii) At approximately midway through the leach (310
hours) the suspension pH was re-adjusted to give
a pH value of 2.0 by the addition of 5N NaOH and
allowed to leach.

Figure 7 illustrates the metal release curves of the respective
tests compared to a normal leach test. These results imply
that when alkaline is added to the leach suspension it precipitates active amorphous hydroxide which may transform to either
crystalline geothite (FeOOH) or to a jarosite complex
(K Fe$_3$ (SO$_4$)$_2$(OH)$_6$) depending on the pH and solution composition
and also the temperature of the medium (20). The graph
indicates that the leach tests which were buffered with 5N NaOH
produced iron compounds which were probably not solubilized, in
this case, by 5N HCL which was used for total iron analysis as
outlined in the analytical section.

A series of tests were also conducted on a variety of pyrite
specimens from different geological origins to confirm that
during the bacterial oxidation of pyrite, the lower pH value of

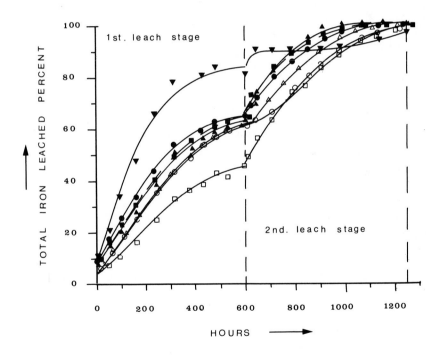

Fig. 8. Two stage bacterial leach profile of pyrite samples from various geological origins. Pulp density 6% (w/v), temperature 35°C, Initial pH 2.0.

the system was limiting. Six samples of different pyrite
specimen as outlined in Table I together with the Tharsis pyrite
at 6% (w/v) pulp density were leached in a two stage test.
Figure 8 illustrates the percentage extraction of the respective
sample with the typical predicted two stage leach profile for
the 6% (w/v) pyrite samples. The leach rates of the respective
pyrite, as expected were slightly different, and probably
caused by geological variations during formation and mineralogical factors such as lattice imperfections, non-stoichiometry,
porosity etc (21). It must be stressed that the percentage
extraction indicated in Figure 8 refers to the total iron present
in the concentrates as represented by the chemical analysis
outlined in Table II . The presence of iron compounds as impurities in the pyrite accounts for the 96 percent extraction of
iron from the Blaenant pyrite sample which contains approximately
4 percent Siderite ($FeCO_3$) identified by optical microscopy in a
previous study (14).

Throughout this particular aspect of the study, the stage
leaching on the respective samples were repeated even after all
the iron in the sample had been monitored in solution. This
procedure was used basically as a check on the iron analysis
and to verify that the mathematical corrections made after each
leach stage, to compensate for loss of sample through analysis
was correct.

It was noted that during this check stage of 600 hours (after
all the iron had been monitored in solution) with no further
appreciable iron release (less than 0.8 percent) the pH profile
drifted slowly from pH 2.0 to approximately 1.3. This indicated
that a sulphur substrate was present which was being slowly
oxidized to form sulphuric acid as expressed by the following
equation.

$$S + 3/2O_2 + H_2O \xrightarrow{\text{Bacteria}} H_2SO_4$$

The leach residue were also analyzed as a check and a mass balance confirmed the complete bacterial oxidation of pyrite.

IV. CONCLUSIONS

It has been shown that bacteria grown on a variety of substrates provide a readily available source of active seed cultures for use in the bacterial oxidation of pyrite. Similar metal release and extraction results were obtained from pyrite using bacteria grown on a variety of substrates.

It has been shown that the optimum cell inoculum is 1×10^9 cells/ml/1% pulp density for the bacterial oxidation of pyrite and varying the inoculum size by a factor of 10,000 produced very little difference in the leach results obtained. This result demonstrated that inoculum size is not critical providing the optimum concentration of cell numbers is not exceeded.

Experiments performed to study the affect of varying the solid to liquid concentration showed that leach rates were proportional to pulp density up to 10% (w/v) at which point the availability of mineral surface was no longer rate limiting.

This study has demonstrated that the relationship between iron release and pH is linear with time during the bacterial leaching of pyrite and the rate of reaction within the pH limits of 2.0 to 1.0 was of zero order. It was found that complete oxidation of pyrite samples could be obtained by con- secutively re-suspending the pyrite in fresh media at pH 2.0 af- ter reaching the limiting pH value of 1.0. An attempt to use an alkaline buffer to maintain a favorable pH of the solution at pH 2.0 resulted in an erratic metal release and iron precipi- tation and demonstrated the necessity to re-suspend at pH 2.0 when the pH became limiting.

It was also shown conclusively that pyrite samples from different geological origins could be leached to completion in the presence of bacteria provided the pH was maintained in the pH range of 2.0 to 1.0. This result demonstrates that previous speculation on the formation of a protective reaction layer on the surface of sulphide particles which accounted for their incomplete oxidation, (or the toxicity of metal in solution) is invalid for iron pyrite. The complete dissolution of pyrite provides a valuable hydrometallurgical process for the concentration of gold and silver usually associated with pyrite. The method envisaged would be to completely bacterially oxidize the pyrite and thus forming a concentrate for conventional treatment by cyanidation.

V. ACKNOWLEDGMENTS

I would like to extend my sincere thanks to my supervisor of studies, Dr. F. D. Pooley, Senior lecturer and Professor J. Platt, Head, both of the Department of Mineral Exploitation, University of Wales, Cardiff, for their help, advice and encouragement throughout this research programme.

Thanks are also extended to Dr. J. R. Lucas, Head, Department of Mining and Minerals Engineering, Virginia Polytechnic Institute and State University, Blacksburg, for his constant interest and encouragement.

The financial support of the Science Research Council is also gratefully acknowledged.

VI. REFERENCES

1. Madsen, B. W., Groves, R. D., Evans, L. G., and Potter, G. M. "Prompt copper recovery from mine strip wastes". *R. I. No. 8012 U. S. Bureau of Mines* (1975).

2. Roman, R. J., and Benner B. R., "The dissolution of copper concentrates," *Min. Sci. Eng.* 5, 1, pp 3-24 (1973).

3. Colmer, A. R., and Hinkle, M. F., "The role of microorganisms in acid mine drainage." *Science 106* pp 253-256 (1947).

4. Corrans, I. J., Harris, B., and Ralph, B. J., "Bacterial leaching and introduction to its application and theory and a study of its mechanism of operation." *J. South African Inst. Min. Met.* March, pp 221-230 (1972).

5. Leathen, W. W., Braley, S. A., and McIntyre, L. D. "The role of bacteria in the formation of acid from certain sulphur constituents associated with bituminous coal." *Appl. Microbiol.*, 1, pp 61-64 (1953).

6. Bryner, L. C., and Anderson, R., "Microorganisms in leaching sulphide minerals." *Ind. and Eng. Chem.*, 49, 10 pp 1721-1724 (1957).

7. Bryner, L. C., and Jameson, A. K., "Microorganism in leaching sulphide minerals." *Appli. Microbiol.*, 6, pp 281-287 (1958).

8. Silverman, M. P., Rogoff, M. H., and Wender, I., "Bacterial oxidation of pyritic materials in coal." *Appl. Microbiol.*, 9, pp 491-496, (1961).

9. Silverman, M. P., Rogoff, M. H., and Wender, I., "Removal of pyritic sulphur from coal by bacterial action." *Fuel.* 42 pp 113-124 (1963).

10. Silverman, M. P., "Mechanism of bacterial pyrite oxidation." *J. Bacteriol.*, Oct., pp 1046-1051 (1967).

11. LeRoux, N. W., North, A. A., and Wilson, J. C., "Bacterial oxidation of pyrite." *10th Inst. Min. Proc. Congress London* (1973).

12. Silver, M. and Torma, A. E., "Oxidation of Metal Sulphides by *Thiobacillus ferrooxidans* grown on different substrates." *Can. J. Microbiol.*, 20, pp 141-147 (1967).

13. Silverman, M. P., and Ehrlich, H. L., "Microbiol formation and degradation of Minerals." *Advan. Appl. Microbiol.*, 6, pp 153-206 (1964).

14. Atkins, A. S., "Bacterial Oxidation of Iron Sulphide with Reference to Coal." *M. Sc. Thesis, University of Wales* (1974).

15. Lawrence R. W., "Bacterial Extraction of Metals from Sulphide Concentrates." *Ph.D. Thesis, University of Wales* (1974).

16. Pinches, A., "The Use of Microorganisms for the Recovery of Mineral Materials." *Ph.D. Thesis, University of Wales* (1972).

17. Silverman, M.P., and Lundgren, D.C., "Studies on the chemoautotrophic iron bacterium *Ferrobacillus ferrooxidans* 1. An improved medium and a harvesting procedure for securing high cell yields". *J. Bacteriol.*, 77, pp 642-647 (1959).

18. Atkins, A.S., "Studies on the Oxidation of Ferrous Sulphide in the Presence of Bacteria". *Ph.D. Thesis, University of Wales* (1976).

19. B.S. 1016, "British Standard method for the analysis and testing of Coal and Coke. Par 6". *Forms of Sulphur in Coal* (1969).

20. McAndrew, R.T., Wang, S.S., and Brown, W.R., "Precipitation of iron compounds from sulphuric acid leach situations". *C.I.M. Bull.*, Jan., pp 101-110 (1975).

21. Prosser, A.P., "Influence of mineralogical factors in the rates of chemical reaction of minerals". *Proc. 9th Commonwealth Min. and Met. Cong.*, 3, pp 59-91 (1969).

BACTERIAL LEACHING OF COPPER SULFIDE ORES

P.N. Rangachari,
V.S. Krishnamachar,
S.G. Patil,
M.N. Sainani, and
H. Balakrishnan

National Chemical Laboratory
Poona, India

Since most of the reserve copper of about 366 million tons is of low grade variety (1% or less copper) bacterial leaching of different Indian ores was studied. Water samples from different mine locations were collected and examination by standard procedures showed the presence of thiobacilli which oxidized sulfur as well as ferrous sulfate. Ten low grade copper ore samples from different mines with copper content of 0.2% to 1.5% were subjected to leaching studies under standard conditions with the six different samples of mine waters. Five percent ground ore (-90 mesh) was used in 9K medium at a pH of 2.0 - 2.5 in shake flask experiments at 28°C. The following observations were made: (i) The mixed flora from the different samples of mine water showed leaching of 40-80% in 90 days. (ii) The different ore samples showed variation in leachability of 20-40%. By repeated subculture, bacterial strains were obtained which can tolerate 25,000 ppm of Cu^{2+}. Repeated subculturing of two of the mixed flora and two standard thiobacillus cultures in the presence of the ore and in the absence of a source of nitrogen in the medium gave rise to a mixed culture which can grow in the absence of added nitrogen and leached about 60% of the copper from the ore in 60 days. Studies on the nitrogen metabolism of these cultures are in progress.

I.   INTRODUCTION

   Colmer and Hinkle (1) isolated the organism *Thiobacillus
ferrooxidans* and earlier Waksman and Joffe (2) had isolated
another organism of the related species, *Thiobacillus thiooxidans*.
Since then the economic and scientific interest in these organisms has steadily grown and their role in the biodegradation
of copper sulfide ores had been recently reviewed by Silverman
and Ehrlich (3), Beck (4), Ehrlich and Fox (5), Zajic (6),
Miller (7), Trudinger (8), Karavaiko, et al. (9), Tuovinen and
Kelly (10, 11), and Dutrizac and MacDonald (12). With respect to
copper, it has been recently estimated that   5-10% of World's
annual production is contributed by the biodegradation process
using low grade ores [Malouf (13)]. Most of the reserve copper
of about 366 million tons in India is of the low grade variety
(1% or less of copper) and in order to supplement the indigenous
production of 61,000 tons per year, the project on bacterial
leaching of ores was undertaken to isolate the organisms and
develop the necessary technology. The present paper is a report
of the work carried out to locate the thiobacillus group of
organisms which could show good leachability of the available
copper sulfide ores.

II.   MATERIALS AND METHODS

A.   Source of the Organism and the Ores

   Mine water samples were collected from different locations
in the country especially copper mining areas like Mosaboni,
Khetri, Amjohr, and the coal mining areas like Assam, Dhanbad,
etc. Low grade copper sulfide ores which are mostly of the
chalcopyrite type were acquired through the courtesy of various
organizations; they were powdered and the material passing
through a -90 mesh was used in the experiments.

B. Culture Technique

Preliminary studies to isolate the organisms were carried out (i) in the sulfur medium used by Waksman and Joffe (2) for the isolation of *Thiobacillus thiooxidans*, and (ii) in the 9K salts medium of Silverman and Lundgren (14) for the study of *Thiobacillus ferrooxidans*. In the case of the former the medium had the following composition: $(NH_4)_2SO_4$ 2.0 g, KH_2PO_4 1.0 g, $MgSO_4 \cdot 7H_2O$ 0.5 g, KCl 0.5 g, $FeSO_4 \cdot 7H_2O$ 0.01 g, sulfur 10.0 g, $Ca_3(PO_4)_2$ 10.0 g, distilled water 1 litre. The initial pH of the medium was between 5.6 to 6.0. 100 ml of the medium was placed in 500 ml Erlenmeyer flask, sterilized as usual and then inoculated with 10 ml of the mine water samples. After incubation in a reciprocatory shaker at 28-30°C, samples were periodically removed and the increase in acidity was titrated against standard sodium hydroxide.

For the isolation of the *ferrooxidans*, the 9K salts medium was employed, inoculated with the mine water samples (10 ml), incubated as indicated above and periodic samples were analysed for the decrease in the ferrous iron content.

Leachability studies had been exclusively carried out in the 9K salts medium using a high concentration of ferrous sulfate. In some experiments the addition of ferrous sulfate had been omitted and instead pyrites of a known composition had been used as an energy source for the growing organisms.

In the experiments without a source of nitrogen the medium of the following composition was employed: $(NH_4)_2SO_4$ 0.15 g, KCl 0.05 g, $MgSO_4 \cdot 7H_2O$ 0.5 g, K_2HPO_4 0.05 g, $Ca(NO_3)_2$ 0.01 g, distilled water 1 litre, pH 2.5. The ammonium sulfate was omitted from the medium and the calcium nitrate was replaced by an equivalent quantity of $CaCl_2$ and 1.0 ml of a 10% $FeSO_4 \cdot 7H_2O$ was added to 100 ml of the medium before inoculation.

All the experiments were uniformly carried out in Erlenmeyer flasks which were shaken in a reciprocatory shaker with 180-200 revolutions per min. Incubation temperature was 28°-30°C.

C. Analytical Methods

Ferrous iron estimation was carried out using orthophenathroline for color development according to Willard (15) and total iron was estimated after reduction of a suitable aliquot with hydroquinone in acetate buffer (15). Copper was estimated by the method of Kuang Lu Cheng and Bray (16) using sodium diethyldithiocarbamate as the complexing agent.

For making media, chemically pure grade materials were employed while for analytical estimations, chemicals of the available highest grade purity were used.

III. RESULTS AND DISCUSSION

For preliminary isolation of the organisms, the increase in acidity as indicated by the increase in sodium hydroxide consumed was used as an indicator. Those samples which showed the highest acidity in the shortest time were designated as samples having the *T. thiooxidans* type of cultures. The rapid disappearance of ferrous iron in the 9K medium indicated the presence of the *T. ferrooxidans* type of organisms. Many of these mixed flora were maintained in the respective media (sulfur or iron medium) for routine subculturing and this was carried out once in a month. They have been with us now for the last four years without much loss in activity. In the following, these mixed flora are designated as Amj II, PCC-B, NML-JI, and Khetri.

The leaching experiments were carried out exclusively in the 9K salts medium using ferrous sulfate as the energy source. 5 g of the ground ore (-90 mesh) were used in 100 ml medium, the initial pH was adjusted to 2.0 to 2.5 with sulfuric acid, the

medium was sterilized and then inoculated with 10 ml of the
actively growing culture (invariably 4-6 day grown mixed culture).
Controls were run without inoculum and since the duration of the
experiments was quite long   controls contained 1% phenol as a
bactericidal agent. After inoculation, zero hour samples were
also removed and suitable aliquots of about 10 ml were withdrawn
periodically. Estimations of ferrous and total iron and of copper
were carried out after centrifugation. The accompanying Table 1
shows the ores from the different sources with the composition of
the few important elements.

Leachability of copper was studied with the mixed culture
Amj II and the results are shown in the Figs. 1 and 2. As is
evident, the leachability varies in a 90 day period between a
minimum of 20% in the case of ore A and a maximum of 70-75% with
the ores B, D, E, and F. In the next series of experiments the
energy source of ferrous sulfate was replaced by pyrite (0.5 g
for every 5 g of ore). The extraction was carried out with a few
of the ores using the mixed flora from the four different sources.
These results are shown in Figs. 3, 4, 5, and 6. One can observe
a clearer picture and a better copper extraction, reaching even

TABLE I *Composition of the Different Ores*

| Ore | % Content of | | | Location of Ore |
	Cu	Fe	S	
A	1.4	3.5	0.48	Malanjkhand
B	0.3	1.7	1.10	Balaghat
C	1.0	3.99	0.50	Malanjkhand
D	0.4	9.25	0.91	Mosaboni
E	0.65	2.73	0.86	Balaghat
F	0.20	2.10	0.34	Rakha
G	1.5	12.9	4.0	Khetri
K	0.7	13.04	1.22	Rakha
L	0.77	12.83	2.06	Surdha
Pyrite	-	33.25	33.00	

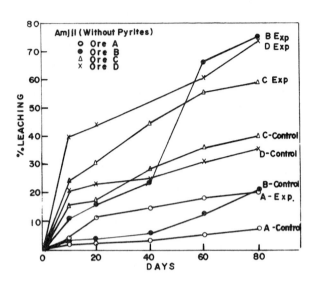

Fig. 1.   Amj II (experiments A-D without pyrites).

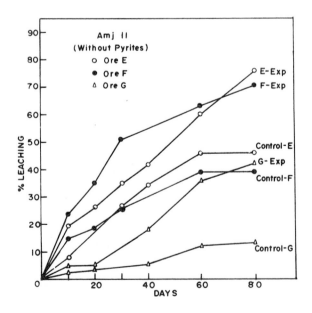

Fig. 2.   Amj II (experiments E-G without pyrites).

up to 90%. These figures also show a variation in extractability
with the different ores in the case of the individual mixed flora.
A comparison of the leachability of the set of ores B, D, and K by
the different samples of organisms shows a wide variation, for
example, ore B varies from 87% with Amj II to about 60% with
Khetri, ore D 70% with PPC-B to 12% with Khetri, and ore K 87%
with Amj II to 57% with Khetri. These types of observations have
been reported in literature [Zajic (6), Sutton and Corrick (17)],
Corrick and Sutton (18) had even indicated that in their experiments only one out of 13 different samples showed leachability.
The results observed could be attributed (i) to the different
activity of the microorganisms present in the mixed flora, and
(ii) to the nature of the ore formation. Petrological studies had
not been carried out with the different ore samples.

In order to select strains fromof the present group of mixed
flora with the highest possible tolerance to copper, the organisms
isolated from the mine water samples and designated Amj II, PCC-B,
NML-JI, and Assam II were inoculated into the 9K salts medium
containing ferrous sulfate as energy source and also varying
amounts of copper as copper sulfate. Initial pH was 2.5. The
flasks were kept on the shaker at 28-30°C and periodically samples
were removed for the estimation of ferrous iron. The decrease in
iron was used as an indication of the growth of the organism.
Starting with 5 ppm of Cu^{2+}, the concentration was gradually increased, the growth in an earlier concentration was used to
inoculate the medium containing a higher concentration of copper;
the process was continued by following the disappearance of
ferrous iron and it had thus been possible to reach a limit of
25,000 ppm. These four elective cultures are now under continuous
subculturing in the presence of 25,000 ppm of Cu^{2+} for about the
last two years. Experiments are under way to study the leaching
capacity of these organisms with the different types of ores.

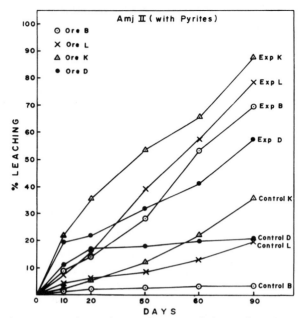

Fig. 3.   Amj II (experiments with pyrites).

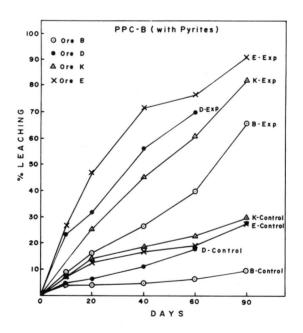

Fig. 4.   PPC-B (experiments with pyrites).

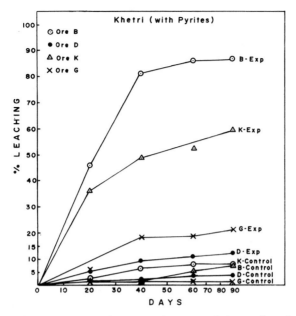

Fig. 5.   Khetri (experiments with pyrites).

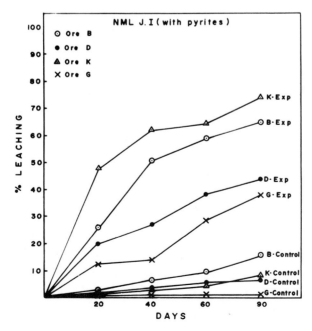

Fig. 6.   NML J.I (experiments with pyrites).

Since the main interest of this project had been to evolve out a technology with low costs, any type of alteration which can be economical was to be investigated. There had been a recent report by Tsuchiya, Trivedi, and Schuler (19) that they could demonstrate an increased leachability of copper when the nitrogen fixer *Beijerinckia lacticogenes* (NCIB 8846) was used along with a *Thiobacillus ferrooxidans*. The nitrogen fixer served to economize the use of the nitrogen source in the culture medium. Having this in mind, leachability experiments were carried out with Amj II, PPC-B, NCIB 8455 (*Thiobacillus ferrooxidans*) and ATCC 13598 (*Thiobacillus ferrooxidans*). Ore C having a copper content of 1% was used as the substrate. The first batch consisted of experimental flasks without the nitrogen source but contained the same nitrogen fixer *B. lacticogenes* (NCIB 8846) and the isolated cultures indicated above. Estimations of copper at the end of 20 days showed an extraction of 20-40% and 40-50% at the end of 40 days. There was leaching of copper but the results did not show any significantly larger extraction of copper as indicated by Tsuchiya, et al.(19). Hence, the use of the nitrogen fixer was discontinued. In order to find out whether this kind of phenomenon was peculiar only to the mixed flora from the mine water, similar experiments without a source of nitrogen were carried out with two standard cultures of the *Thiobacillus* sp. and it was found that these also showed good and equal leachability of copper viz. 30-40% in 20 days and 50-60% in 50 days. These experiments were continued by inoculating fresh media from a previous batch after it had been grown for 30 days. Growth was always followed by estimation of the decrease in the ferrous iron. Repeated subculturing helped to dilute even the traces of nitrogen and 20 such transfers have been so far carried out with the results indicated in Table II. The growth was recognized as indicated by the periodic estimations of ferrous and ferric iron. Copper estimations were also carried out.

TABLE II

Culture	Average % Cu^{2+} Leached Days of Incubation		
	15	30	50
Amj II	35-40	45-55	50-65
PPC-B	30-45	40-55	50-60
NCIB 8455	25-35	40-50	50-60
ATCC 13598	30-40	45-55	55-65

Since the growth of the organism was indicated both by the disappearance of iron and the accumulation of copper as compared to the controls, there was reason to believe that the *Thiobacillus* group of organisms could survive possibly without a source of nitrogen and fix nitrogen as well [Zimmerley, et al. (20); MacIntosh (21)]. This was indicated by the growth of the standard cultures. But it is possible that some minimal amount of nitrogen could have been provided in these standard cultures by the carry-over of dead cells; this amount should then be very small since only about 10 ml of inoculum were used. There is a possibility of some dissolved ammonia from the surroundings contributing to the nitrogen requirement. There is already such a report in the literature [Tuovinen, et al. (22)]. In the case of the mixed flora from the mine water samples, these could be factors contributing to the source of nitrogen for the growth; there could also be a nitrogen fixer in the mixed group of organisms which could provide sufficient nitrogen by fixation under the experimental conditions. A more detailed study including the nitrogen metabolism of these cultures is in progress to elucidate the observed results.

IV. ACKNOWLEDGMENT

One of the authors (P.N.R.) is thankful to Dr. L.E. Murr, Symposium Director, New Mexico Tech. for financial assistance. The authors wish to express their grateful thanks to Mr. S.R. Modak, Sri M.D. Deshpande and Mrs. U. Puntambekar for technical assistance in the early phases of this work and to Dr. V. Jagannathan for his continued interest. Our thanks are also due to National Metallurgical Laboratory, Jamshedpur, Central Mining Research Station, Dhanbad and Regional Research Laboratory, Bhubaneswar for the supply of mine water and ore samples.

V. REFERENCES

1. Colmer, A.R., and Hinkle, M.E., *Science*, 106, 253 (1947).

2. Waksman, S.A., and Joffe, J.S., *J. Bacteriol.*, 7, 239 (1922).

3. Silverman, M.P., and Ehrlich, H.L., *Adv. App. Microbiol.*, 6, 153 (1964).

4. Beck, J.V., *Biotechnol. Bioeng.*, 9, 487 (1967).

5. Ehrlich, H.L., and Fox, S.I., *Biotechnol. Bioeng.*, 9, 471 (1967).

6. Zajic, J.E., "Microbial Biogeochemistry", p. 345. Academic Press, New York, 1969.

7. Miller, J.D.A., "Microbial Aspects of Metallurgy", p. 202. Elsevier, New York, 1970.

8. Trudinger, P.A., *Min. Sci. Engng.*, 3, 13 (1971).

9. Karavaiko, G.I., Kuznetsov, S.I., and Golomzik, A.E., "Role of Microorganism in Leaching of Metals and Ores", p. 248. Nauka, Moscow, 1972.

10. Tuovinen, O.H., and Kelly, D.P., *Z. Allg. Mikrobiol.*, 12, 34 (1972).

11. Tuovinen, O.H., and Kelly, D.P., *Int. Metall. Rev.*, 19, 21 (1974).

12. Dutrizac, J.E., and MacDonald, R.J.C., *Miner. Sci. Eng.*, 6, 59 (1974).

13. Malouf, E.E., *Min. Eng.*, 23, 43 (1971).

14. Silverman, M.P., and Lundgren, D.G., *J. Bacterial.*, 70, 642 (1959).

15. Willard, H.H., *Ind. Eng. Chem. Anal. Edn.*, 10, 13 (1938).

16. Kuang Lu Cheng and Bray, R.H., *Anal. Chem.*, 25, 655 (1953).

17. Sutton, J.A., and Corrick, J.D., "U.S. Bur. Mines Rept. Invest.", 5839, p. 16 (1961).

18. Corrick, J.D., and Sutton, J.A., "U.S. Bur. Mines Rept. Invest.", 6714, p. 16 (1965).

19. Tsuchiya, H.M., Trivedi, N.C., and Schuler, M.L., *Biotechnol. Bioeng.*, 16, 991 (1974).

20. Zimmerley, S.R., Wilson, D.G., and Prater, D.G., U.S. Patent 2,829,964 (1958).

21. MacIntosh, M.E., *J. Gen. Microbiol.*, 66, i (1971).

22. Tuovinen, O.H., Niemela, S.I., and Gyllenberg, H.G., *Biotechnol. Bioeng.*, 13, 517 (1971).

CONTINUOUS BIOLOGICAL LEACHING OF CHALCOPYRITE CONCENTRATES:
DEMONSTRATION AND ECONOMIC ANALYSIS

R.O. McElroy
A. Bruynesteyn

B.C. Research
Vancouver, B.C.
Canada

Results of a two year study on biological leaching of chalcopyrite concentrate are presented. Technical and economic aspects of the process are discussed.

Continuous biological leaching of chalcopyrite has been demonstrated using a 30 l single stage reactor; steady state copper leaching rates obtained were sufficient to produce electrowinning grade leach solutions in 50 hours. Chalcopyrite in leach residues was recovered by flotation, reground and releached to obtain an overall extraction of >96%. Locked cycle flotation and leaching tests showed no buildup of unrecoverable copper; process losses are due to the copper content of jarosite (waste) and chalcopyrite lost in flotation.

Leach solution grade can be controlled at levels in the range of 20->50 g/l copper. Satisfactory cathode copper has been produced by direct electrolysis of leach solution and by LIX-electrowinning technology.

Barren liquor from electrowinning or LIX raffinate can be recycled to the process after partial neutralization with limestone. Solid wastes including gypsum and jarosite are suitable for disposal with mill tailing or in a separate impoundment.

On the basis of the experimental results, flowsheets and equipment lists have been prepared for a leaching plant to treat 200 tpd of chalcopyrite concentrate.

Projected capital and operating costs are presented with an analysis of suitable applications of the method.

Modifications of the process to recover gold and silver from leach residues are described.

I. INTRODUCTION

B.C. Research has been involved in technical studies on microbiological leaching since 1958. In the early phases of these studies the most significant development was a laboratory procedure which allowed rapid determination of the amenability of ore samples to microbiological leaching (1). Careful optimization of all the parameters affecting the performance of bacteria finally resulted in a set of conditions under which very rapid leaching of ore minerals occurred. As a natural outgrowth of these experiments on low grade ores and waste rock, leaching tests on sulphide mineral concentrates were initiated. Results of the early tests suggested that if certain problems, particularly the incomplete leaching of chalcopyrite, could be overcome, then biological leaching of sulphide concentrates might be an economically and environmentally attractive method for treatment of copper and other sulphide concentrates (2).

A program was developed for a two year study to determine the technical and economic feasibility of biological leaching as a process for copper production from chalcopyrite concentrates. Funding for the program was obtained from a consortium of eight mining-, engineering-, and copper consuming companies. This paper presents the highlights of that program.

II.   TEST PROCEDURES

A.   Leaching

The strain of leaching bacteria *(Thiobacillus ferrooxidans)* used in this study was obtained in 1958 from the Britannia Mine near Vancouver and has been maintained in laboratory culture since that time (3).

Continuous leaching tests were conducted in an air sparged, turbine agitated tank shown schematically in Figure 1.

The total volume of the tank was 30 ℓ, but the effective (slurry) volume was maintained at 22 ℓ. (Freeboard was required to prevent overflow of foam).

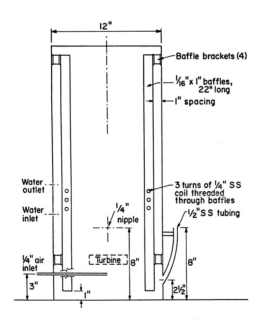

Fig. 1.   Leaching tank (stainless steel construction through-out) with turbine agitator, baffles, heating/cooling coil, and air sparger.

Dry copper concentrate, ground to 100% -400 mesh, was introduced continuously to the leaching tank by a BIF "Omega" variable speed disc feeder. Water and/or stripped and neutralized recyled leach solution were added by metering pumps. Air containing 0.1% CO_2 was introduced through the sparger located below the center of the turbine. Airflow was adjusted to maintain dissolved oxygen levels of >80% of air saturation in the slurry. Leached slurry was removed from the tank as overflow.

Concentrate was prepared for leaching by wet ball milling for 3 hours in a 50 cm X 50 cm (inside dimensions) batch ball mill charged with approximately 130 kg of 1.25 cm dia steel balls, 20 kg of concentrate, and 10 ℓ of water. The ground material was air dried at ~40°C after settling and decantation of excess water. Dried concentrate was crushed and blended with nutrient salts (ammonium sulphate, and dipotassium hydrogen phosphate at levels of 20 and 3 g/kg of concentrate, respectively) before being placed in the feeder bin.

Concentrate feed rates in the range of 100 - 400 g/hr were used, depending on concentrate leachability, tank performance, and the pulp density being tested. The liquid feed rate was normally adjusted to maintain ~500 ml/hr slurry overflow.

The pH of the leach slurry was maintained in the range 1.7-2.2. Fine ground calcium carbonate was used either mixed with concentrate or as batch additions to neutralize excess acid; on rare occasions, technical sulphuric acid was added to lower the slurry pH.

B. Source of Concentrates

Results presented in this paper were obtained in tests on chalcopyrite concentrate (23.8% Cu, 31.6% Fe) from the Craigmont Mines Ltd. operation near Merritt, B.C. During the study, concentrates from four other mines were successfully tested in continuous operation.

C. Flotation Testing

Initial flotation tests were done in a Denver D-2 bench cell operated at 1800 rpm, with the aeration rate adjusted to maintain adequate froth level. Bulk flotation of leach residue was accomplished in a 2 X 85 ℓ cell Galigher unit. Locked cycle flotation tests were done in the bench scale unit, using leach residue thickened to 17% solids. Each cycle consisted of four stages of flotation; the first stage concentrate was considered to be the recycled product, while the second, third and fourth concentrates were retained for addition in the appropriate stage of the succeeding cycle. No flotation reagents were added in these tests.

D. Analytical Methods

Solutions from leaching and copper recovery tests were analyzed by atomic absorption spectrophotometry; solids were digested by wet chemical methods before analysis.

The composition of the jarosite (acid soluble) fraction of leach residue was determined by treatment of filtered solids with cold ($0^{o}C$) 1N hydrochloric acid solution and sulphur dioxide. Under these conditions, jarosite dissolves readily and ferric ions are rapidly reduced by sulphur dioxide (sparged into solution) so that no significant amount of copper is released by reaction of ferric ions with residual chalcopyrite.

III. RESULTS AND DISCUSSION

A. Continuous Leaching of Chalcopyrite Concentrate

Previous studies (1, 2, 4) have established the feasibility of microbiological leaching of chalcopyrite concentrate in batch operation. Under optimum conditions, biological leaching rates (copper release rates) of 500-700 mg/ℓ/hr are readily obtained after a short incubation period. For industrial applications, however, continuous leaching is desirable to facilitate plant operation and eliminate the incubation or "lag" time associated with batch fermentation.

The leaching tank was operated initially in batch culture; when satisfactory performance was established (i.e., batch leaching rates >400 mg/ℓ/hr), feeding of concentrate, nutrients and water was started. No significant problems were encountered in this shift from batch to continuous operation and operational runs of several weeks duration were achieved. Mechanical problems in the solids feeder were the most frequent cause of upsets in the system, but re-establishment of continuous operation did not pose major problems. After brief shutdowns (1-4 hours) steadystate operation resumed almost immediately on start up. Extended shutdowns or (most commonly) extended periods with no addition of concentrate, resulted in increased lag time before re-establishment of satisfactory leaching rates.

Under the initial operating conditions (10% pulp density), continuous leaching rates were consistently in the range of 200-300 mg/ℓ/hr, i.e. signficantly lower than rates for the same concentrates in batch operation. Review of the test conditions revealed that the reactor was being operated under substrate-limiting conditions. Short circuiting of the single stage reactor was reducing the amount of leachable material to the point that the rate was limited by the amount of feed present. Subsequently, the pulp density was increased to 40%. When steady state operation was established, continuous leaching rates of ≥400 mg/ℓ/hr were obtained. The tank performance over an operating period of 6 days is shown in Figure 2.

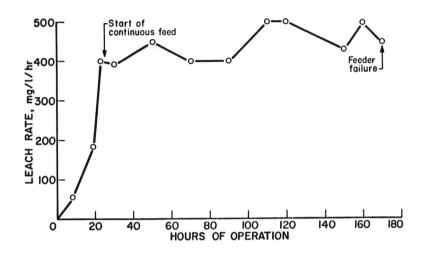

Fig. 2. Copper leaching rate in continuous operation at 40% pulp density.

This operational period was equivalent to 2.4 residence times, and there was no indication that - barring mechanical failures - the high rate could not be maintained indefinitely. During this period, dissolved copper levels of 35-50 g/ℓ were obtained.

It should be emphasized that, in a pilot or commercial leaching system, operation at high pulp densities will not be necessary because leaching will be done in a series of four or more tanks. In such a system the residence time of all particles in the system will be approximately equal (i.e., negligible short circuiting).

B. Recovery of Chalcopyrite from Leach Residue

One of the major problems to be overcome in the leaching of chalcopyrite concentrates was the fact that they generally only leach to about 50% extraction under biological leaching conditions. In an attempt to overcome this problem tests were run to determine if the chalcopyrite content of leach residues could be recovered by froth flotation.

When aerated in a flotation cell, without addition of any reagents, leached slurry forms a brittle but manageable froth in which chalcopyrite is preferentially concentrated. The results of a four stage, four-cycle flotation test on leach residue are presented in Table 1.

TABLE I   *Results of Locked Cycle Flotation Tests on Residue from Continuous Leaching of Chalcopyrite Concentrate*

Cycle	Stage	% Cu in Concentrate	% Cu in Final Tailing	% Cu Recovery in Flotation	Overall % Cu Recovery (Leaching and Flotation)
1	1	24.4			
	2	16.2			
	3	7.3			
	4	5.0	4.1	94.3	97.0
2	1	23.2			
	2	15.5			
	3	8.0			
	4	5.1	4.0	95.1	97.5
3	1	22.9			
	2	15.6			
	3	7.7			
	4	5.3	4.2	94.1	96.9
4	1	23.4			
	2	15.0			
	3	7.9			
	4	4.7	4.3	94.4	97.1

These data show that in cycles 2-4, the grade of the first stage concentrate (for recycle) stabilized at approximately 23% copper, only slightly less than the original feed grade of 23.8% copper. Middling grades were quite consistent throughout the four cycles. The fourth stage produced very little concentration.

Of the copper lost in tailing, ~60% was unrecovered chalcopyrite and ~40% was present in the acid soluble (jarosite) fraction.

C. Leaching of Recycled Chalcopyrite

A series of four batch-leaching/flotation tests were conducted to confirm that chalcopyrite recovered from leach residue and reground, could be recycled to extinction within the process. After each test, the first stage flotation concentrate was mixed with an appropriate amount of the original concentrate, ground, and leached in the 30 ℓ tank. The second stage flotation concentrate was reserved for addition to the leach residue from the subsequent test. Results are presented in Table II.

These results indicate that recycling of chalcopyrite is feasible; in tests 2-4 copper extractions and leaching rates were better than for the fresh feed alone. Equipment and time limitations necessitated use of only two stages of flotation, and grinding in small (30 cm ID) mills was less effective than in the 50 cm ID mill normally used. These factors account for the relatively low recoveries (compared to the 4-stage locked cycle test), and the relatively low extraction values.

TABLE II   *Recycling of Unleached Chalcopyrite*

Batch Test No.	Initial Tank Conditions		% Cu in Recycle Feed	Extraction %
	% Solids	% Cu in Solids		
1	17.8[a]	22.9[a]	–	33.0
2	18.0	22.4	19.2	39.1
3	16.8	23.8	20.9	44.0
4	17.8	22.6	20.7	40.0

Batch Test No.	Maximum Leach Rate (mg/ℓ/hr)	2-Stage Flotation Recovery %	Total Recovery (Leaching and Flotation) %
1	283	79.7	86
2	445	83.0	90
3	415	84.4	91
4	430	83.4	90

[a]*Ground concentrate plus inoculum.*

Following these controlled tests, residue from continuous runs was routinely collected and treated by 2-stage batch flotation in a pilot scale (2 X 85 ℓ cells) Galigher unit. These concentrates were reground with fresh feed and recycled to leaching. Control over the concentrate grade and recovery was less precise than in the small scale batch tests, but satisfactory leaching performance was obtained over several weeks of operation. There was no evidence of a buildup of unleachable material or of any deterioration in leaching rates.

D. Copper Recovery from Leach Liquor

The copper content of leach solution can be maintained at levels up to 70 g/ℓ by adjustment of operating variables including pulp density and recycle of leach liquor. In this study, the continuous leaching system was normally adjusted to produce pregnant solution containing 25 to 30 g/ℓ of copper at a pH of 1.5-1.8. The iron content of the pregnant solution depended upon the concentrate being tested, but was generally in the range of 4 to 7 g/ℓ, entirely as ferric salts.

Pregnant solution was separated from leach residue by decantation after flocculation and settling. Of the flocculants tested, "Jaguar MDD" (Stein Hall Ltd.) was found to be most satisfactory at a dosage of approximately 2 lb/ton of residue. Two methods of copper recovery were investigated: direct electrowinning and LIX-electrowinning using LIX 65N and LIX 70.

The LIX laboratory studies were conducted by General Mills Chemicals, Inc. Use of LIX 65N on a leach solution containing 28 g/ℓ copper produced raffinate containing 2.1 g/ℓ copper (92.5% recovery); use of LIX 70 on a leach solution containing 30 g/ℓ copper produced a raffinate containing 0.1 g/ℓ copper (99.7% recovery). Iron rejection in these tests was good; only 0.01 g/ℓ of iron was transferred from the leach solution through the organic phase to the electrolyte.

Cathode copper was produced from the stripping electrolyte; analyses are not available, but it is assumed that the quality of the metal would be comparable to that produced by operating plants.

The solutions provided from these tests were produced by decantation without flocculant treatment, and some sludge formation was observed in the LIX system due to fine solids present in the leach solution. Flocculation and/or a polishing filtration would probably be required in practice.

Direct electrowinning of copper from leach solutions was investigated in batch tests and by continuous operation in a laboratory cell equipped with lead anodes and an aluminum starter sheet. Laboratory tests on direct electrowinning indicated that, due to the relatively low copper and free acid content of the solution, low current densities are required to produce satisfactory deposits. To maintain cathode quality while operating the cell at 6 g/ℓ copper, the maximum feasible current density was 8 amps/sq ft; operating at 20 g/ℓ, the maximum feasible current density was approximately 15 amps/sq ft. In a typical test, cell parameters were as follows:

Current density	8 amps/sq ft
Cell voltage	2.2-2.4 v
Current efficiency	90%
Cell temperature	30-34°C
Agitation	none
Solution	28 g/ℓ Cu^{++}, 2 g/ℓ Fe
Feed	"
Terminal solution	6 g/ℓ Cu^{++}, 2 g/ℓ Fe

Chemical analyses of copper produced in a representative batch test on solution from a chalcopyrite concentrate are presented in Table III.

TABLE III   *Analyses of Laboratory Produced*
Cathode Copper Samples

	Element	Concentration (%)
Above Detection Limit	Zinc	0.0037
	Iron	0.0002
	Calcium	0.0004
Below Detection Limit	Nickel	<0.0001
	Lead	<0.0005
	Aluminum	<0.001
	Chromium	<0.0001
	Tin	<0.0005
	Arsenic	<0.0001
	Sulfur	<0.002

E.   Neutralization and Recycle of Spent Electrolyte

Partial neutralization (to pH 2-2.5) of spent electrolyte containing 6 g/ℓ Cu and ~30 g/ℓ H_2SO_4 was done by addition of fine (-200 mesh) limestone under strong agitation. Provided that the pH was not raised above 2.5, a free settling, fast filtering gypsum ($CaSO_4 \cdot 2H_2O$) product was produced. In the laboratory, a single displacement water wash using 10% of the volume of the electrolyte produced a filter cake containing <0.1% copper. Recycling of the filtrate had no deleterious effect on leaching.

F.   Recovery of Precious Metals from Leach Residue

Development of techniques for recovery of gold and silver from leach residues was not an initial objective of the program. However, when it was discovered (5) that addition of silver compounds to fine ground chalcopyrite at a level of 50 mg of silver/kg of concentrate increased the normal single stage extraction from approximately 50% to over 80%, it became desirable to investigate possible recovery methods for silver.

Analyses of leach residues from tests with added silver indicated that the silver was concentrated in the acid (HCl) soluble fraction; presumably as argento-jarosite. (i.e., only silver present in the sulfide matrix of the concentrate would be recovered by flotation of the leach residue). Recovery of silver from dilute (approximately 50 mg/ℓ) solutions containing large amounts of iron and copper was not considered practicable, so acid leaching of biological leach residues under reducing conditions was investigated as a method for converting the silver to an insoluble form which could be recovered (with residual chalcopyrite) for subsequent copper-silver recovery by conventional smelting.

Treatment of leach residue with spent electrolyte (containing 6 g/ℓ Cu and ~36 g/ℓ H_4SO_4) and SO_2 (1 atmosphere pressure) for ten minutes followed by filtration, produced an acid insoluble residue containing >99% of the silver added to the leach slurry. This residue, consisting mainly of chalcopyrite, pyrite and siliceous material, contained 19.2% copper, compared to the original concentrate grade of 22% copper. However, since 80% of the copper originally present was extracted by leaching, the weight of concentrate to be smelted for copper and precious metal recovery was only 23% of the original weight. Therefore, the original silver content of the concentrate would be increased by a factor of 4.3, thereby increasing the proportion of original silver which would be payable under normal smelter contracts. This may be advantageous since a representative smelter deduction for silver is 2.2 oz/ton (75 g/metric ton)of concentrate.

The reducing acid leach has not been tested to determine its effect on the gold content of leach residue, but the chemical properties of gold suggest that it would be concentrated with silver in the insoluble fraction.

G. Molybdenum Content of Leach Residue

Molybdenite (MoS_2) in copper concentrate is - for all practical purposes - unaltered by the leaching procedures used in this study. In flotation of leach residue molybdenite is concentrated with chalcopyrite, so - in closed circuit - molybdenite would build up as a circulating load. This study did not include any assessment of recovery methods for molybdenite.

IV. FLOWSHEET DEVELOPMENT

Results of the laboratory study were used to develop the flowsheet presented in Fig. 3. Fresh concentrate from storage (unit 1) moves via a feeder (unit 2) to a ball mill feed tank (unit 3) which also receives recycled concentrate from residue flotation; pH is controlled by addition of ground limestone (units 3, 4). Slurry from the feed tank passes through a ball mill (unit 5) to hydrocyclone classification (unit 7); cyclone overflow (containing -10 micron solids) passes to a leach feed thickener (unit 8).

Thickener underflow passes via a feed tank (unit 9) to aerated, agitated leaching tanks (unit 10) arranged in series. Leached slurry passes to a thickener (unit 12); underflow passes to flotation (unit 11) and overflow to the pregnant solution tank.

Flotation tailing is treated with flocculant and passed to a thickener (unit 14). Tailing thickener underflow passes to washing (countercurrent decantation (CCD) or filtration (unit 15)) and disposal; overflow passes to the pregnant solution tank. Flotation concentrate is also thickened (unit 13),

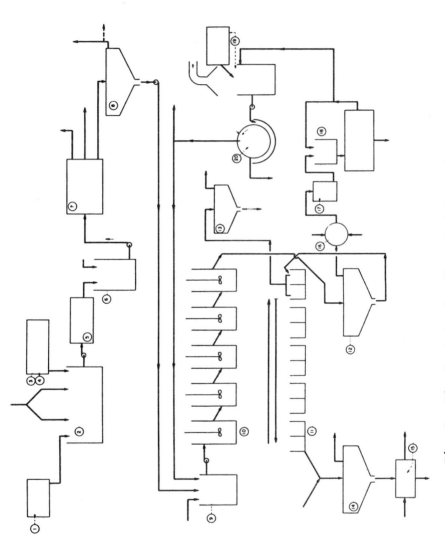

Fig. 3. Flowsheet for leaching of chalcopyrite concentrate.

with overflow to the pregnant solution tank and underflow to
the ball mill feed tank.

Pregnant solution is filtered (unit 17) and passed to the
tankhouse (unit 18) for electrolysis. Spent electrolyte is neutralized to pH 2.5 with limestone (unit 19), filtered (unit
20) and returned to the head of the leaching or milling
circuits.

From this flowsheet, an equipment list and specifications
were prepared for a plant to treat 200 tons/day of chalcopyrite
concentrate containing 25% copper. Engineering and manufactur-
ing firms (particularly Dravo Corp., and Greey Mixing Ltd.)
provided assistance in sizing and specifications for the
ball mill, hydrocyclone unit, leaching tanks and aeration
equipment. (The grinding/classification circuit is the only
unit operation in the flowsheet which was not investigated
in the laboratory program).

V. ECONOMIC ANALYSIS

A. Capital Cost

On the basis of the flowsheet, equipment list and specifications developed from test data, Dravo Corporation prepared a capital cost estimate for a plant to treat 200 tons/day
of chalcopyrite concentrate. The estimate included engineering, procurement, management, construction and profit, but
excluded the cost of a level site, insurance and taxes. A
summary of this estimate (1973 U.S. dollars) is presented in
Table IV.

B. Direct Operating Cost

Estimated direct operating costs for the 200 ton/day
facility (based on 1973 British Columbia labour-, power- and
material costs in 1973 U.S. dollars) are summarized in Table V.

TABLE IV   Summary of Capital Cost Estimates

| Facility | Estimated Capital Cost | |
	Facility ($)	$/Annual Net Ton Cu[a]
Leach plant	9.4×10^6	560
Electrowinning plant	4.2×10^6	250
Total	13.6×10^6	810

[a] Assuming 350 operating days/year and 96% copper recovery.

TABLE V   Summary of Estimated Direct Operating Costs

| Operation | Costs | | % of Total | Cumulative % of Total |
	$/net T Cu[a]	¢ of lb Cu[a]		
Labour	41.83	2.09	28.6	28.6
Power (10 mills/kwh)	41.20	2.06	28.2	56.8
Maintenance (5% of fixed capital)	40.47	2.02	27.7	84.5
Limestone ($5.00/T)	15.00	0.75	10.2	94.7
Flocculants	4.17	0.21	2.8	97.5
Nutrients	2.47	0.12	1.7	99.3
Grinding Balls (cast steel)	1.20	0.06	0.7	100
Total	146.34	7.31	100	

TABLE VI   Copper Extraction and Oxygen Consumption in Biological Leaching of "Sulphide" Copper Minerals

Mineral	Biological Copper Extraction[a] (%)	Stoichiometric Oxygen Requirement (kg/kgCu)
Chalcopyrite, $CuFeS_2$	45–55	2.14
Bornite, $CuFeS_4$	70–80	0.93
Covellite, CuS	90–100	1.01
Chalcocite, Cu_2S	90–100	0.63

[a] Normal value for -400 mesh material under optimized conditions.

C. Overall Costs

Using the estimates presented, with suitable modifications
for particular circumstances, allows an estimation of overall
operating costs per unit of copper produced. To take a simplified example, assume that the total investment for a 200 tpd
plant (fixed capital of $13.6 x 10^6 + working capital of
$13.6 x 10^5) is to be repaid by annuity at 10% interest over
a 15 year period. On this basis, financing cost per pound of
copper produced is 5.95 cents. Adding the estimated direct
operating cost (7.31 cents, Table V) produces a total treatment
cost 13.26 cents per pound of copper.

In 1973, smelter charges per pound of copper in concentrate
were quoted in the range of 8.5-14.5 cents per pound, so (estimated) biological leaching costs appeared to be at the high
end of the range. However, this comparison does not include any
allowance for possible savings on freight costs, or elimination
of penalties for problems elements (e.g. arsenic, antimony,
mercury).

It should also be noted that - compared to other "sulfide"
copper minerals - chalcopyrite is the most refractory in terms
of response to biological leaching.

As shown in Table VI, chalcopyrite has the lowest (single
stage) copper extraction value and the highest oxygen consumption of the common "sulfide" copper minerals. Unfortunately,
chalcopyrite is the most important ore mineral of copper. Never-
theless, many operations produce copper concentrates with substantial quantities of bornite and chalcocite. For these operations, capital and operating costs would be substantially lower
than for treatment of chalcopyrite.

Assuming that an appropriate recovery method can be developed, the concentration of molybdenite in leach residue may also represent an economic advantage for biological leaching.

The 200 tons/day plant flowsheet includes 5 series of leaching tanks so - with some increase in unit cost - optimum sized leaching units could be used to treat as little as 40 tons/day of concentrate.

The factors noted above, and other site specific considerations, may have a substantial effect on the economics of an individual project. Examination of present smelter schedules and escalated costs for copper produced from biological leaching of chalcopyrite suggests that the relative cost comparison developed in 1973 is still applicable.

VI. APPLICATIONS OF CONCENTRATE LEACHING

This experimental study has demonstrated the technical feasibility of biological leaching for processing of chalcopyrite concentrates. Results of the economic analysis, however, suggest that the process may be only marginally competitive with conventional smelting unless or until restrictions on atmospheric emissions force a substantial increase in smelting charges. Also, the process is power intensive, which may decrease its competitiveness in areas of relatively high power costs. (As an historical aside, it is interesting to note that in mid-1973, when the initial cost analysis was done, a power cost of 10 mills/KWH was regarded as conservatively high).

On the basis of available data, the process does have a number of competitive advantages including:

- negligible atmospheric emissisons
- production of refined copper (i.e. potentially lower transport cost and increased market flexibility).

- feasibility for small scale minesite operations (minimum ~40 tpd of concentrate).
- reduced cost for concentrates containing appreciable bornite, chalcocite, etc., regardless of grade or pyrite content.
- potentially increased returns for silver, gold and (possibly) molybdenite in concentrates.
- production of dilute sulfuric acid suitable for leaching of oxide ore and/or mill tailing and/or acid consuming waste dumps.

These features suggest that biological leaching should be considered as an option where one or more of the following factors apply:

- severe restrictions on atmospheric emissions from smelting.
- high transport costs for concentrate.
- local market for refined copper.
- local use for byproduct acid.
- concentrates contain bornite, chalcocite, etc. (especially if pyrite content is high and/or grade is low).
- concentrates contains significant amounts of molybdenite.
- high precious metal prices and/or high smelter deductions.

A combination of any of these factors and the feasibility of small scale operation may make biological leaching an attractive option for minesite production of copper.

VII. ACKNOWLEDGEMENTS

Financial support from the following companies is gratefully acknowledged: Amax Exploration Inc., Bagdad Copper Corp., Canadian General Electric, Dravo Corp., Kupferbergbau GmBh, Phelps Dodge Corp., Placer Development Ltd., Rio Algom Mines Ltd.

Cost estimates, drawings and other technical services were provided by Dravo Corp; Mr. R.A. Danielle was particularly helpful in arranging for these services.

Greey Mixing Co. Ltd. provided data on aeration equipment and power requirements.

General Mills Chemicals Ltd. performed the LIX tests.

B.C. Research staff members who contributed extensively to the study included: Dr. D.W. Duncan, Mr. D.R. Watt, Ms. H. Kurtz and Ms. M. Lewis.

VIII. REFERENCES

1. Duncan, D.W., Trussell, P.C., and Walden, C.C. Leaching of chalcopyrite with *Thiobacillus ferrooxidans:* effect of surfactants and shaking. *Appl. Microbiol. 12; 122* (1964).

2. Brynesteyn, A., and Duncan, D.W. Microbiological leaching of sulfide concentrates. *Canadian Metallurgical Quarterly 10; 57.* (1971).

3. Razzell, W.E., and Trusell, P.C. Isolation and properties of an iron oxidizing *Thiobacillus. Bacteriol. 85; 595.* (1963).

4. Duncan, D.W. and McGowan, C.J. Rapid bacteriological metal extraction method. Canadian Patent 869, 470 (April 27, 1971); U.S.

5. McElroy, R.O., and Duncan, D.W. Copper extraction by rapid bacteriological process. U.S. Patent 3,856,913 (Dec. 24, 1974). 1974.

EXAMINATION OF A COPPER ORE AFTER LEACHING WITH BACTERIA

N.W. Le Roux
and
K.B. Mehta

Warren Spring Laboratory
Stevenage
Herts SG1 2BX
England

Low-grade ore has been bacterially leached on many occasions but only limited attention has been given to the state of the ore and the bacterial population after leaching. In this report the state of a partially leached copper ore is described after it had been subjected to four years bacterial leaching in a column. The ore of particle size -38 mm initially contained 0.68% copper as chalcopyrite. The column was sawn open without disturbing the ore bed. Samples of rock and precipitates were taken at 0.3 m intervals, transferred aseptically into modified Silverman Lundgren medium and growth of bacterial monitored. Representative samples of rock were analysed for copper and total iron and two visually different precipitates on the rocks were examined by X-ray fluorescence and X-ray diffraction. Some precipitates were also analysed for copper. The effect of coatings of precipitated salts on leaching rate was also considered. The results showed that over the four-year period leaching had occurred throughout the three metre ore bed although there was approxmiately a decrease in the activity/numbers of Thiobacillus ferrooxidans *from top to bottom of the bed. There was evidence of blockage of rock surfaces by precipitated iron compounds and non penetration of leach liquors to the inside of the lumps of rock.*

I. INTRODUCTION

Low grade ore has been leached in dumps and columns for many years but only limited attention has been given to the state of the ore and bacterial activity after leaching. Mineralogical examination of leached ore has been carried out and reported by Duncan, et al. (1) and Bruynesteyn, et al. (2). The latter reported clay layer formation in the test dumps. Bhappu, et al.(3) carried out bacterial activity measurements in the laboratory on core samples taken from different depths in a leach dump.

It is considered that the decline in leaching rate during bacterial leaching (Fig. 1) could be due to changes in the physical state of the ore. At Warren Spring Laboratory, ore from the mine waste dumps at the Old Parys and Mona Copper Mines, Angelsey, North Wales was bacterially leached in shake flasks using finely ground material and in columns using larger particles (-6.3 mm and -38 mm). The largest column 0.3 m diameter and 3.0 m long, using -38 mm ore, was leached over a period of 4 years before dismantling. This provided a rare opportunity to examine the leached residue, and the results of this examination are the subject of this paper. The factors considered were the presence of bacteria through the ore bed, the degree of channelling, rock breakage, the extent and nature of precipitates and the metal or mineral gradient through the ore bed.

II. MATERIAL

The ore was obtained from mixed mine waste at Parys Mountain, Anglesey, N. Wales. It was used at -38 mm particle size.

Chemical Analysis

Copper	0.68%
Zinc	0.064%
Lead	0.12%
Silver	0.0018%

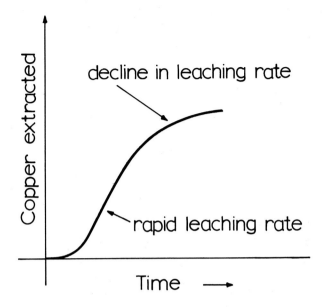

Fig. 1. Bacterial Leaching Curve

A. Mineralogy

A mineralogical analysis of the particular sample used showed that the bulk of the material was quartz-rich and included some shale fragments. A sample crushed to pass 44 mesh (355 μ) gave 10.88% "sinks" when separated in a heavy liquid of specific gravity 3.3. The "sinks" contained most of the sulphides in the ore sample and consisted of pyrite, 60-70%, and chalcopyrite, 10-20%, with minor hematite and sphalerite.

III. EXPERIMENTAL

For the visual examination of the ore bed the column was dismantled and opened without disturbing the ore. This was achieved by tightly packing the end space of the column, gently lowering the column to the ground and then opening an area of 1.8 m^2 along the length of the column (Fig. 2) by using a high-speed electric saw. Visual observations were made for liquor channelling in the ore bed, breakage of rocks, precipitation on the rocks and photographs were taken.

A.   Sampling

During the four-year leaching period the liquor flow through the ore bed was downward and this could have created a chemical and bacterial activity gradient. To quantify this the ore bed was sampled at 0.3 m intervals.

B.   Bacterial Examination

To avoid contamination, samples were taken, immediately the column was cut open. To obtain rock samples, the uppermost layer of the ore bed, which could have been contaminated on sectioning,

Fig. 2.   Sectioned column - exposing the ore bed.

was removed and ten rock samples were taken from the centre of the bed using a sterile spatula. Three samples of precipitates on the column wall (from the top, middle and bottom of the column along its length) and one sample of scrapings from the column wall where there was no visible precipitate were taken to see if bacteria were associated with the precipitates and wall.

C. Mineralogical and Chemical Examination

Selected rock samples for mineralogical examination and grab samples of about 0.6 kg for leaching tests were taken from ten sample spots. The rest of the ore was divided into ten portions at 0.3 m intervals and from each portion a representative sample, for chemical analysis, was prepared by crushing, grinding and riffling.

D. Bacterial Tests

The ten rock samples were placed aseptically in separate 500 ml wide-necked flasks containing 150 ml of suitable sterile medium. This was a modification of Silverman and Lundgren medium (4) using half the amount of ferrous iron, i.e. 4.5 gℓ^{-1} Fe^{2+} and a pH of 2.1. The three precipitate samples and one scraping sample were also placed aseptically in separate 150 ml flasks containing 50 ml of the same sterile medium.

All the samples were incubated in the dark at 30°C as stationary cultures and the bacterial numbers were estimated daily.

E. Mineralogical Examination and Chemical Analysis

Selected pieces of ore were cut open and polished surfaces prepared for examination under reflected light microscopy.

Each of the ten representative ground samples were analysed for copper and total iron. Two visually different precipitates on the rock were analysed by XRF and XRD to ascertain

their composition and identity. Precipitates coating rocks from
the top, middle and bottom of the column were analysed for copper.

F. Leaching Tests

During the four years of leaching it was suspected that the
ore had become coated with precipitated salts. To investigate
the effect that these coatings might have had on leaching rates
samples were needed which were both coated and clean. The first
attempt to obtain non-coated leached ore was by washing the -38 mm
leached ore with 3 M HCl but at this strength the acid attacked
the ore.

Therefore, it was decided to crush the -38 mm leached ore to
expose clean surfaces for leaching. As leaching rates for untreated crushed Parys ore was already available for -6.35 mm particle size, a 3 kg sample of -38 mm treated ore was crushed to
this size and leached by percolation in a PVC column (63.5 mm I.D.)
using distilled water acidified to pH 2.0.

At weekly intervals the pH of the solution was adjusted to
2.0 and the solution analysed for iron and copper.

IV. RESULTS

Visual observations of the distribution of precipitate in the
ore bed indicated that the leach liquors had not percolated
through well-defined channels. Also there was no visual evidence
that the rock had broken or crumbled during the four years of
leaching. The ore was mainly covered with a yellow, light-brown
precipitate, but some rocks were covered with a brown lacquer-like
precipitate (Fig. 3).

A. Bacterial Tests

The results of testing for the distribution of bacteria in
the ore bed showed that *Thiobacillus ferrooxidans* was present

Fig. 3.   Precipitate covering the rock.

throughout the ore bed, associated with rocks as well as precipitates (Table I).   Bacteria were isolated more readily from the
samples taken from the top of the column than the bottom (Table 1).
These tests also showed that *T. ferrooxidans* was associated with
the iron-coloured precipitates on the walls of the column but not
with the bare wall itself (Table II).

B.   Mineralogical Examination

Mineralogical examination showed that there were no differences between the composition or textures of the surfaces of the
pieces of rock and ore taken from different parts of the column,
although most of the material was partially coated with various
coloured precipitates.

The main coating was a yellow, fine-grained powder the amount
of which varied from sparse to moderate.   Occasionally resinous

TABLE I   Isolation of T. ferrooxidans *from Rock Samples*

Sample Position	Incubation (Days)									
	7	8	9	12	14	15	19	21	23	26
10 Top of Bed	±	±	++	++++	++++	++++	++++	++++	++++	++++
9	+	+	+++	++++	++++	++++	++++	++++	++++	++++
8	–	±	+++	++++	++++	++++	++++	++++	++++	++++
7	–	–	–	++	++++	++++	++++	++++	++++	++++
6	–	–	–	+	+++	++++	++++	++++	++++	++++
5	–	–	–	±	+	++	++++	++++	++++	++++
4	–	–	–	±	+	++	++++	++++	++++	++++
3	–	–	–	±	++	++++	++++	++++	++++	++++
2	–	–	–	–	–	–	–	±	+	+++
1 Bottom of Bed	–	–	–	–	–	±	+++	+++	++++	++++

TABLE II   Isolation of T. ferrooxidans *from
Precipitates on the Wall of the Column*

Sample Position	Days									
	7	8	9	12	14	15	19	21	23	26
Top (Precipitate)	–	–	+	+++	+++	++++	++++	++++	++++	++++
Middle (Precipitate)	±	±	±	±	+	+++	+++	++++	++++	++++
Bottom (Precipitate)	–	–	–	–	±	+	+++	++++	++++	++++
Middle Scraping (No precipitate)	–	–	–	–	–	–	–	–	–	–

Approximate bacterial counts obtained from isolation test
++++> 10^3/ml; +++ $2 \times 10^2 - 10^3$/ml; ++ 4×10^2/ml; + 5-40 ml; ±
less than 5/ml.

C. Mineralogical Examination

Mineralogical examination showed that there were no differences between the composition or textures of the surfaces of the pieces of rock and ore taken from different parts of the column, although most of the material was partially coated with various coloured precipitates.

The main coating was a yellow, fine-grained powder the amount of which varied from sparse to moderate. Occasionally resinous or vitreous types of coatings underlay the yellow powdery deposits. These coatings were usually black, dark brown, bluish-grey or deep red in colour, often less than 100 μ thick and sometimes with a blistered appearance. In places these resinous coatings had cracked to reveal a layered structure.

Microscopic examination of the ore surface revealed that in places the sulphide minerals appeared not to have been greatly altered. Elsewhere cavities had been formed where the sulphides had been partly removed. In these cases the remnant sulphides usually had highly irregular black-coated surfaces. However, the polished sections of the rocks showed that the sulphides disseminated in the interior of the lumps usually appeared unaffected but sulphides occurring in veinlets sometimes showed signs of partial dissolution. It was also noted that the remnant sulphides in the cracks and veinlets were encased in the iron precipitates.

D. Chemical Analysis

The amount of copper and iron left with the ore after four years of leaching is shown in Table III.

The analysis by XRF showed that in the yellow and dark brown lacquer precipitates the major element was iron and the minor element was copper with trace amounts of As, Zn, Pb, Ba, Sn, Zr, Rb and Sr.

TABLE III   Chemical Analysis of Rock Samples

Sample Position	% Copper	% Total Iron
10 Top of Bed	0.76	11.4
9	0.48	11.5
8	0.53	11.7
7	0.65	11.9
6	0.58	11.4
5	0.45	11.4
4	0.46	11.3
3	0.44	10.9
2	0.54	12.0
1 Bottom of Bed	0.46	11.4
Average of Column (Excluding Sample 10)	0.51 + 0.071	11.49

From the above results the average copper extracted in four years was 25%.

The analysis by XRD showed that both coloured precipitates were mainly amorphous but some natrojarosite ($NaFe_3(SO_4)_2(OH)_6$) was present in both. The dark brown lacquer precipitate had some serpentine ($Mg_3Si_2O_5(OH)_4$) present. The copper content of the precipitates from the top, middle and bottom of the column is given in Table IV.

E. Leaching Test

Leaching rates for copper extracted from crushed uncoated leached ore (-6.35 mm), prepared from the large column residue was 1.06% Cu/28 days for the first 91 days but it dropped to 0.33% Cu/28 days for the next 73 days. The copper leaching rates previously obtained for crusted-unleached ore (-6.35 mm) were 0.53% Cu/28 days and 0.36% Cu/28 days.

TABLE IV   *Chemical Analysis of Precipitate Samples*

	% Copper
Top of Bed	0.45
Middle of Bed	0.41
Bottom of Bed	0.38

V. DISCUSSION

The distribution of bacteria in the leached ore bed was assessed by measuring the rate of oxidation of a solution of ferrous iron to which had been added equal amounts of ore plus associated bacteria.

Quick oxidation of the ferrous iron accompanied by good growth is often interpreted as an indicator of high bacterial numbers in the inoculum. However, it is possible that an equivalent oxidation rate could result from fewer, but more active, bacteria being added. In this study there was a very consistent trend for an increase in bacterial activity or numbers from the top to the bottom of the leached bed of ore. The greater activity and/or numbers of bacteria at the top of the ore bed might have been due to the more ready availability of oxygen at the top which was open to the atmosphere compared with the bottom which had only a narrow opening. The amount of ferrous iron available at the top was also greater.

The copper remaining in the top of the ore bed was 0.76% which was higher than the 0.68% Cu in the head sample. This high copper content in the top of the ore bed could have been due to higher copper in the precipitates from the top of the bed (Table IV). This could have been due to recycling of leach liquor to the top of the ore bed without a complete copper recovery stage. In order to determine if there was a copper metal gradient down the ore bed the top sample 0.76% Cu (Table III) was excluded from the

regression line analysis. The remaining nine samples gave a slope
of 0.0098 $\pm$ 0.00214 for 95% confidence limit, indicating a zero
metal gradient. The mineralogical examination also showed no
obvious metal concentration gradient throughout the ore bed.

The residual copper left (Table IV) after four years of leaching indicated that the average copper extracted in four years was
25%. This compared reasonabl with low extraction rates obtained
by Bruynesteyn (5) for 38 mm non-porous iron. However, the same
author showed (2) that extraction rates of 64% in 12 weeks were obtained with porous 51 mm ore. The low copper extractions were due
to:

(1)   unleached sulphide minerals in the interior of the rock;
      this indicating low rock porosity
(ii) the blinding of the rock surfaces and solution penetra-
      tion barrier in the veinlets and cracks due to iron
      salts precipitation

Visual observation and mineralogical analysis showed that the
rocks had not crumbled and most of them were coated with a yellowy
iron-type precipitate but some were also coated with a brown resinous precipitate. Physically the two precipitates looked very
different but chemically both had the same elements and were both
amorphous. The yellow precipitate was powdery and while it could
"blind" the ore it would allow diffusion of fluid between the particles. The brown resinous precipitate looked impervious but it
was brittle and would allow penetration of fluids if it cracked.

When the effect of iron precipitates on the leaching rate was
studied, leaching of crushed, uncoated leached ore (6.35 mm) gave
an initial rate of 1.06% Cu/28 days. This was higher than the
best leaching rate of 0.53% Cu/28 days previously obtained for
crushed unleached ore of the same size (-6.35 mm).

Further leaching of the crushed leached ore after the initial
91 days gave a rate of 0.33% Cu/28 days which was comparable with

the average leaching rate of 0.36% Cu/28 days for crushed unleached ore. The higher initial leaching rate of 1.06% Cu/28 days for
crushed leached ore could be attributed to: -

(1) releasing of trapped copper on crushing. This trapped
     copper could have consisted of copper co-precipitated
     with iron salts or blocked by an iron salt barrier.

(2) production of significantly higher amount of smaller
     particle size material in the crushed leached ore than
     in the crushed unleached ore during comminution.

The smaller particle size could have resulted from physical/
chemical changes in the ore during four years of leaching (e.g.
porosity changes).

The fact that copper had coprecipitated with iron salts was
substantiated by the copper found in the precipitates (Table IV)
but the suggestion of undissolved copper in (1) above is purely
speculative. The suggestion that smaller particle sizes were
produced on crushing the leached ore was supported indirectly by
the fact that a higher initial leaching rate was maintained for
91 days and it is unlikely that this could have been due only to
precipitated and undissolved copper which should have come into
solution in a much shorter time.

VI. CONCLUSIONS

The results showed that over the four-year period, leaching
had occurred throughout most of the three meter ore bed, although
there was apparently a decrease in bacterial activity from top to
bottom. The results indicated two reasons for the low extraction
of copper over the period. Firstly there was evidence of blockage of rock surfaces and mineralized veins with precipitated iron
compounds which would hinder contact with leach liquor. Secondly
since sulphides in the interior of lumps had not reacted it was clear
that leach liquors had not penetrated rock surfaces due to low porosity.

The higher initial leaching rate for the crushed residue could have been due to

(i)   soluble copper in the ore

(ii) extraction from fines produced during crushing.

The sample of ore used in the above tests showed poor leaching characteristics. However, literature reports and field visits have indicated that the Parys Mountain ore has been leaching for years and some of the ore particles are in a friable state. So it could be concluded that the Parys Mountain waste dumps are inhomogeneous and to assess whether the dumps of ore would be amenable to bacterial leaching, tests would have to be carried out on varied samples and at a larger scale.

VII. REFERENCES

1. Duncan, D.W., and Drummond, A.D., *Can. J. Earth Sci.*, 10 (4), 476, (1973).

2. Bruynesteyn,A., and Cooper, J.R., "Solution Mining", Proceedings of a symposium 103rd AIME Annual Meeting, Dallas, Texas, Feb. 25-27, 1974, p. 268.   The Ass. Inst. Min. Metall. and Pet. Eng., Inc., New York, 1974.

3. Bhappu, R.B., Johnson, P.M., Brierley, J.A., and Reynolds, D.M., *Trans. Soc. Min. Eng. AIME*, 244, 307 (1969).

4. Silverman, M.P., and Lundgren, D.G., *J. Bacteriol.*, 77, 642, (1959).

5. Bruynesteyn, A.,and Duncan, D.W., in Solution Mining", Proceedings of a Symposium 103rd AIME Annual Meeting, Dallas, Texas, Feb. 25-27, p. 324 (1974).   The Ass. Inst. Min. Metall. Pet. Eng., Inc., New York, 1974.

MICROBIAL LEACHING OF COPPER AT
AMBIENT AND ELEVATED TEMPERATURES

James A. Brierley

New Mexico Institute of Mining and Technology
Socorro, NM 87801, U.S.A.

Corale L. Brierley

New Mexico Bureau of Mines and Mineral Resources
Socorro, NM 87801, U.S.A.

This report compares the leaching ability of Thiobacillus ferrooxidans, *a microbe commonly associated with copper leaching operations at ambient temperature, and a high-temperature* Sulfolobus-*like microbe, an organism recently determined to be associated with metals extraction at 60°C. The mineral substrates used for leaching included mine waste and concentrate from the Kennecott Copper Corporation, Chino Mines Division, NM, ore from the Phelps-Dodge Corporation, Tyrone Division, AZ, and concentrate from the Phelps-Dodge Corporation, Ajo Division, AZ. The bio-leaching experiments were run in stationary batch reactors. Four nutrient conditions were evaluated for their effect on bio-leaching of copper minerals at room temperature and 60°C: 1) basic leach solution only; 2) basic leach solution supplemented with 0.02% yeast extract; 3) basic leach solution with 36 mM ferrous iron; and 4) a combination of basic leach solution, 0.02% yeast extract and 36 mM ferrous iron. Soluble copper concentration was monitored after 30 and 60-days leaching. The yeast extract did not greatly affect copper leaching by either* T. ferrooxidans *or the* Sulfolobus-*like microbe except with the Ajo concentrate which was leached more effectively. The addition of ferrous iron enhanced the bio-leaching of copper. Apparent increased bio-leaching at 60°C was attributed to temperature affect rather than enhanced bio-activity by the thermophile except with the Ajo concentrate, indicating greater ability by the Sulfolobus-like microbe for extracting copper from a mineral substrate rich in chalcopyrite.*

I. INTRODUCTION

Recent reports have discussed the use of an acidophilic thermophilic microbe, which grows at 60°C, in leaching of molybdenum ores and concentrates (1, 2) and copper concentrates, ores and mine waste (2, 3). The characteristics of this microbe, which oxidizes ferrous iron, sulfur and sulfides, have been previously described (4). Recent work (5) has shown that this microbe shares many characteristics with the bacteria of the genus *Sulfolobus*. This *Sulfolobus*-like microbe can catalyze the leaching of metals in conditions where temperatures exceed the upper range (30° to 40°C) at which *Thiobacillus ferrooxidans* is capable of growing and catalyzing biogenic metal extractions (6). It is possible that both microbes may be present in an environment in which a temperature gradient exists. Such an environment may be a leach dump with temperatures ranging from the ambient air temperature at the surface to higher than 80°C with the dump (7).

This paper compares bench-scale leaching data obtained from the use of *Thiobacillus ferrooxidans* at ambient temperatures with the extraction data obtained at 60°C using the *Sulfolobus*-like microbe. The extractive capabilities of the two microbes were tested using various nutrient conditions with differing grades and mineralogical qualities of copper sulfide.

II. MATERIALS AND METHODS

A. Culture Medium

The leach solution (8) consisted of the following components (grams per liter): $(NH_4)_2SO_4$, 1.0; K_2HPO_4, 0.1; $Al_2(SO_4)_3 \cdot 18H_2O$,

4.0; $MgSO_4 \cdot 7H_2O$, 3.0; $Ca(NO_3)_2 \cdot 4H_2O$, 0.1; $MnSO_4 \cdot H_2O$, 0.05; Na_2SO_4, 0.05; KCl, 0.05. The pH was adjusted to 2.5 with 1 N H_2SO_4.

In some experiments, 100 ml quantities of sterile leach solution were supplemented with 36 mM ferrous iron. A solution of 25 g of $FeSO_4 \cdot 7H_2O$ in 95 ml of distilled water and 5 ml of 1 N H_2SO_4 was sterilized at 121°C, 18 psi for 15 minutes. Four ml of the ferrous iron solution was added to 96 ml of the leach solution.

Where indicated, leach solutions were supplemented with 0.02% yeast extract (Difco). Two ml of a sterile 1% solution were added to 98 ml of leach solution.

B.   Mircobes

A thermophilic chemolithotroph, a microbe which oxidizes reduced sulfur and iron compounds at temperatures between 45° to 70° C (4), was used for batch reactor experiments at 60°C. This microbe was maintained in leach solution supplemented with 0.02% yeast extract using elemental sulfur as an energy source. These batch cultures were used as inoculum for mineral substrate experiments. The inoculum for ambient temperature experiments was American Type Culture Collection strain 19859, *Thiobacillus ferrooxidans*, which was prepared by culturing the organism in leach solution with 36 mM ferrous iron.

C.   Substrates

Table 1 summarizes the types, particle sizes, concentrations, mineralogy and sources of minerals used in the study. These substrates were analyzed for opaque minerals by reflected-light microscopy. The minerals were placed in 250 ml Erlenmeyer flasks for batch reactor testing. Sterilization of the minerals was completed by dry heat at 160°C for 2 hours; sterile nutrient solutions were then added.

TABLE I Mineral Substrates
Used for Microbial Leaching

Substrate	Mineralogy	% Cu	Particle size mesh	Pulp Density g/100 ml
Kennecott Copper Corp. (Chino Division) NM	Chalcocite Bornite Pyrite	0.31±0.01	-16+50	10
Waste				
Phelps-Dodge Corp. (Tyrone Division) NM	Chalcocite Covellite Bornite Chalcopyrite Pyrite	0.45±0.04	-48+65	10
Ore				
Phelps-Dodge Corp. (Ajo Division) AZ	Chalcopyrite (in excess of 50%) Covellite Bornite Pyrite Sphalerite Molybdenite	29.0±0.8	-65+200	1.0
Concentrate				
Kennecott Copper Corp. (Chino Division) NM	Chalcocite Chalcopyrite Covellite Bornite Pyrite Molybdenite	26.8±0.3	-65+200	1.0
Concentrate				

480

D.   Experimental Conditions

All experiments for ambient temperature bio-leaching were conducted at laboratory temperature. The high temperature bio-leaching experiments were performed in hot-air ovens or water baths at $60° \pm 1°C$.

The bio-leaching experiments were run in stationary, batch reactors with 100 ml of leach solution (8). The amounts of waste, ore and concentrate used per reactor appear in Table I. Four nutrient conditions were evaluated for their effect on bio-leaching of copper minerals: 1) basic leach solution only; 2) basic leach solution supplemented with 0.02% yeast extract (YE); 3) basic leach solution with 36 mM ferrous iron (Fe); and 4) a combination of basic leach solution, 0.02% yeast extract and 36 mM ferrous iron. Ten flasks for each mineral and each nutritional condition were inoculated with 1 ml of an actively growing iron culture of *T. ferrooxidans* and incubated at ambient temperature. Similarly, 10 flasks were prepared for each experimental condition and inoculated with 1 ml of an actively growing sulfur culture of the thermophilic *Sulfolobus*-like microbe. Ten uninoculated control flasks were also incubated for each experimental condition at room temperature and 60°C.

E.   Analysis

The flasks were sampled after leaching periods of 30 and 60 days. Soluble copper concentration was determined using standard techniques for a Perkin-Elmer model 303 atomic absorption spectrophotometer. The mean value and standard deviation for extracted copper were established for each experimental condition.

III.   RESULTS

A.   Copper Mine-Waste and Ore Leaching

Following 60 days of incubation, the microbial leaching of

copper from the Chino Mine waste in unsupplemented solution (Table II) was negligible when using *T. ferrooxidans* at room temperature. About 8% of the copper was readily solubilized in the acidic leach solution as indicated by the control conditions with no microorganisms. At 60°C up to 22% copper was extracted in the control condition and 36% with the thermophilic microbe after 60 days without supplementation. The addition of 0.02% yeast extract depressed leaching in control reactors at room temperature and 60°C. The addition of ferrous iron increased the biological leaching at room temperature by *T. ferrooxidans* and at 60°C by the thermophilic microbe. At room temperature the maximum biological effect was achieved during the first 30 days of leaching as reflected by the data in Table II; at 60 days no further copper had been extracted. The addition of ferrous iron did not greatly enhance copper extraction in room temperature control reactors during the 60-day leach period. At 60°C with ferrous iron, the thermophilic microbe increased copper extraction only during the initial 30-day leach period. The percent copper extracted in 60°C control reactors was greatly increased by the addition of ferrous iron, and by 60 days the percent copper in 60°C control reactors equaled the percent copper in reactors inoculated with the thermophilic microbe. The effect of supplementing room temperature and 60°C control and inoculated reactors with both yeast extract and ferrous iron was no different than using the iron only. The yeast extract appeared to slightly depress copper extraction at 60°C as was noted when yeast extract was the sole supplement.

Table III presents data on the extraction of copper from ore obtained from Phelps-Dodge Corp. When *T. ferrooxidans* was used to extract copper without supplementation, the percent copper extracted after 30 days was no different than percent copper extracted in the control reactor. The addition of 0.02% yeast extract to the leach solution slightly increased the effectiveness

TABLE II *Microbial Leaching of Chino Mine*
Waste at Ambient Temperature and 60°C

		Percent Copper Leached			
		30 Days		60 Days	
Experimental Set-up		Room Temp.	60°C	Room Temp.	60°C
Waste	Inoc.	7.7+1.8	21.6+ 3.6	8.0+0.7	36.0+14.6
	Control	7.3+0.7	14.7+ 3.1	8.0+0.7	22.3+ 9.4
Waste + YE	Inoc.	6.7+0.6	17.8+ 7.3	7.6+1.1	26.0+ 4.5
	Control	4.8+0.5	10.5+ 4.6	4.8+0.4	17.8+ 7.1
Waste + Fe	Inoc.	27.9+2.6	55.1+ 8.6	20.3+1.2	100
	Control	10.6+1.4	46.7+ 6.6	9.9+0.9	97.0+ 4.2
Waste+Fe+YE	Inoc.	28.5+2.9	57.6+13.8	22.7+1.7	80.2+28.1
	Control	9.1+1.1	35.4+ 6.3	9.1+1.0	87.0+15.1

of *T. ferrooxidans* for copper extraction; however, copper extraction in the control reactor was depressed. Only slightly more copper was leached by 60 days. The effect of bio-leaching was more pronounced at 60°C with 78% copper leached in 30 days compared with 57% for the uninoculated control without supplementation. At 60 days and 60°C, 98% copper was leached in the presence of the high-temperature microbe and 87% without the microbe. Similar results were noted with the addition of yeast extract at 60°C. The addition of ferrous iron increased microbial leaching. The effect was most noticeable during the first 30 days of leaching at room temperature. After 30 days, the leaching of copper was complete at 60°C for the inoculated condition and complete by 60 days for the uninoculated condition. Adding yeast extract and iron did not affect the biological leaching, but did depress leaching without microbes at room temperature and 60°C at 30 days; control leaching was depressed at room temperature up to 60 days.

TABLE III *Microbial Leaching of*
Phelps-Dodge Copper Sulfide Ore
at Ambient Temperature and 60°C

		Percent Copper Leached			
		30 Days		60 Days	
Experimental Set-up		Room Temp.	60°C	Room Temp.	60°C
PD	Inoc.	36.8+2.5	78.6+26.6	40.0+3.3	98.8+ 2.7
	Control	36.5+2.1	57.4+17.8	34.0+1.4	87.0+17.8
PD + YE	Inoc.	42.0+4.0	74.2+26.7	44.0+3.3	96.8+ 7.2
	Control	31.9+1.4	51.4+18.6	26.3+1.1	86.2+22.2
PD + Fe	Inoc.	81.5+3.9	93.5+16.4	66.6+1.6	96.8+ 7.2
	Control	59.6+2.2	85.2+20.7	53.6+3.9	97.5+ 5.0
PD + Fe + YE	Inoc.	82.2+4.8	92.0+11.5	64.3+3.0	97.0+ 6.7
	Control	47.3+1.4	74.4+24.5	46.5+1.2	100[1]

[1] *Two reactors used and results averaged*

B.   Copper Concentrate Leaching

Table IV presents data for microbial leaching of concentrates. Biological leaching of the Kennecott concentrate without added yeast extract or ferrous iron was very slight at 60°C, since the percent copper leached was nearly the same with or without the thermophilic microbe. Less copper was leached at room temperature; however, the affect of *T. ferrooxidans* was greater since more copper was extracted in the presence of this organism. The addition of yeast extract only served to depress leaching both with and without microbes at 60°C and room temperature incubation conditions. The addition of ferrous iron enhanced copper leaching, especially in the presence of *T. ferrooxidans*. Leaching with added ferrous iron was greater at 60°C than at room temperature with 99.8% extraction achieved in 60 days in inoculated batch reactors. Yeast extract in combination with ferrous iron depressed the amount of biologically leached copper.

TABLE IV *Microbial Leaching of Copper Concentrates at Ambient Temperature and 60°C*

		Percent Copper Leached			
		30 Days		60 Days	
Experimental Set-up		Room Temp.	60°C	Room Temp.	60°C
Kenn.	Inoc.	26.9+2.1	30.5+ 2.9	27.8+2.1	47.5+ 6.4
	Control	22.1+0.9	29.3+ 2.3	18.4+0.4	43.4+ 9.2
Kenn. + YE	Inoc.	25.6+1.2	27.2+ 2.8	25.7+1.8	38.4+ 9.4
	Control	17.7+0.4	25.5+ 2.0	14.0+0.3	34.8+ 7.2
Kenn. + Fe	Inoc.	42.4+1.5	52.1+ 9.2	43.1+0.9	99.8+ 0.6
	Control	27.3+0.9	45.8+ 9.1	24.0+1.1	65.6+16.5
Kenn. + Fe + YE	Inoc.	35.4+0.9	56.6+14.8	37.6+1.1	90.2+15.2
	Control	23.8+1.7	43.1+ 5.3	21.5+0.8	69.4+16.9
Ajo	Inoc.	6.2+1.0	8.0+ 3.3	8.0+1.5	51.3+14.8
	Control	2.3+0.1	5.3+ 1.3	2.4+0.1	7.5+ 2.1
Ajo + YE	Inoc.	6.2+0.6	29.2+ 6.6	9.8+1.0	81.7+19.0
	Control	1.5+0.1	3.7+ 1.2	1.4+0.1	3.7+ 1.6
Ajo + Fe	Inoc.	8.1+0.6	42.5+18.0	9.6+0.8	58.9+17.3
	Control	5.2+0.5	29.9+11.0	6.2+0.7	36.2+12.3
Ajo + Fe + YE	Inoc.	10.9+0.7	39.5+11.0	12.7+1.0	89.7+12.6
	Control	3.9+0.3	18.8+ 5.8	4.9+0.4	43.1+10.7

The percent copper microbially extracted from the Ajo concentrate (Table IV) was considerably less than the percent copper extracted from the Kennecott concentrate. The thermophilic microbe had leached 51.5% of the copper in 60 days with only 7.5% leached in the 60°C uninoculated control. Only 8.0% copper was leached in the presence of *T. ferrooxidans* with 2.4% leached in the uninoculated control. The addition of yeast extract greatly enhanced the microbial leaching at 60°C with 29.2% and 81.7% copper extracted by 30 and 60 days, respectively. Yeast extract did not significantly affect leaching by *T. ferrooxidans*. Yeast extract slightly decreased the percent copper extracted in uninoculated control reactors at 30 and 60 days at room temperature. Addition of ferrous iron resulted in increased thermophilic mi-

crobial leaching at 30 days, but this increase was not noted at
60 days. Ferrous iron enhanced copper leaching in uninoculated
controls at 60°C. Ferrous iron addition did not significantly
influence copper leaching by *T. ferrooxidans* at room temperature.
The combination of ferrous iron and yeast extract slightly increased the extraction of copper by *T. ferrooxidans* and the thermophilic isolate after 60 days of incubation.

C.   Comparison of Leaching at Room Temperature and 60°C

Table V shows the maximum percent copper extracted in 30 and
60 days by *T. ferrooxidans* and the thermophilic *Sulfolobus*-like
microbe. The percent copper extracted by the corresponding uninoculated controls has been subtracted; therefore, all values
represent the actual percent extracted as a result of microbial
activity. These data show that the percent copper extracted in
30 days from Chino Mine waste was nearly the same for both microbes; this was also noted at 60 days with each microbe extracting about 13.6% copper. *T. ferrooxidans* was able to extract
about 12% more copper from the Phelps-Dodge ore than the high
temperature microbe in 30 days. However, for a 60-day period the
percent copper extracted by each microbe was nearly the same.
Both *T. ferrooxidans* and the thermophilic microbe were only able
to extract 13-15% of the total copper present in a concentrate
obtained from the Chino Operation in 30 days. However, after 60
days of leaching, the thermophile had extracted 34% of the total
copper present in the concentrate - nearly twice the amount extracted by *T. ferrooxidans*. The chalcopyrite concentrate from
the Ajo Division of Phelps-Dodge Corp. was not easily leached by
T. ferrooxidans with only 7% of the total copper microbially extracted in 30 days and 8.4% in 60 days. The thermophilic microbe
was more efficient at extracting this copper with 25% extracted
in 30 days and 78%, or nearly 10 times the amount extracted by
T. ferrooxidans in 60 days.

*TABLE V   Maximum Percent Copper
Extracted by Microbial Action[1]*

Substrate	Temp. (°C)	Percent Cu	
		30 Days	*60 Days*
Chino waste	RT[2]	19.4	13.6
	60	22.2	13.7
P. D. ore	RT	34.9	17.8
	60	22.8	11.8
Chino concentrate	RT	15.1	19.1
	60	13.5	34.2
Ajo concentrate	RT	7.0	8.4
	60	25.5	78.0

[1] *The percent copper extracted in uninoculated control reactors
has been subtracted.*

[2] *Room temperature.*

IV.   DISCUSSION AND CONCLUSION

Microbial leaching was dependent on environmental conditions
and mineralogy of the samples leached.

When unsupplemented leach solution was used, the thermophilic microbe was better able to extract copper from the waste material, ores and concentrates tested, with the exception of the
Kennecott concentrate which was somewhat more susceptible to the
leaching ability of *T. ferrooxidans.*

Supplementing batch reactors with ferrous iron greatly enhanced copper extraction from all test substrates. *T. ferrooxidans* leached considerably more copper from test substrates when
supplemented with ferrous iron. When the leach solution was supplemented with iron at 60°C, copper extraction increased substantially in both inoculated and control reactors. The thermophilic
microbe did not contribute substantially to copper extraction
from the mine waste and ore. Rapid spontaneous oxidation of iron

occurred at 60°C in the uninoculated controls providing an oxidant for the solubilization of the copper sulfides. However, the
thermophilic microbe did enhance the extraction of copper from
both concentrates tested at 60 days of incubation.

The affect of added organic material in the form of yeast
extract on bio-leaching of copper was evaluated since it has been
shown that yeast extract stimulated the growth of the *Sulfolobus*-
like microbe and increased the isolate's ability to extract molybdenum from molybdenite (1). In hydrometallurgical extraction
of copper, both *T. ferrooxidans* and the thermophilic microbe may
be used for leaching. In this event it would be necessary to establish whether an additive (*e.g.*, organic matter) which stimulated the leaching by one microbe would positively or negatively
affect the leaching ability of another microbe. When 0.02% yeast
extract was added to the leach solution or leach solution plus
ferrous iron, biological and non-biological extraction of copper
was not significantly affected. Only slight depression of biological and non-biological copper leaching occurred with the
Kennecott concentrate. Only biological extraction of copper from
the Ajo concentrate was enhanced by the addition of yeast extract.
The basis of the effect of yeast extract was not explored. It is
evident that appropriate research has to be performed prior to adding
organic matter to leaching operations in order to predict the
results.

The thermophilic *Sulfolobus*-like microbe and *T. ferrooxidans*
had differing capabilities for copper extraction for the substrates tested. The response was probably dependent on the mineralogy of the substrate. The thermophilic microbe, growing at
60°C without iron supplementation, extracted more copper from the
Chino mine waste, the Phelps-Dodge ore, and Ajo concentrate than
did *T. ferrooxidans* at room temperature. This may have been the
result of differences in microbial mineral interaction.

Both the thermophilic microbe and *T. ferrooxidans* will extract copper from the same types of substrates and will function in the same habitat with a temperature gradient. *T. ferrooxidans* function optimally in the cooler (20°-40°C), oxygen-rich zones of leach dumps and *in-situ* operations while the thermophilic microbe grows at higher temperatures (45°-70°C) and can function in an oxygen-poor region (1). Recent reports (9, 10) have indicated that a *Thiobacillus*-like microbe is capable of iron and pyrite oxidation at 50° to 55°C. This type of microbe may also be capable of leaching various metallic sulfides and occupy a high-temperature niche in leach dump environments thereby supplementing the function of the mesophilic *T. ferrooxidans*.

The apparent increased microbial leaching of copper at 60°C observed in Tables II, III and IV, was not due to the thermophilic microbe itself, but rather was due to the increased temperature. Copper extraction in the uninoculated controls was also increased. When the differences between the inoculated and uninoculated conditions were determined, the copper extracted in the presence of the respective microbes was nearly the same with the Ajo concentrate being the notable exception (Table V). It was apparent from this data that the thermophilic *Sulfolobus*-like microbe has a greater ability for extracting copper from a mineral substrate rich in chalcopyrite.

V. ACKNOWLEDGMENTS

This research was supported by U.S. Bureau of Mines Grant No.: G0133131 and in part by N. S. F. Institutional Grant No.: GU-3533. The authors wish to express their gratitude to the students assisting the project and Joseph Taggart for the ore microscopy.

VI. REFERENCES

1. Brierley, C.L., *J. Less-Common Met.*, 36, 237 (1974).
2. Brierley, C.L., and Murr, L.E., *Science*, 179, 488 (1973).

3. Brierley, C.L., *Dev. in Ind. Microbiol.*, 18, 273 (1976).

4. Brierley, C.L., and Brierley, J.A., *Can. J. Microbiol.*, 19, 183 (1973).

5. Langworthy, T.L., *J. Bacteriol.*, 130, 1326 (1977).

6. Tuovinen, O.H., and Kelly, D.P., *Zeitschrift fur Allg. Microbiol.*, 12, 311 (1972).

7. Beck, J.V., *Biotechnol. Bioeng.*, 9, 487 (1967).

8. Bryner, L.C., and Anderson, L., *Ind. and Eng. Chem.*, 49, 1721 (1957).

9. LeRoux, N.W., Wakerley, D.S., and Hunt, S.D., *J. Gen. Microbiol.*, 100, 197 (1977).

10. Brierley, J.A., and LeRoux, N.W., in "Conference Bacterial Leaching 1977," (W. Schwartz, Ed.), p. 55. Verlag Chemie, New York, 1977.

THE USE OF LARGE-SCALE TEST FACILITIES IN STUDIES OF THE ROLE OF MICROORGANISMS IN COMMERCIAL LEACHING OPERATIONS

L.E. Murr and James A. Brierley

New Mexico Institute of Mining and Technology
Socorro, New Mexico, U.S.A.

Although temperature profiles have been identified in a few commercial leach dumps (such as Kennecott Copper Corporation's Midas test dump at Bingham Canyon, Utah), there is no correlation on this large scale with bacterial activity (defined quantitatively as the most probable number, MPN, or number of Thiobacillus ferrooxidans *per gram or cc of sample) or oxygen consumption. Recent experiments on unitized copper-bearing waste bodies weighing approximately 1.7×10^5 kg in insulated stainless steel (dewartype) containers 3.07 m in diameter and 12.3 m in height have allowed vertical temperature profiles in these ore bodies to be correlated with bacterial population (T.* ferrooxidans *most probably number, MPN) in the ore body, and oxygen distribution and consumption. It is suggested from these experiments that bacterial catalysis is somewhat critical within a range above about 10^5 organisms/g sample in the waste body. Oxygen consumption initially occurs when MPN reaches approximately 10^7 cells/g; independent of marked temperature influence. While the temperature profiles observed for two different waste bodies show varying degrees of correspondence with the mean ambient temperature, there is overwhelming evidence that the reaction kinetics (mainly exothermic pyrite oxidation) develop intrinsic temperatures within the ore body since temperatures have been observed as high as $59°C$. This high temperature seems to be correlated with a marked decline in the* Thiobacillus ferrooxidans *population, and a concomitant indication that a high-temperature microorganism*

491

exists. One waste body (containing Kennecott Corp. Chino Mines type 1 waste) has been stabilized at temperatures above 52°C. That is, a temperature plateau has been established between 50 and 60°C, and this is suggestive of bacterial control of the temperature of the waste body. The implications of bacterial control of the waste-body temperature and the use of high-temperature microorganisms to establish higher temperature plateaus and enhanced reaction kinetics is discussed in relation to commercial leaching operations.

I. INTRODUCTION

It is now well established on the laboratory scale that bacterial catalysis is essential for efficient oxidation to occur in the solubilization of copper during the leaching of porphyry copper deposits or copper-bearing waste; principally as chalcopyrite. This is illustrated for example in Chapters 6, 13, and 14. Bacteria have been observed in the effluent solutions in the commercial leaching of copper in-situ or in leach dumps and heaps (1,2). It is likely, in fact, that in any successful or reasonably successful leaching operation, bacteria must be present in sufficient numbers to efficiently catalyze the oxidation reactions ($Fe^{2+} \rightarrow Fe^{3+}$ and sulfides to sulfates). Judging from laboratory studies, these numbers appear to be in the range of 10^5 to 10^8 microorganisms per gram sample.

Bhappu, et al. (2) carried out bacterial activity measurements in the laboratory on core samples taken from different depths in a leach dump. The results of the study by Bhappu, et al. (2) suggested that sufficient bacterial activity for sustained catalysis only existed near the dump surface.

Madsen and Groves (3), on leaching a low-grade chalcopyrite waste in a large column (containing 7.9×10^3 kg of waste) for several years recently concluded that enhanced leaching and leaching rates observed following aeration by blowing air up through the waste body were due to enhanced bacterial catalysis, although no direct measurements of bacterial activity or bacterial population concentration were made. Similarly, Cathles and Apps (4)

have, in developing a model of the dump leaching process that
incorporates oxygen balance, heat balance, and air convection,
tacitly stated that oxygen, the primary oxidant in the leaching
process, is transported into the dump by means of air convection
and oxidizes ferrous ions through bacterial catalysis. This model
does not include bacterial activity directly and makes no assump-
tions about microbial populations necessary to sustain a certain
level of catalysis. The air convection is promoted by the heat
generated by the oxidation of the sulfides present, including of
course pyrite.

The generation of heat in a dump is also an important para-
meter because it not only presumably promotes air convection, but
it is an integral thermodynamic variable affecting the reaction
kinetics. It also markedly influences bacterial activity since
T. ferrooxidans optimize their metabolism within a fairly narrow
temperature range (20-30°C) (5). Consequently, when dump temperatures exceed roughly 35°C, the bacterial activity due to *T. ferrooxidans* will begin to markedly decline. It is unknown whether
bacterial activity is controlled by exothermic reactions or
whether the temperature is moderated by some mechanism which
limits catalysis and thereby limits exothermic heating to maintain
near optimum conditions for microbial metabolism.

Although there is an abundance of laboratory data which
clearly demonstrates the effect of aeration and oxygen levels, etc.
on bacterial activity, there is no comparable data (that is
quantitative data) for an actual leach dump. Except for the
temperature profiles for the Kennecott Midas dump (4), there is
also no systematic indication of temperatures within a dump, and
there is no data up to the present time which demonstrates a clear
connection between bacterial activity, oxygen consumption, and
temperature in a large waste body undergoing leaching.

In the work to be described herein, nearly 200 tons of waste
rock ($\sim 1.7 \times 10^5$ kg) were leached in insulated stainless steel

containers which allowed the temperature and oxygen levels at
locations within the waste column to be determined, and samples
to be extracted from these locations for determinations of bacterial
populations. Since the waste-body temperatures were not
markedly influenced by ambient temperature fluctuations, these
experiments represented a unitized, equivalent thermodynamic segment
of a large waste dump from which meaningful results could be
obtained relating to a large-scale dump or heap leaching system.

II. EXPERIMENTAL TECHNIQUES

The experimental facility employed in the investigations to
be described consisted of an array of four stainless steel
liquid-oxygen storage tanks modified to accommodate roughly
2×10^5 kg of waste rock. Two of these tanks constituting the
John D. Sullivan Center for In-Situ Mining Research (field-test
facility) were converted for leaching two distinctly different
porphyry copper wastes. Each tank measured roughly 12.3 m in
overall height and 3.1 m inside diameter. A 0.3 m thick perlite
insulating layer separated the inner stainless steel liner from
the thicker, outer carbon steel jacket. Prior to loading the
waste rock, the base of each tank was built up using concrete
overcoated with 4 layers of a heavy epoxy paint. A drain was cut
into the forward end of the sloping tank bottom and a square
schedule 80 PVC pipe array was fabricated to allow for pressurized
air convection through the waste body. A 1 m high quartzite bed
(+2 in) was placed upon the PVC air distributor. Thermocouples
(and in one waste body experiment an array of moisture blocks)
were installed at the top of the gravel bed, and waste rock hand
loaded upon the instrumentation to a depth of 0.3 - 0.5 m before
elevator loading through the upper end of the tank.

Side access ports were cut into the tanks at various locations
from the top of the gravel bed through which additional
instrumentation imbedded in the waste rock could be monitored.

A 5 cm diameter access tube was fabricated at each port level for
the periodic withdrawal of solid samples for bacterial analysis.
The waste rock was loaded to each access level, instrumentation
installed, the rock hand loaded over the instruments, and elevator
loading resumed. Four such ports were cut into each tank, each at
90 degrees along a diameter from the previous one. Funnels or
similar collectors, including lysimeters (porous cup devices which
could be evacuated to draw solution from the location in the waste
body, then pressurized to expel the solution sample collected;
consisting of 100 ml maximum volume) were installed at each access
port location in each of the experimental waste bodies, and at
another location below the access levels in the case of one waste
body.

Figure 1 shows an overall view of the leach tank facility
while Fig. 2 illustrates, schematically, the solution and air
distribution systems design. While solution flow rates were
varied somewhat, the majority of the leaching experiments utilized
a rate of 2.6 ℓ/min for both waste bodies. Air flow rates were
also varied, with a maximum flow rate of 70 ℓ/min. While air
flow was normally maintained continuous, solution (lixiviant) was
normally applied in periods of 24 to 48 hrs (flushes) with rest
periods between flushes ranging from 2 to 6 weeks.

Figure 3 shows a more extensive view of the overall height of
the leach tanks [Fig. 3(a)] and a comparative schematic view showing the general disposition of the access ports and associated
instrumentation and sampling features. Figure 4 shows a simple
schematic view of the solution circuit [Fig. 4(a)] and a close-up
view of the cementation system (launder) and holding tanks shown
in Figs. 1 and 3(a). One of the two experimental waste bodies to
be considered in this chapter consisted of a Kennecott Chino Mines
Division (CMD) Santa Rita waste (~-4 in). The total copper
content was 0.36% with non-sulfide copper accounting for 0.14% and
the balance of the sulfide copper disseminated primarily as

Fig. 1.   *Overall view of leach-tank test facility showing tank array.*

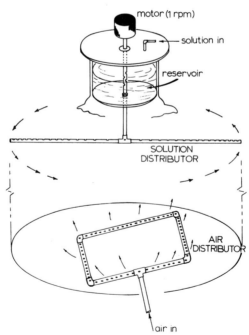

Fig. 2.   *Schematic view of solution and air distribution systems.*

(a)

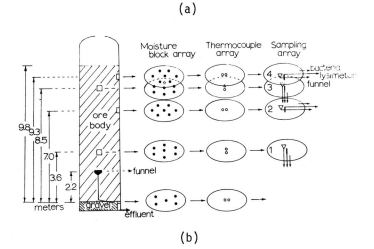

(b)

Fig. 3. Experimental leach tank design features. (a) End view of leach tanks showing one of several air compressors (ac), one of two cementation systems (c), and the instrumentation trailer (i), (b) Schematic view illustrating sampling and instrumentation arrays in one of the experimental waste bodies.

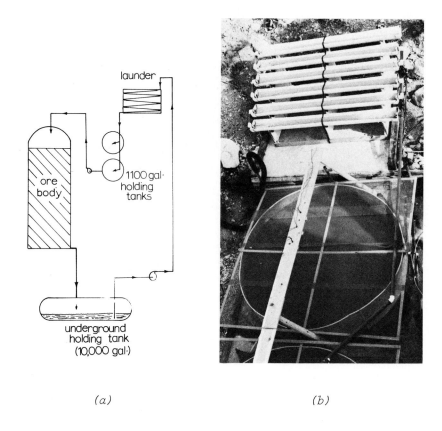

(a)   (b)

Fig. 4. Solution circuit schematic (a) and close-up view of one of two cementation systems consisting of a launder and solution holding tanks (b). The iron scrap which normally fills the six PVC pipe half-sections (0.38 m diameter) has been removed in the photograph.

chalcocite; with some chalcopyrite. The pyrite/copper sulfide ratio was determined to be roughly 10.6. The rock matrix was a quartz monzonite requiring little initial neutralization.

The second waste body consisted of a Duval-Sierrita rock type (~-6 in) with a total copper content of 0.34% of which only

0.03% was non-sulfide (acid soluble) copper, and the balance primarily chalcopyrite. The pyrite/chalcopyrite ratio was roughly 4, and the rock matrix contained approximately 2% carbonate compounds (mainly calcite) which required a considerable neutralization period.

Table I shows for comparison the analyses of the two different rock types used in these experiments. It is readily observed that, as outlined above, these waste bodies represented rather extreme considerations in rock type and concomitant leachabilities. The leaching procedure was also different for these two waste rock types. In the case of the Kennecott waste, CMD tailings water (pH 2.6) containing roughly 0.4 g/ℓ Cu, about 1 g/ℓ total Fe, and approximately 10^5 *Thiobacillus ferrooxidans* cells/cc of solution (MPN) (6) was used to saturate the waste body following dry body permeability measurements described previously (7). The waste body bacterial population (*T. ferrooxidans*) by comparison, was approximately 10^3 cells/g sample prior to the solution application. Following the drain down of the tank (7), the waste body was aerated for a period of time before solution was applied as previously described.

By comparison, the Duval waste was continuously flushed for a period of roughly 5 months using well water containing no initial iron (less than 50 ppm total Fe) and no detectable copper. Solution onfluent flow rates ranged from 2 to 4.4 ℓ/min during this initial neutralization period. Onfluent pH was also varied from 1 to 2.5 (with H_2SO_4). During this neutralization period, aeration varied between 0 and 35 ℓ/min. The solution bacterial population initially measured zero (no detectable *T. ferrooxidans*). The initial waste body bacterial population also measured between zero and 10 cells/g just prior to the initial solution application.

Once the waste bodies were saturated and solution cycles with aeration rest periods began, the effluent solutions were

TABLE I   Waste Body Analysis and Properties

Parameter	Kennecott (Santa Rita) Waste	Duval (Sierrita) Waste
Quartz (wt. %)	68	29
K-feldspar (wt. %)	2	38
Biotite (wt. %)	2	5
Dickite (wt. %)	---	5
Total Cu (wt. %)	0.36	0.34
Non-Sulfide Cu:		
(% of Total)	0.22	0.03
Chalcocite (wt. %)	~1.0	---
Chalcopyrite (wt. %)	<0.1	0.8
Pyrite (wt. %)	---	3.3
Pyrite/Non-Sulfide Cu	10.6	4.2
Carbonate (wt. %)	<0.1	2.0
Total SiO_2 (wt. %)	75	71.4
Total Fe (wt. %)	5.6	2.7
Dry body bacteria:		
(cells/g)	10^3	<10
Max. rock size	~4 in	~6 in
Total wt. in kg	1.6×10^5	1.7×10^5
Dry body permeabilities, (in darcys):		
Top surface	0.37	2.6
Port #1	0.31	2.4
Port #2	0.20	2.0
Port #3	0.21	1.7
Port #4	0.12	0.8
Wet body permeabilities, (in darcys):		
Top surface	0.41	1.1
Port #1	0.29	1.0
Port #2	0.22	1.0
Port #3	0.31	1.1
Port #4	0.13	0.6

analyzed at close intervals (2-6 hrs) during flushes for copper and iron (Fe^{2+} and Fe^{3+}) using atomic absorption spectroscopy and standard titration methods. Oxygen concentrations referenced to air (~21%) were determined at the waste body access port locations [Fig. 3(b)] using a long probe inserted into the bacterial analysis ports, using a commercial oxygen meter. pH and Eh measurements were made of the solution effluent of both waste bodies and within the waste rock using solutions withdrawn through the porous-cup lysimeters and samples withdrawn through the funnels embedded within the waste body. Solid samples withdrawn periodically from the waste body were utilized in MPN determinations of *T. ferrooxidans* populations, and rock samples were also periodically examined in a scanning electron microscope (Hitachi Perkin-Elmer HHS-2R). These results were frequently compared with more controlled laboratory experimental results involving systematic studies of bacterial catalysis in low-grade waste rock systems as discussed in greater detail in Chapter 6, and in other work (9, 10).

A most-probable-number method (6) was used to estimate the concentration of *T. ferrooxidans* in solid (waste) and liquid samples. The 9K medium (8) was used in the procedure which involved incubation of culture tubes at room temperature (~25°C); and the results were considered positive or negative for growth following 21 days incubation.

III. RESULTS

A. Kennecott Waste Body Test Data

Figure 5 illustrates the ambient air temperature features at the test site, the waste body temperature profiles, the corresponding oxygen profiles and bacterial populations within the waste body, and the total oxygen consumption by the waste body. It is clear that a correlation exists between the waste body temperatures, oxygen profiles and consumption, and bacterial population. While there is some relationship of ambient air temperatures to the waste body temperature profiles, the waste body temperatures are observed not to be strictly controlled by the ambient air temperature. In addition, the maximum waste body temperatures exceed the average maximum ambient air temperatures by nearly a factor 2.

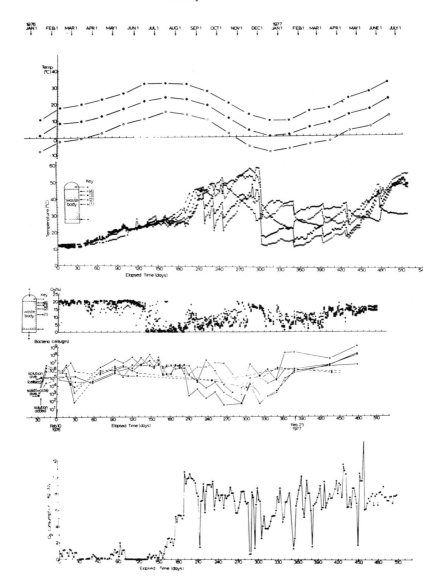

*Fig. 5. Kennecott waste body leaching data. From top: am-
bient air temperatures at the test facility during the experimental period showing average monthly maximum, mean, and minimum
temperatures, waste body temperature profile, oxygen profiles
(expressed as O₂% at each port location), bacterial (T. ferrooxidans) population in the solution effluent and onfluent and the
waste body profile, and the daily oxygen consumption of the waste
body in kg/day (bottom). The zero point on the abscissa is
coincidental in all graphs.*

It can be observed in Fig. 5 that when the waste body began to heat up rapidly, the oxygen consumption also rose sharply. In addition, when the waste body attained its maximum temperature profile, the *T. ferrooxidans* population declined markedly. In fact, in the mid-section of the waste body where temperatures ranged from about 55 to 59°C [at ports 1 and 2, respectively in Fig. 3(b) and as noted in the data key of Fig. 5], Fig. 5 illustrates a concomitant decline in *T. ferrooxidans* MPN to less than 10^2 cells/g which is much less than the same level as that observed in the dry waste body prior to leaching. It must be cautioned that because of the limitations in the experimental technique, MPN measurements are subject to considerable error, probably at least a factor 10. Nonetheless, the trends established over long periods of time utilizing measurements made in the same way, utilizing the same procedures, can be recognized (in Fig. 5) to have established some significant trends. Similarly, the oxygen levels recorded at each access port shown in Fig. 5 are also subject to some error, but these data also illustrate a correlatable trend, particularly with the temperature profile, which can be taken to be an accurate representation of the temperatures within the waste body.

It is of interest to note in Fig. 5 that when the temperature maximum is drastically quenched by a solution flush near the ambient minima after an elapsed time of approximately 300 days, the *T. ferrooxidans* population begins to increase again as the waste body temperatures stabilize in a range below 40°C. Just before the second increase in temperature at roughly 400 days elapsed time, the *T. ferrooxidans* population is observed to be maximum again. This clearly establishes a connection between the waste body temperature profile and the bacterial population numbers. During the period of quenching the first waste body temperature maximum (300-330 days), there is also an increase in the oxygen levels and a concomitant decline (for a short period)

in the oxygen consumption. To some extent this also corresponds
to the trough in the bacterial population data (between roughly
270-330 days elapsed time).

It can be noted in Fig. 5 that the waste body temperatures
are effectively quenched with each solution application (flush),
with a marked quenching near the upper portion of the waste body,
especially after the initial heating (>210 days elapsed time).
Figure 6 illustrates in more detail the sequencing of the solution
flushes, and shows the copper recovery as well as the aeration
schedules associated with the flushes. The bottom graph in Fig.
6 also illustrates the average iron content of the solution
effluent during the first 16 flushes, and shows the preponderance
of Fe^{3+} as a result of the bacterial catalyzed oxidation of the
Fe^{2+}, prominent in the solution onfluent following cementation
[Fig. 4(a)]. This response is typical of bacterial leaching
studies performed in the laboratory (11).

Figure 7 illustrates the details of several later (more
current) solution effluent (flush) analyses. The oxidation efficiency illustrated in Fig. 6 (bottom) is presented in more detail
in the solution flush data of Fig. 7 which clearly shows the
Fe^{3+}/Fe^{2+} ratio to be between 5 and 10 on the average. Solution
effluent Eh and pH characteristics during the 2 and 4 day flush
cycles are also illustrated along with solution onfluent and
effluent temperatures, and copper solubilized in Fig. 7.

B. Duval Waste Body Test Data

As is readily apparent from Table I, the Duval (Sierrita)
waste body represented a considerably different mineral regime by
comparison with the Kennecott waste body. Because of the large
proportion of acid-consuming minerals (primarily carbonates) in
the host rock, the Duval waste body required a considerable
neutralization period. At the start of the Duval test, the dry
body and wet body permeabilities were also determined as previously

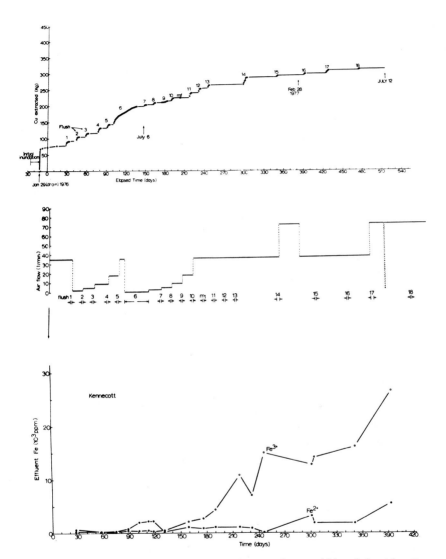

Fig. 6. Solution flush (leach) cycles utilized in the Kennecott waste body leach test. Top: Cumulative copper extraction with each flush. (mf) denotes a mini-flush for solution test purposes. The initial extraction as a result of the inundation and drain down represents primarily non-sulfide copper dissolution; Middle: Aeration schedule (m_1 corresponds to mf in top graph; Bottom: Average effluent iron for the first 16 flushes (iron in solution). Onfluent iron in solution corresponding to (c) would essentially reverse Fe^{2+} and Fe^{3+} after an elapsed time of 150 days, i.e., Fe^{2+} was dominant in the solution onfluent.

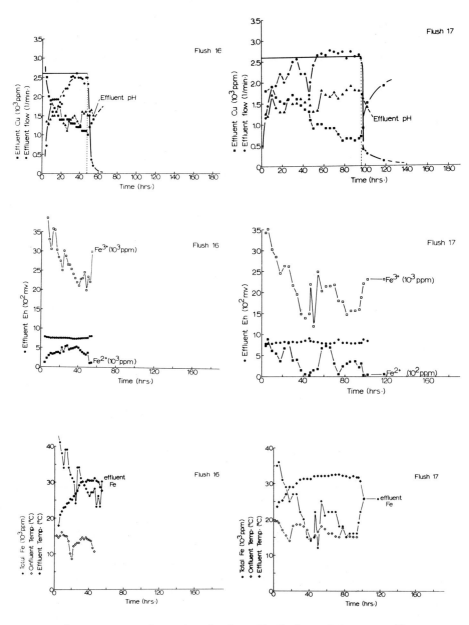

Fig. 7.   Examples of solution flush data (corresponding to Fig. 6).

reported (7); which were considerably different from the Kennecott waste body permeabilities as noted from Table I. Following the initial inundation (saturation) of the waste body with fresh well water acidified (with H_2SO_4) as described above, as well as smaller column inundation to act as scaling comparison tests (7), the pH of the effluent was monitored continuously. After approximately 140 days of essentially continuous solution application, the effluent pH was reduced to approximately 2.5, and the waste body began to consume oxygen. This feature is illustrated in Fig. 8. Figure 8 also shows the very close correlation of the scaling tests, indicating that at least acid neutralization of a waste body seems to be a feature which can be very successfully predicted even for a very large waste body, so long as the rock size and solution application is geometrically scaled to the larger regime as previously outlined (7).

It can be observed from Fig. 5 (top) that coincident with the onset of oxidation [oxygen consumption in Fig. 8(b)], the ambient temperature was beginning to increase, and the waste body temperatures were also beginning to increase correspondingly. When the rate of O_2 consumption had reached a range of roughly 4.5 kg/day [Fig. 8(b)], the waste body temperatures were above 30°C (Fig. 9). This corresponds to the behavior of the Kennecott waste body as shown in Fig. 5. It can be observed, correspondingly in Fig. 9, that when oxygen was being consumed in the upper regions of the waste body, the temperature was slightly higher there, and the bacterial population was higher. This feature was also apparent in the Kennecott waste body (Fig. 5).

An important feature to note with regard to the O_2 consumption of both the Kennecott and Duval tests (Figs. 5, 8, and 9, respectively), is that although the bacterial population was initially very high, and increased in the solid (waste) to roughly 10^7 cells/g, oxidation was not appreciable until this level of bacteria were present. Correspondingly, the same level

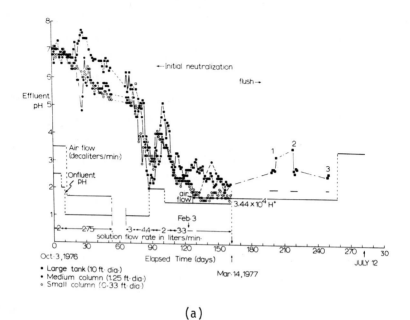

(a)

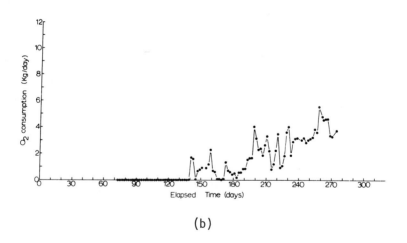

(b)

Fig. 8.   Initial acid neutralization and scaling of H^+ consumption of the Duval waste body (a), and the corresponding waste body oxygen consumption (rate of consumption) (b).   Subsequent solution flushes and aeration rates are also shown following apparent neutralization in (a).

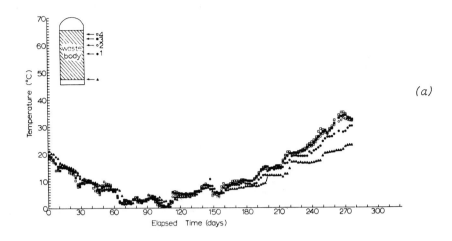

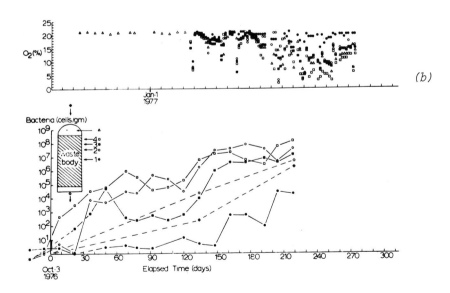

Fig. 9. Duval waste body leaching data. (a) Waste body temperature profile, (b) Oxygen levels within the waste body, and Bacterial (T. ferrooxidans) population in the solution effluent and onfluent and the waste body profile.

was attained in the upper three access levels of the Duval waste
body when oxygen levels dropped there, and the total O_2 consumption began to increase. It is interesting to note that while the
waste body temperature at this stage was nearly 25°C in the Kennecott waste body, the temperature had only reached approximately
10°C in the Duval waste body. This would seem to be indicative of
the fact that the bacteria play a significant role in oxidation in
a waste body. Although in both the Kennecott and the Duval experiments, the initial increase in temperature of the waste bodies
is observed to coincide with the annual ambient increase in
temperature, it might be that exothermic reactions (the result of
bacterially and chemically catalyzed oxidation) are contributing
markedly, and that the waste body heating is due in large part to
the presence of bacteria and the ferric iron produced as a result
of this activity.

IV.   DISCUSSION

With reference to the last points made regarding the test
results above, the merits of large column leach tests become
immediately obvious. In laboratory tests such as the many discussed in the previous chapters of this volume, and many previous
studies, the test temperature is established externally (that is,
external to the leaching regime or waste body in the case of low-
grade waste leaching), and usually maintained constant. By
comparison, leach dump and large heap temperatures are dictated by
exothermic reactions, the ambient temperature, or both. Oxidation
reactions are not strictly determined by temperature although this
is obviously a principal variable in the reaction kinetics. Large
column tests are able to provide direct thermodynamic data relating to commercial waste body leaching and temperature drive of
the system whereas small scale tests are not. However, with the
large columns, it is difficult to establish abiotic controls in
order to precisely (and unambiguously) define the contributing
role of the bacteria.

One of the significant observations described above is the onset of oxygen consumption which, considering drastically different leaching parameters and waste body characteristics, including temperature and bacterial population, occurs in an identical manner. That is, after a sufficient period of solution cycling, which is also very different by comparison, oxygen consumption, and correspondingly sulfide oxidation (Cu_2S, $CuFeS_2$, FeS_2), begins when the $T.\ ferrooxidans$ population has reached approximately 10^7 cells/g (compare Figs 5, 6, 8, and 9). Since, in the case of the Kennecott column, the initial solution contained 10^5 cells/cc, it appears that there will be no bacterial catalysis of any consequence until the MPN exceeds roughly 10^5 cells/g.

Figure 6(c) also shows that when sulfide oxidation (O_2 consumption) begins in the Kennecott waste body, the ratio Fe^{3+}/Fe^{2+} begins to exceed 1 and increases as the temperature increases and the O_2 consumption increases (compare Figs. 5 and 6). As a consequence, it appears that the onset of bacterial catalysis, and O_2 consumption, is related somehow to iron in solution (Fe^{2+}). It is well known, in this regard, that new leach dumps require a period of time (sometimes referred to as an incubation period) before leaching will begin, and that in many cases, mine tail water is needed or the period extends for periods beyond 0.5 yrs., or leaching never really occurs. There appear to be two factors involved. The solution either does not contain sufficient numbers of bacteria (primarily $T.\ ferrooxidans$), or the chemistry is not sufficient to promote bacterial growth and activity. By bacterial activity we refer to metabolic function, which is influenced by available CO_2, O_2, nutrients, etc., as well as iron in solution. It may be that in a leach dump, the critical factor is ferrous iron in solution (which depends upon solution pH, copper in solution, etc.). Bacterial activity also depends strongly upon pH, independent of ferrous iron content. The ferric iron concentration may also be an important factor in determining bacterial activity as it acts as a competitive inhibitor of microbial ferrous iron oxidation (see Chap. 2).

Some evidence of the importance of these factors in a large waste body is provided in the data for the Duval column. In Fig. 9(b), the *T. ferrooxidans* population is essentially nil at the start of solution application. However, in the upper few meters of the waste body, the population rises rapidly, increasing by 10^5 in only 60 days. During this time, the temperature is decreasing. The solution chemistry was observed to change, but not noticeably with regard to ferrous iron and copper in solution. However, the neutralization front was moving down the column of waste rock, and the pH was slowly declining. The slow buildup of bacterial population near the waste body mid-section [10^4 cells/g compared with $>10^6$ cells/g in the upper sections after 210 days elapsed time in Fig. 9(b)], is indicative, we feel, of remaining, unneutralized rock portions which raise the local pH. This portion, as observed in the O_2 level data, is not consuming oxygen, and it is also the coolest part of the waste body [Fig. 9(a)]. This kind of data, and data correlation, is not possible in laboratory tests, and is highly representative of leach dump segments which are thermodynamically insulated regimes.

It is of interest to note that in both the Kennecott and Duval waste body test data of bacterial population, the numbers in solution are less, by roughly 10^2, than those recorded in the solid. As discussed in Chap. 6, it may not be necessary for bacteria to attach in order to catalyze the oxidation reactions, particularly the oxidation of iron sulphides, although the conversion might be more efficient as a result of attachment, and as a consequence the leaching rate might be enhanced by attachment. Unfortunately, due to the difficulty in removing rock specimens intact from the waste column, and in specimen preparation (particularly drying) it has not been possible to observe bacterial attachment in the waste body up to the present time. The difficulties encountered in drying the specimens, and in masking the microorganisms in surface layers and slimes during dry down, might be avoided in future attempts utilizing the critical-point drying technique.

The question of attachment and the catalytic function which might be associated with attachment is an interesting one, and one which can have important implications in the processes occurring within a leach dump. Correspondingly, there are operational features of a leach dump which can have a significant effect upon the attachment, or the events which occur near the rock surfaces. In the present experiments, solution application (flushes) were interrupted by rest periods during which the waste bodies are aerated. Although the schedules of these cycles were dictated somewhat by a specific experimental format (14), the advantages of leach cycles (including rest periods) are now documented (1). Figure 10 illustrates the changes which can occur in a region of a waste body during such rest periods and even at various locations in a dump during continuous solution application if channeling of solution away from a region were to occur. As the region is aerated [either directly or by thermal convection (4)], moisture held in capillaries is reduced until surface tension effects cause a solution film to cover the rock particles [Fig. 10(b)]. Under this condition, bacteria are forced to come close to the surface, and the microbial ecosystem at the rock surface is drastically changed because concentrations are increased, pH may change significantly, and other factors may be altered. We observe, in fact, that such rest periods are necessary for significant heating to occur in the waste body. Certainly this is due in large part to heat transfer and thermal transport of the solution cooling or quenching the waste body as shown in Fig. 5, but the features illustrated in Fig. 10(a) and (b) could also contribute in a significant way.

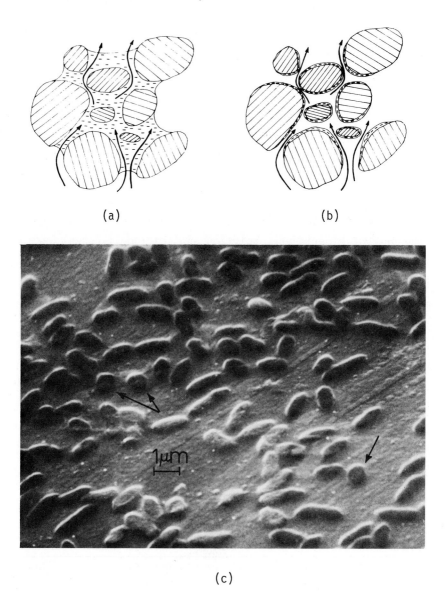

(a)   (b)

(c)

Fig. 10. Variations in local waste body moisture content
which could influence leaching reactions during aeration. (a)
Solution saturated regime (solution-filled capillaries),(b) Solution films covering rock (particle) surfaces. The arrows indicate
air convection, the dashes are bacteria in solution, (c) Scanning
electron micrograph showing rod-like T. ferrooxidans on a pyrite
surface. Note spherical-shaped organisms (arrows). (Courtesy
V.K. Berry)

Figure 10(c) illustrates the appearance of the *T. ferrooxidans* attached to a pyrite surface region following shake-flask leaching at 28°C for 2 weeks in the laboratory. Observations of attachment by *T. ferrooxidans* are not common (see Chap. 6) even in laboratory experiments, and have not been made for rock samples extracted either from the large column tests reported here, or commercial leach dumps.

One of the most significant features indicated in the Kennecott test data, particularly the waste body temperature profiles, is a prominent correlation between *T. ferrooxidans* population and temperature after the onset of O_2 consumption. Most striking is the prominent decline in *T. ferrooxidans* population (by as much as 10^5) when the first temperature maximum occurs. During this short, high temperature plateau (between an elapsed time of 280-300 days) sampling from the waste body indicated the presence of a high-temperature thermophilic iron oxidizing bacterium having a rod shape (12), similar in some respects to the rod-shaped thermophilic *Thiobacillus*-like bacterium recently described by Brierley and LeRoux (13). Similar examination of waste rock particles cultured at 55°C during the second high temperature plateau shown in Fig. 5 indicated not only an active rod-shaped organism similar in appearance to *T. ferrooxidans* [as shown in Fig. 10(c)], but also a spherically shaped microbe similar to those shown in Fig. 10(c). This microbe is morphologically similar to *Sulfolobus* but growth and iron oxidation does not occur at 60°C. Preliminary examination has shown that in port areas 1 and 2, which are shown in Fig. 5 to have differed by roughly 5°C, there were more spherical microbes at the higher temperature location (port #2) than those were at the lower temperature port (port #1).

While there is insufficient data at this writing, there is some evidence that a second microbe, a thermophile, exists naturally within the Kennecott waste body or in the mine tail water (12) since the test was never intentionally inoculated. Obviously a thermophile of some kind exists since the O_2 consumption continues, and even increases, as the *T. ferrooxidans* population declines, and the temperature increases.

On examining the temperature profiles of the Kennecott waste body (Fig. 5) in comparison with the *T. ferrooxidans* population, we are struck by the possibility that because of the distinct temperature plateaus, and the distinct differences in functional organisms within these regimes, the bacteria might actually be regulating the temperature. It has been mentioned in previous chapters, for example, that the leaching process involving bacteria seems to function like a complex feedback mechanism. It has been suggested nearly two decades ago that temperature increase in leach dumps may be associated with bacterial oxidation of iron and sulfides (15). It is unclear whether bacteria function as a result of reactions occurring or whether the reactions occur in a more direct response to bacteria (or bacterial catalysis). In other words, since bacteria do catalyze the exothermic reactions within a waste body, they could, via a feedback mechanism, regulate the rate of catalysis in order to regulate the temperature to a range where their ability to catalyze (as a result of metabolic optimization) would be maximum.

It might then be possible, through some external manipulation, to alter the optimum conditions for the operation of a feedback mechanism associated with a particular microorganism, and to force the reaction control to another regime and another microorganism capable of functioning as a regulator of the higher-temperature feedback regime. At this stage, the higher temperature microbe (a thermophile) would regulate the reactions (and the temperature). Such a process would not necessarily involve additional oxygen consumption to speed up the reactions to create exothermic heat because the regime is insulated and little heat might be lost, and the fact that the thermophile would perhaps limit catalysis in order to maintain an optimum temperature for metabolism, thereby stabilizing O_2 consumption. This is in fact observed in Fig. 5.

The implications of such processes occurring are of course both exciting and important because even if a thermophile did not increase the rate of catalysis (which it probably does not) the higher temperature regime it would regulate could markedly increase the rate of copper going into solution (roughly doubling with each 10°C temperature). In other words, if bacteria limit dump temperature via a feedback mechanism, then advancing the temperature regime by inoculating with a higher-temperature microbe could conceivably increase copper leach dump operation very markedly. However, one cannot discount the process of "natural selection" whereby the microbes are occupying a thermal niche provided by the insulated-tank environment rather than controlling the temperature profile.

There is evidence of temperature regulation within existing leach dumps. For example, the maximum temperature recorded in Kennecott's Midas test dump was 54°C (6). By comparison, other field studies have indicated that other low-grade copper waste dump temperatures may reach 80°C (14). The reason for this temperature difference could well be a difference in the leach-dump microbial ecosystem, with the higher-temperature dumps containing microbes having a higher-temperature regime for optimum metabolism. These concepts should receive intensive efforts to verify them because the consequences could have enormous practical significance.

V. SUMMARY AND CONCLUSIONS

The current status of large-scale leaching studies on two different copper-bearing waste bodies has been outlined in this chapter. The results have shown that large scale testing can provide valuable information about variations in leaching parameters within a waste-ore body which cannot be obtained from laboratory or other small-scale leaching experiments. Direct correlations between temperature profiles (vertical profiles) having a relationship to existing leach dump temperature profiles and O_2 consumption, O_2 levels, bacterial populations and leach solution chemistry and related data over long periods of elapsed time have been obtained for the first time, and have shown very convincingly that there is indeed a relationship. We have observed that as the waste body temperature reaches certain temperature plateaus, a reaction regime is defined which is related to the function of specific microorganisms. It is suggested that commercial leach dumps may be temperature limited because of the absence of microbes or microflora in the microbial ecosystem which would be able to regulate the exothermic reactions at higher-temperature plateaus.

There is a real need to investigate the ecosystem of existing leach dumps or proposed dumps to determine the nature of the microorganisms present, and perhaps the degree of synergism which might be involved in the attendant bacterial leaching processes. It is also important to determine whether in fact a higher temperature plateau can be achieved in a waste body by inoculation with a higher-temperature thermophile than those which are determined to exist naturally. This experiment will in fact be performed in the Kennecott waste body test, and the results will be reported at some later time.

Finally, it must be concluded that while the use of large-scale testing such as that reported herein may not provide significant guidance for the development of new operating parameters for existing leach dumps, such testing can have a significant impact on new dump design and parameterization. Much of this data cannot be obtained from lower-scale testing as pointed out above.

VI. ACKNOWLEDGMENT

This research was supported in part by the Kennecott Copper
Corporation Metal Mining Division and by the National Science
Foundation (RANN) under Grant No. AER-76-03758 and AER-76-03758-
A01. The authors thank Drs. Carl Popp and W.J. Schlitt for pro-
viding some of the analytical data in Table I, Dr. M.H. Wilkening
for supplying the ambient air temperature data, and to many
colleagues connected with the John D. Sullivan Center for In-Situ
Mining Research for other important contributions related to
the design phases and the testing programs. These include D.A.
Reese, Drs. L.M. Cathles and John Apps, Dr. V.K. Berry, Mr. P-C.
Hsu, Mack Stallcup, Stewart Ingham, William Herrington, Steve
Lockwood, Barbara Lopez, Sara MacMillan, and Al Hansen (now
deceased).

Special recognition must be given to Dennis Bloss, who is now
deceased. Without his dedication throughout the initial design
and testing phases, much of the data might not be available. The
help and guidance provided by the NSF Project Oversight Committee
was especially important in the early phases of the Duval test.
These members included Milton Wadsworth, Jim Grunig, Rees Groves,
Larry Cathles, Ken Temple, and John Apps. Discussions with
Nathaniel Arbiter and Ed Malouf have also added some stimulus to
our perspective in writing this manuscript. Finally, the help of
Lamar Kempton and the TERA staff at New Mexico Tech must be
acknowledged because without it the design modifications might not
have been possible. The provision of waste rock by Duval Corp. is
also gratefully acknowledged.

VII. REFERENCES

1. Karavaiko, G.I., Kuznetsov, S.I., and Golonizik, A.I., "The
    Bacterial Leaching of Metals from Ores," (English translation
    by W. Burns), Technology Ltd, Stonehouse, Gloster, England,
    1977.

2. Bhappu, R.B., Johnson, P.H., Brierley, J.A., and Reynolds, D.H., *Trans. Soc. Min. Engrs. AIME*, 244, 307 (1969).

3. Madsen, B.W. and Groves, R.D., in "Extractive Metallurgy of Copper," (J.C. Yannopoulous and J.C. Agarwal, Eds.), Vol. II, Chap. 47, p. 926. AIME, New York, 1976.

4. Cathles, L.M. and Apps, J.A., *Met. Trans.*, 6B, 617 (1975).

5. Bryner, L.C., Walker, R.B., and Palmer, R., *Trans. Soc. Min. Engrs. AIME*, 238, 56 (1967).

6. Collins, C.H., "Microbiological Methods," Plenum Press, New York, 1967.

7. Murr, L.E., Cathles, L.M., Reese, D.A., Hiskey, J.B., Popp, C.J., Brierley, J.A., Bloss, D., Berry, V.K., Schlitt, W.J., and Hsu, P-C., in "Proceedings of American Nuclear Topical Meeting: Energy and Mineral Resource Recovery", April 12-14, Golden, Colorado, 1977 (to be published in 1977; also to be published in *In-Situ*, 1978).

8. Silverman, M.P and Lundgren, D.G., *J. Bacteriol.*, 77, 642 (1959).

9. Murr, L.E. and Berry, V.K., *Hydromet.*, 2, 11 (1976).

10. Berry, V.K. and Murr, L.E., in "Scanning Electron Microscopy/ 1977", (O. Johari, Ed.), Vol. I, p. 137, 1977.

11. Dutrizac, J.E. and MacDonald, R.J.C., *Mineral Sci. Engr.*, 6, 59 (1974).

12. Brierley, J.A. and Lockwood, S.J., *FEMS Micro. Letters*, in press (1977).

13. Brierley, J.A. and LeRoux, N.W., in "Conference Bacterial Leaching 1977," (W. Schwartz, Ed.), p. 55. Verlag Chemie, New York, 1977.

14. Beck, J.V., *Biotechnol. Bioeng.*, 9, 487 (1967).

15. Lyalikova, N.N., *Microbiology*, 29, 281 (1960).

Index

A
B
C 8
D 9
E 0
F 1
G 2
H 3
I 4
J 5